Applied Simulation and Optimization 2

Miguel Mujica Mota · Idalia Flores De La Mota
Editors

Applied Simulation and Optimization 2

New Applications in Logistics, Industrial and Aeronautical Practice

Editors
Miguel Mujica Mota
School of Technology
Amsterdam University of Applied Sciences
Amsterdam, North Holland
The Netherlands

Idalia Flores De La Mota
Facultad de Ingeniería
Universidad Nacional Autónoma de México
Mexico City
Mexico

ISBN 978-3-319-85754-1 ISBN 978-3-319-55810-3 (eBook)
DOI 10.1007/978-3-319-55810-3

Softcover reprint of the hardcover 1st edition 2017

Printed on acid-free paper

This Springer imprint is published by Springer Nature
The registered company is Springer International Publishing AG
The registered company address is: Gewerbestrasse 11, 6330 Cham, Switzerland

To my Parents and Christina

Foreword

Simulation allows to reproduce complex behaviors of existing real-world and hypothetical future systems as well as to carry out experiments on virtual environments to improve understanding, to develop new concepts, and to test technologies in a cost-effective way and safe environment.

Experience gained in the last years indicates that the extensive use of simulation constitutes an invaluable methodology to cope with most critical scenarios and threats, and it is evident that the future will introduce further challenges.

These scenarios typically need to be studied by using the theory of complex systems, characterized by many highly correlated elements, strongly not linear components and emergent behavior. In order to succeed in these scenarios, it becomes necessary to develop new solutions (i.e., technologies, systems, new doctrines, operational plans, etc.). These innovative solutions require to be immersed within the specific operational scenario which means to be able to conduct detailed experimentation in realistic conditions. Obviously, the adoption of a comprehensive approach stresses even more these aspects and creates additional challenges for being able to acquire such experience. It results evident that this knowledge should be available in advance to support decision process; therefore, simulation is probably the only suitable approach and the critical technology for succeeding in this area.

This means that in order to anticipate innovation and transformation, it is necessary to simulate the new solution within virtual environments. Therefore, the new challenging scenarios and the innovative solutions introduce the necessity to create, maintain, and extend knowledge and experience.

Indeed, simulation is a science to study real world based on facts learned through experiments and observations carried out on a virtual environment. The concept is quite revolutionary, providing the opportunity to conduct "a priori" experimentation to know a real-system behavior even before it is realized or put in place. Obviously, models and simulators are able to reproduce real systems just based on approximations that should be regulated based on the specific nature of their specific purpose; sometimes, very simple models could be effective, while in other cases

Big Data is required. In any case, M&S (modeling and simulation) potential is very impressive even if it is still subjected to limitations in terms of fidelity, usability, and maintainability as well as of the Simuland: the representation of the real system under analysis based on our knowledge, assumptions, hypotheses, and available data, as it was defined by John McLeod, founder of the Society for Modeling and Simulation International (SCS).

After more than 50 years, these concepts are still valid and M&S represents a strategic asset further reinforced by data abundance and enabling technologies as well by advances in methodologies and techniques. Indeed, simulation science enables scientific and quantitative analysis and operational exercises within synthetic environments reproducing real scenarios, future challenges, new solutions, and systems.

So, it is evident that M&S is a cornerstone in modern optimization of complex systems, especially when facing cases with many entities and interactions, high degrees of stochasticity and complex behaviors.

The present volume addresses this topic and presents cases, where it becomes clear how M&S could serve optimization in addressing very challenging environments from public transportations to retail and from mechatronic systems to decision-making.

The introduction of new tools and techniques is very important to guarantee a strategic advantage in designing, engineering, managing, and operating modern systems, and this book presents them in application to real problems providing clear understanding of their potential, guidelines for their reuse as well as proofs their validation. This is exactly how the scientific method was defined when it was introduced over four centuries ago that requires to be able to test, experiment and to repeat experiences in order to check their validity; in this sense, this book brings a value for the reader for being able to acquire capabilities and to understand how to put them at work.

Simulation is a continuous evolving world, and new techniques and methods continuously emerge to get benefits of upcoming developments and to face emerging challenges; based on this consideration is easy to understand how developing and presenting case studies is critical community. This is very important for young researchers because it allows them to develop new skills and also to being aware of experiences and real developments to direct their future researches and projects. Therefore, the contents proposed in this book are also very useful for experienced simulationists to further extend their knowledge with reliable data, references, and case studies.

Each day real world situations are the result of more and more interconnected fields areas, and the resulting complex systems are characterized by a impressive number of variables, high uncertainty, and challenging dynamics and behaviors; as soon as we are able to develop reliable models of a system, we are used to discover that we need to develop new ones to address a more extended case where additional precision or comprehensive approach are required to address the new needs of users and decision makers. This is a loop but also a great opportunity for simulation considering its flexibility and capabilities. However, it requires to develop real transdisciplinary capacities among researches and scientists as well as to develop the user community in being able to use and trust simulation science; this book presents, defends, and promotes advances in simulation science applied to optimization and related tools and techniques.

In my opinion, such achievement is not surprising knowing since very long time the authors and their enthusiasm and capabilities in developing scientific researches coupled to real cases; personally I consider this the most proper approach to research, keeping it strictly connected with development to create solutions to the most challenging problems, that usually are the real one: industry, social life, and transportations are environments able to propose big problems, and simulation is an elegant solution to face and solve them. I really appreciated the opportunity to provide them with few of my thoughts for this valuable contribution to simulation advances.

Agostino G. Bruzzone
President of the Simulation Team and of International
Master Program in Industrial Plant Engineering
and Technologies of the University of Genoa

Contents

Contributors

Luis Alvarez-Icaza Universidad Nacional Autónoma de México, Coyoacan, Mexico; Ciudad Universitaria, Ciudad de México, Mexico

Alenka Baggia Faculty of Organizational Sciences, University of Maribor, Kranj, Slovenia

Guillermo Becerra CONACYT - Universidad de Quintana Roo, Chetumal, Q Roo, Mexico

Gail W. DePuy Department of Industrial Engineering, University of Louisville, Louisville, KY, USA

D. Delahaye Ecole Nationale de L'Aviation Civile, Toulouse, France

Roland Deroo Service Technique de l'Aviation Civile, Bonneuil-sur-Marne, France

Gerald W. Evans Department of Industrial Engineering, University of Louisville, Louisville, KY, USA

Idalia Flores De La Mota Universidad Nacional Autónoma de México, Coyoacan, Mexico; Ciudad Universitaria, Ciudad de México, Mexico; Facultad de Ingeniería, Deparatmento de Ingeniería de Sistemas, Universidad Nacional Autónoma de México, Mexico City, Mexico

Alexandre Gama Service Technique de l'Aviation Civile, Bonneuil-sur-Marne, France

Aman Gupta Department of Decision Sciences, Embry-Riddle Aeronautical University - Worldwide, Daytona Beach, FL, USA

Wulfrano Gómez Gallardo Facultad de Ingeniería, Universidad Nacional Autónoma de México, Mexico City, Mexico

Aída Huerta-Barrientos Facultad de Ingeniería, Deparatmento de Ingeniería de Sistemas, Universidad Nacional Autónoma de México, Mexico City, Mexico

Deogratias Kibira Department of Industrial and System Engineering, Morgan State University, Baltimore, MD, USA

Alexander Kurzhanskiy California PATH, Richmond, USA

Jennie Lioris École des Ponts-ParisTech, Champs-sur-Marne, France

Jose Luis Mendoza-Soto CINVESTAV, Gustavo A. Madero, Ciudad de Mexico, Mexico

M. Mujica Mota Aviation Academy, Amsterdam University of Applied Sciences, Amsterdam, The Netherlands

Blaž Rodič Faculty of Information Studies, Novo Mesto, Slovenia

P. Scala Aviation Academy, Amsterdam University of Applied Sciences, Amsterdam, The Netherlands

Esther Segura Pérez Facultad de Ingeniería, Universidad Nacional Autónoma de México, Mexico City, Mexico

Guodong Shao Engineering Laboratory, National Institute of Standards and Technology (NIST), Gaithersburg, USA

Daniel Tello Gaete Facultad de Ciencias de la Ingeniería, Universidad Austral de Chile, Valdivia, Chile

Pravin Varaiya University of California, Berkeley, Berkeley, USA

Ann Wellens Facultad de Ingeniería, Universidad Nacional Autónoma de México, Mexico City, Mexico

Part I
Novel Tools and Techniques in Simulation Optimization

A Conceptual Framework for Assessing Congestion and Its Impacts

Jennie Lioris, Alexander Kurzhanskiy and Pravin Varaiya

Abstract In urban areas, intersections are the main constraints on road capacity while traffic flows do not necessarily directly conform to the speed-flow relationship. It is rather the signal timing and the interplay between the clearing rate of each intersection which determines the formation and duration of congestion. Junctions often differ in their design and throughput. General conclusions on the relationship between vehicle speed and traffic flows on a junction link are rarely possible. Well-adapted models are required for a comprehensive study of the behaviour of each intersection as well for the interactions between junctions. This chapter assesses the potential benefits of adaptive traffic plans for improved network management strategies, under varying traffic conditions. Queueing analysis in association with advanced simulation techniques reveal congestion mitigation actions when the pre-timed actuation plan is replaced by the max pressure feedback control. The case of unpredicted local demand fluctuation is studied, where uncontrolled congestion is progressively propagated to the entire network under the open-loop policy. Travel-time variability is measured under both plans and within all traffic schemes while frequency of stop-and-go actions are also encountered. Reliability of predictable trip durations is a major factor to be considered when ensuring "on time" arrivals and the related costs when the time is converted into benefits.

Keywords Traffic responsive signal · Adaptive control · Pre-timed control · Max-pressure practical policy · Discrete event simulation · Performance evaluation · Queueing network model

J. Lioris
École des Ponts-ParisTech, Champs-sur-Marne, France
e-mail: jennie.lioris@cermics.enpc.fr

A. Kurzhanskiy
California PATH, Richmond, USA
e-mail: akurzhan@berkeley.edu

P. Varaiya (✉)
University of California, Berkeley, Berkeley, USA
e-mail: varaiya@berkeley.edu

M. Mujica Mota and I. Flores De La Mota (eds.), *Applied Simulation and Optimization 2*, DOI 10.1007/978-3-319-55810-3_1

1 Introduction

In contrast to freeways, urban traffic is distinguished by the existence of junctions and/or roundabouts involving conflicting traffic streams and thus interrupting vehicle flows. Intersections play a major role in determining the quality and volume of traffic in arterial networks by arbitrating conflicting movements in order to allow users to share the same road space sequentially.

Traffic control and signal coordination contribute to improve travel conditions by reducing frequent vehicle stops and queue lengths. More precisely, signal coordination may contribute to an optimised use of the current infrastructure, by establishing platoon type vehicle departures.

Automobile dependent cities often associated with large traffic volumes, tend to imply poor road performance especially when varying travel schemes occur forcing the related transportation structure towards heavy traffic or even congestion states. Moreover, spatial complexity is characterised by even more complex journeys difficult to be predicted and consequently controlled. Activity changes influence spatial distribution and consequently complex travel patterns are manifested and congestion is possible.

Classical congestion management policies maximise the ability of urban areas to deal with current and expected demand. Such flow-based management policies, associate capacities to road links expressed in flow and density. Under that scope, network performance is increased when higher density and flows are reached.

Alternatively, cost-congestion approaches involve an "economically optimal" traffic level for each road and tend to measure the congestion cost incurring when traffic exceeds the "optimum" levels by taking into consideration the related road demand and supply.

Many traditional traffic managements aiming to increase road capacity and thus to mitigate congestion impacts by improving traffic operations while others seek to involve road infrastructure and/or to shift roadway demand to public transportation. Although such approaches are suited for particular congestion types such as bottlenecks they can deliver long-lasting results when they are paired with other policies controlling the newly created capacity. Non-recurrent congestion caused by unplanned events influencing the system behaviour which frequently becomes unpredictable, and may cause extreme congestion conditions and/or become system-wide. A vehicle breakdown may create bottlenecks, prohibiting transit in a part of road or obliging other vehicles to deviate and thus varying demand patterns may be caused. Similar effects can occur from other events such as crashes, bad weather, work zones etc.

Congestion influences both travel times (indicators concerning mostly policy makers) and the reliability of the predicted travel conditions (indicators interesting to road users). There is no single congestion metric which is appropriate for all purposes. Consequently, quantitative and qualitative metrics should be provided when measuring congestion such as queue lengths and related duration, variance of travel times etc.

This work explores the effectiveness of the currently available road infrastructure management under open and close loop signal plans within various traffic contexts. The outline of this paper is as follows:

Section 2 introduces the problem to resolve while it describes the two principal control categories determining traffic signals: the open loop policy defined by the Fixed-time control and the feedback Max-Pressure and Max-Pressure practical plans. Section 3 presents the adopted mathematical approach for studying the traffic control problem while it briefly introduces the principal notions of Discrete event systems (DES) and suggests various methods for the study of DES. Section 4 discusses appropriate metrics for performance appraisal, according to the needs of the study. Section 5 presents the case study (network, associated demand and other data) and develops the notion of stability. Section 6 proceeds to the performance appraisal of the two control categories and verifies the expected system stability under both Fixed-Time and Max-Pressure plans. Section 7 compares Fixed-Time and Max-Pressure control policies in terms of queue lengths and the related probabilities, average total travel times, delay measurements etc. Section 8 studies the network behaviour under demand variation while Sect. 9 is devoted to the system reaction when a non-recurrent congestion occurs for a limited period at a particular intersection. Section 10 reveals the simulation advantages, where a detailed analysis allows a complete reconstruction of the simulated scenario. Consequently, any system observation can be deeply examined and consequently justified. An illustrative example is discussed.

2 Traffic Control: Problem Statement and Timing Plans

This section presents the dealing problem in order to understand the insights resulting from the framework on traffic management as discussed in the next sections. Moreover, an open loop control scheme and versions of Max-Pressure feedback algorithm are briefly introduced. A much more extensive study as well all theoretical properties of Max-Pressure control are presented in [1] where stability guarantees are also provided.

2.1 Problem Formulation

Let us consider an intersection n. A *phase* (i, j) indicates a permitted movement from an incoming link i of node n towards an outgoing link j. A *stage* U_n indicates a set of simultaneously compatible phases of node n and is represented by a binary matrix such that:

$$U_n(i,j) = \begin{cases} 1, & \text{if phase } (i,j) \text{ is actuated} \\ 0, & \text{otherwise} \end{cases}$$

The set of all considered intersection stages is denoted by $\mathcal{U}$. For simplicity, the optimisation horizon is divided into intervals or *cycles* of fixed width, each one comprised of T time periods. Let $L < T$ denote the idle time, that is the time period during which no phase is actuated and occurring within two different stage switches. This time corresponds to amber lights, pedestrian movements etc. Hence, the total available actuation period per cycle is $T - L$. If $q(t)$ is the array of all queue lengths at time t then the system state $X(t)$ is defined by $X(t) = q(t)$. A stabilising time plan maintains the mean queue length bounded. At time t a stabilising stage $u(t) = U$, with $U \in \mathcal{U}$ and a cycle proportion λ_u have to be defined such that $\sum_{u \in \mathcal{U}} \lambda_u T + L = T$.

2.2 *Signal Plans*

2.2.1 Fixed Time Policy (FT)

A *pre-timed* or *fixed time* control is a pre-calculated periodic sequence, $\lambda = \{\lambda_U \geq 0,\ U \in \mathcal{U}\}$, actuating each *stage* $u(t) = U,\ U \in \mathcal{U}$ during a fixed time period $\lambda_U T$ and such that $\sum_U \lambda_U = 1 - L/T$ within every cycle. The involved cycle proportions $\{\lambda_U\}$ are the major parameters to be defined. Let d be the average demand vector, R the turn probability matrix, $f = [I - R']^{-1} d$ the average flow vector, S the saturation flow rate matrix and $R(l,m)f_l$ the average required rate of turns to satisfy the demand. The FT timing λ accommodates the demand when the following equalities are satisfied

$$\sum_{l,m} \lambda_i C(l,m) U_i(l,m) > R(l,m) f_l \text{ and } \forall l, m \tag{1}$$

$$\sum_i \lambda_i = 1 - L/T, \text{ with } \lambda_i > 0 \text{ and } \forall l, m \tag{2}$$

In [1] is discussed how Eq. (1) is sufficient for the system stability. As (1) shows, in order to design a stable FT scheme, the knowledge of the demand vector d, the turning probability matrix R and the saturation flow rates C are required. When (1) can be satisfied infinitely many λ are feasible. An "optimum" vector λ can be obtained when maximising the minimum *excess capacity*

$$\min_{l,m} \sum_i \lambda_i C(l,m) U_i(l,m) - R(l,m) f_l. \tag{3}$$

2.2.2 Offsets

The term offset defines the time relationship, expressed in either seconds or as a percent of the cycle length, between coordinated phases at subsequent traffic signals. The offset is dependent on the offset reference point, which is defined as that point within a cycle in which the local controller's offset is measured relative to the master clock. The master clock is the background timing mechanism within the controller logic to which each controller is referenced during coordinated operations. This point in time (midnight in some controllers, user defined in others) is used to establish common reference points between every intersection. Each signalized intersection will therefore have an offset point referenced to the master clock and thus each will have a relative offset to each other. It is through this association that the coordinated phase is aligned between intersections to create a relationship for synchronized movements.

Remark In the following, for the purpose of simplification, the term FT is employed for defining the Fixed-Time Offset Policy (FT-Offs).

2.2.3 Max-Pressure and Adaptive Max-Pressure Plans (MP-AMP)

Max-Pressure is a feedback control policy selecting a stage to actuate as a function of the upstream and downstream queue lengths related to the intersection. Two remarkable properties distinguish this version of MP:

- no knowledge of average demand is necessary
- Max-Pressure is stable whenever the demand can be stabilised

At any time, the MP control selects the stage U^* involving the max gain w, that is

$$U^*(q)(t) = \text{argmax}\{w(q(t), U), U \in \mathcal{U}\}, \textit{ MP stage} \tag{4}$$

with

$$w(q(t), U) = \sum_{(l,m)} \varsigma(l, m)(t) S \circ U(l, m)(t) \tag{5}$$

where if $r_{(m,p)}$ denotes the probability of queue $q_{(m,p)}$ to be selected when a vehicle is located on link m, then

$$\varsigma(l, m)(t) = \begin{cases} q_{(l,m)}(t) - \sum_{p \in \mathcal{O}(m)} r_{(m,p)} q_{(m,p)}(t), \\ \qquad\qquad \text{if } q_{(l,m)}(t) > 0, \\ 0, \text{ otherwise.} \end{cases} \tag{6}$$

A priori, the MP policy does not require any computation actuation duration since it will automatically select the stage requiring green time. More precisely, supposing that too little green time is allocated to the currently employed MP stage. Since the control is going to be revised before the end of the actuation duration of the present MP stage, the feedback policy will determine the stage requiring green time for the next period. Thus, if the current stage still requires actuation, MP control will extend its actuation by selecting it for the next period. On the other hand, if large actuation durations are considered, then it is probable that within the new control revision a different MP stage will be decided. At any decision moment, the MP stage depends on the actual intersection state. As discussed in [1] if there exists a stabilising FT control then MP algorithm will also stabilise the system. Moreover, if matrix R can be consistently estimated it may be employed in (4) without affecting the stability property.

Let $a(l, i)(t)$ denote the vehicle arrivals at time t associated with phase (l, i), where (l, i) is a possible turn from link. The estimated turn ratio value $\hat{r}(l, m)$ of phase (l, m), can be computed as follows

$$\hat{r}(l, m) = \frac{\sum_t a(l, m)(t)}{\sum_k \sum_t a(l, k)(t)} \tag{7}$$

Employing relation (7) in Eq. (5), the *adaptive Max-Pressure* stage is defined.

2.2.4 Max-Pressure Practical Control (MP-Pract)

Aiming at limiting frequent stage switches, the previously presented MP scheme is parametrised and the Max-Pressure Practical algorithm is introduced. Thus, the MP stage is evaluated as frequently as desired but the new taken decision is applied only if it exerts significantly larger pressure

$$\max_U w(U, q(t)) \geq (1 + \eta) w(U^*, q(t)). \tag{8}$$

Parameter η is related to the desired degree of stage switches.

3 Towards a Realistic Model: DES Versus Traditional Approaches

3.1 *Discrete event formalism*

From a formal point of view Discrete Event Systems (DES) are complex dynamic systems whose state variables can take both discrete and continuous values but the state space is discrete or at least it contains several discrete variables. However, the state transition mechanism is *event driven*, that is changes on the system state are due to the occurrence of *asynchronous* events in contrast with Continuous Variable Dynamic Systems (CVDS) where the state space is continuous and time step is driving all state transitions. Thus, traditional modelling through differential or partial differential equations is not anymore appropriate for studying DES. Furthermore, the notion of time interferes when appraising the system performance. More precisely, responses to questions such as *the time spent by a system on a particular state* or *how fast a system can reach a given state, at what time a particular event will occur* etc. are often required in a DES study.

The traffic control problem introduced in Sect. 2 falls into the category of DES. Obviously *asynchronous, random* behaviour is involved (transport demand, vehicle arrival at intersections, incidences, etc.). Moreover, frequent *synchronisation* features are taking place. For example, multiple constraints have to be satisfied for a vehicle to cross an intersection. *Concurrence and parallelism* actions are also common. Since only compatible movements may be simultaneously actuated the presence of traffic lights excludes all vehicles to cross intersections at the same time.

3.2 *Formalising DES*

As previously discussed, in CVDS systems since time derivatives can naturally be defined, differential equation based-models can be used for describing the system dynamics, where the state equation and the initial conditions are stated in the form of

$$\dot{x}(t) = f(x(t), u(t), t), \; x(t_0) = x_0 \tag{9}$$

Since DES based dynamic systems are *event-driven*, the notion of time is not an independent variable. Hereafter, some usual approaches for the study of DES are briefly presented.

3.2.1 Discrete Event Simulations

Broadly speaking, the system evolution is virtually reproduced according to the modelled conceptual framework. The notion of time is frequently involved and "future" events are stored in an event pile until all necessary conditions are satisfied for their realisation. No analytical problem formalisation exists between entries and exits (observations) and consequently detailed statistical analysis plays an important role to the system comprehension. Any kind of methodological statistical tool can be associated and this approach can only be limited by the complexity of the program modelling and the implementation duration.

3.2.2 Perturbation Analysis (PA)

This approach related to optimisation can be considered as a technique for computing the sensitivity of involved variables regarding some parameters. A nominal trajectory is required (obtained for example by simulation) and trajectory modifications are obtained by variations of the considered parameter. A type of *stochastic gradient* can be obtained (by finite differences) which may be used in optimisation the related iterative algorithm.

3.2.3 Petri Nets (P/T Nets)

A Petri net is a graphical mathematical language employed for distributed systems, specifically for describing synchronisation and concurrent phenomena. Temporal P/T nets allow to take into consideration performance evaluation aspects and form concise and efficacious means for the study of dynamic systems. However, when the related system complexity is increased an hierarchical approach is highly recommended.

3.2.4 Dynamic Algebraic Models: Diode Algebra

When considering a *quantitative* aspect, certain DES class systems can be described by mathematical models similar to the ones utilised by the Optimal Control Theory. Nevertheless, *concurrent* behaviour arbitrated for example by priority and scheduling policies should be excluded in order to limit *synchronisation* phenomena.

3.3 Why Simulations?

When dealing with stabilising traffic signal plans, substantiated responses should be provided to a pattern of issues such as:

- which set of simultaneously compatible movements should be actuated at a given time?
- which green duration should be allocated to the selected set of movements?
- how delays of queued vehicles can be controlled?
- when adaptive controllers, how frequent the decision related to the selection of the actuated stage should be defined? Etc.

A model and all the resources of control theory are necessary in order to predict the system performances. Obviously, we are handling a quite complex spatiotemporal decision making problem for which it is almost impossible to write a mathematical model precisely describing its evolution. So what could the answers be and how do we know if the correct ones are being confronted?

Gaining direct experience on a *trial and error* basis would imply raised "risks" especially when poor decision making. A reliable simulation tool could accurately provide deep comprehension when exploring the many aspects of the manifested system behaviour according to the applied strategy, reproducing precisely all stochastic and unforeseen features frequently presented in road networks. Moreover simulation is the only way to reproduce various scenarios with a single factor modified at each implementation. This is a fundamental property when searching optimal traffic policies.

3.4 Addressing Traffic Control Decisions

The employed approach for the study of traffic signal coordination involves a two a level decision problem. The *dimensioning part* accounts for multiple components referring to:

- the network model (topology, link characteristics, turn pockets, probability laws defining mean travel times, etc.)
- demand model (geometry and intensity, diversions etc.)
- vehicle routing (Origin-Destination (OD) matrix, trajectory paths, turning probabilities etc.)
- operating mode determining the desired control policy to be employed
- initial state of the system, etc.

The *real time management* stands for all the control algorithms ruling the system. Thus, for each decision type a whole set of different controls can be modelled and evaluated. Consequently, optimisation of this part quantifying the derived benefits of the selected policy for each decision category is required.

Let us consider the case of the decision related to an optimal selection of the control defining intersection timing plans. Urban Traffic Control is a method for coordinating traffic signals by employing timing plans loaded on a central computer. These schemes are using the same cycle after cycle and the green splits on each approach remain unchanged. This method is frequently employed in Town Centre locations and on ring roads aiming to ensure that the linking from one junction to the next remains constant throughout the day. However, if no traffic was being detected on an approach then, depending on the plan, it can be possible to skip a stage and move to the next stage early to reinforce a busier approach. Different schemes of Urban Traffic Control, such as offset transitioning algorithms for coordinating traffic signals while improving *early return to green phenomena* can be conceived and tried in order to appraise the policy results within its ability to improve the network behaviour under various traffic patterns.

When adaptive (distributed) traffic control strategies need to be modelled, performed and optimised a deep understanding of the network behaviour within the same traffic context is required.

Furthermore, centralised versus decentralised policies can be compared as well combinations of signal plans applied to specific road regions can be considered. Additionally, involving multiple traffic conditions such as varying road access, temporary incidents limiting or prohibiting vehicle movements, demand fluctuations etc. the impact of the integrated strategy to the scope and scale of the involved challenge can be quantified. Measuring the relationship between the related control effectiveness and implied costs of each strategy could provide the required robust *benefit-cost* assessments in order to ensure that the advantages of the strategy plan justify the related costs and preserve an adequate network performance.

For all the previously referred cases the *dimensioning part* remains fixed while different control schemes are applied. Thus, each signal policy performance can be quantified under an identical traffic environment. If the control behaviour under varying traffic conditions needs to be evaluated, then for a given configuration of the *real time management* various traffic contexts within the *dimensioning part* can be considered (demand variation, different types of congestion etc.).

4 Performance Measurement and Observation Bias

Congestion measurement has a crucial impact on the congestion management and can be carried out for several levels and various purposes. Popular congestion metrics are based on speed, road access, delay, costs etc. Each such measure will raise a different congestion statement and may imply a completely different policy approach. At a *micro level* road decision makers require metrics addressing operational issues on network links such as average speed versus rated speed, traffic density versus traffic capacity, speed/flow relationships on network links etc. Thus, by detecting bottlenecks specific link performance can be compared to overall network performance. Even though, these measures are difficult to aggregate and do not directly express road users needs. On the other hand, measuring speeds on specific links is not necessarily a representative indicator for a deep understanding especially in dense arterials where congestion and consequently delays are generated by intersections and specific road access points. Measuring delays is important for an efficacious road control, where issues of how large vehicle volumes impact travel times, are required. Moreover, drivers are concerned by *trip-based* metrics. There are not necessarily *better* congestion metrics than others. However, it may exists a better matching between the selected indicators and the desired outcomes. Thus, a congestion metric should be selected not simply because it is available but mostly because it contributes to quantify a specific purpose.

Furthermore, several techniques are available for measuring congestion from raw data. Point detection using loops, video detection, radars, etc. Vehicle-based detections employing probe-vehicles, cell phones, satellite information etc. Independently from the information means, what traffic managers observe and what indicators are communicated should be one thing. The risk of bias is mostly implied when it comes to interpret and consequently quantify what one sees without considering accuracy measurements. In this study, indicators based on average travel times (trips or specific links) and delays are going to be considered for measuring the network performance within each traffic context.

5 From Theory to Application

5.1 *Study Area*

The operating region is a section of the Huntington-Colorado arterial near the I-210 freeway in Los-Angeles. The network as depicted in Fig. 1 is viewed as a directed graph comprised of a set of 16 signalised intersections, 76 links of which 22 are entry links and 24 are exit ones while 179 different turn movements are possible. All events take place at nodes while vehicles travel on edges. With each internal network link, a random travel time is associated based on the mean value of the related free flow speed. However, the realised vehicle travel time depends upon the current state of the link.

5.2 *Demand: Intensity and Geometry*

Vehicles enter the network at entry links in a Poisson stream with specified demand rates. The elapsed time between successive arrivals at link i follows an exponential law of parameter λ_i. Moreover, vehicles join exit or internal links according to turning probabilities. Distinct vehicle queues are associated with any entry or internal link according to the number of possible movements from the related link.

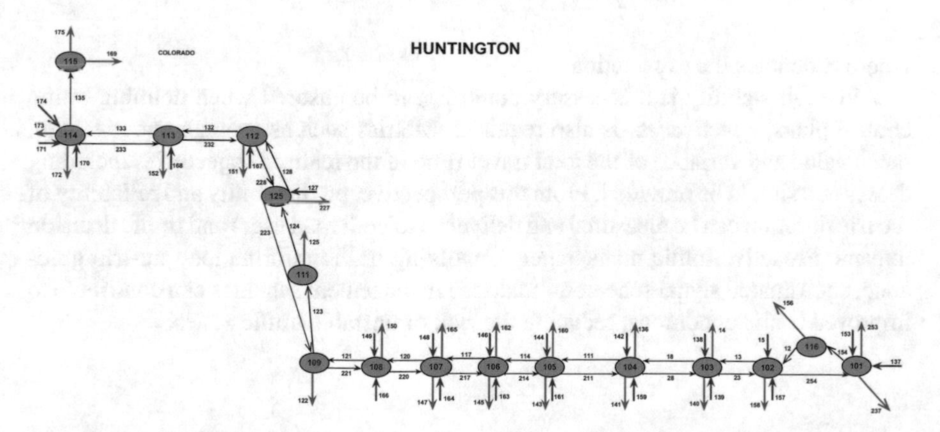

Fig. 1 Directed network graph

Vehicles join the queue corresponding to the desired turn. At any time, a control policy operates a stage for a duration of time. The operation of a movement causes the corresponding queue to be served. The service rate, called the saturation flow rate, is pre-specified and depends on the geometry of the intersection. When a non-empty queue is served vehicles move towards the downstream queue and join it by the completion of the realised travel time. Each link has a fixed finite storage limit depending on the link geometry. When the downstream link reaches its max vehicle storage limit, the upstream queue is blocked even if the control permits the movement.

5.3 System Stability

Consider d the vector of demand rates such that

$$d_l = \begin{cases} > 0 \text{ if } l \text{ entry link} \\ = 0, \text{ otherwise.} \end{cases}$$

and $R(l, m)$ the probability of a vehicle located at link l to turn towards link m, with $R = \{R(l, m)\}$. The vector of mean link flow values $f = \{f_l\}$ satisfies the conservation law $f = R'f + d$, where R' denotes the transpose of R. Thus, $f = [I - R']^{-1} d$ and $R(l, m)f_l$ is the mean rate of turns from link l towards link m. A timing policy able to accommodate demand d and consequently maintains vehicle queues stable operates phase (l, m) at a rate at least equal to $R(l, m)f_l$ in which case the following condition holds true

$$\sup_T T^{-1} \sum_{l,m} \sum_{t=1}^{T} \mathbf{E}q(l, m)(t) < \infty \quad (10)$$

where $\mathbf{E}$ denotes the expectation.

Although stability is a necessary condition to be ensured when defining traffic control plans, *effectiveness* is also required. Metrics such as average queue length, mean value and variance of the total travel time of the realised trajectories including delays etc. should be provided. From this perspective, predictability and reliability of the trip duration can be measured and delivered to both road users and traffic decision makers. Proactive traffic management, involving traffic information, pre-trip guidance, coordinated signal schemes, incidence management schemes can contribute to improved traffic conditions, reducing the risk of unstable traffic zones.

6 Managing Traffic Demand

6.1 *Stability Under Pre-timed Policy*

A Fixed-Time plan $\lambda = \{\lambda(n)\}$ where n is the intersection index is obtained by solving the LP problem maximising the excess capacity as defined in (3) under constraints (2). The considered demand vector d generating 14,344 vehicles per hour is feasible for (2). The expected stability is verified as depicted in Fig. 2 where the sum of all the network queues $\sum_{l,m} q(l, m)(t)$ remains bounded for any value of t during the 3 h simulation duration. The average queue length values 189 and is indicated in Fig. 2 by the red dashed line.

6.2 *Stability Under MP Schemes*

Figure 3 depicts the network behaviour when MP policies define the timing plans and in which the number of control revision varies between four and ten times per cycle. In the same plot the average length of all the sum of all the network queues is also indicated. As expected the network is stable for any MP version. Moreover, one observes that the queue length is inversely proportional to the number of control decisions per cycle, theoretical property discussed in [1]. Thus, as the frequency of control decisions is increased the average queue size decreases. The average total network queue value is 147 when four control revisions per cycle while this number

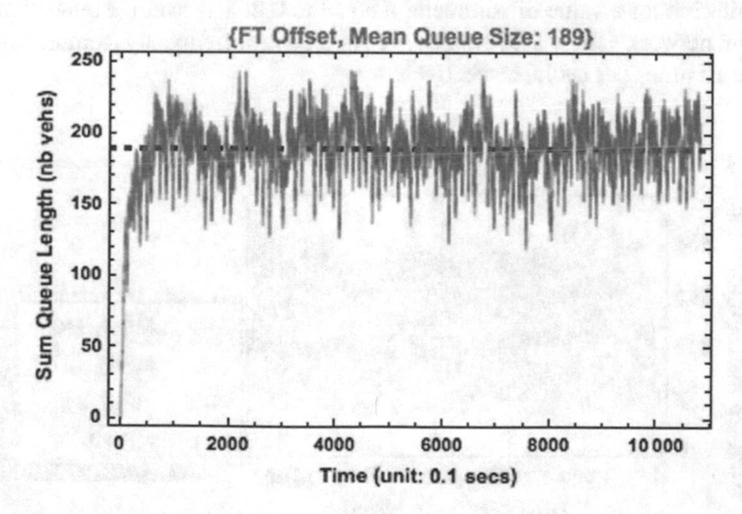

Fig. 2 Sum of all queues under pre-timed plan

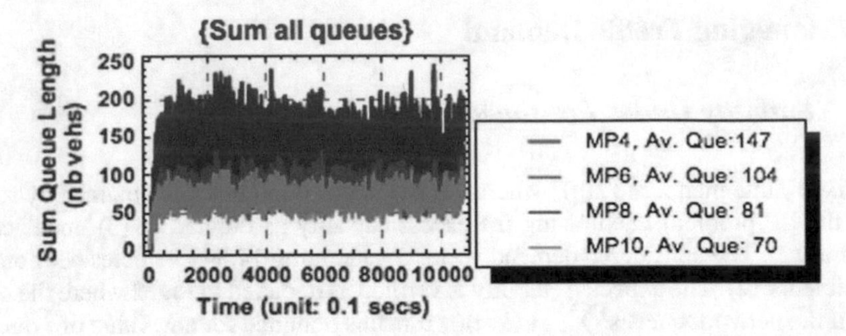

Fig. 3 Sum of all network queues under MP: 4–6–8–10 cycle revisions

becomes equal to 104 under six control decision updates in order to fall down to eighty one and seventy when eight and ten control updates.

6.3 *Stability Under MP-Pract Plans*

Changing stages incurs loss time due to the red clearance period applied between the two consecutive stage switches. Instead different versions of the MP-pract control policy are going to be employed for three different values of the η parameter.

Hence, Fig. 4 illustrates the evolution of the sum of the network queues when the control is revised four (red curve), six (cyan plot), eight (purple plot) and ten times (yellow curve) per cycle. However, the new taken decision is applied under the MP-pract criterion for a value of parameter η equal to 0.2. The average queue length of the entire network values 149, 104, 81, 70 for a control frequency examination of 4, 6, 8 and 10 times per cycle, respectively.

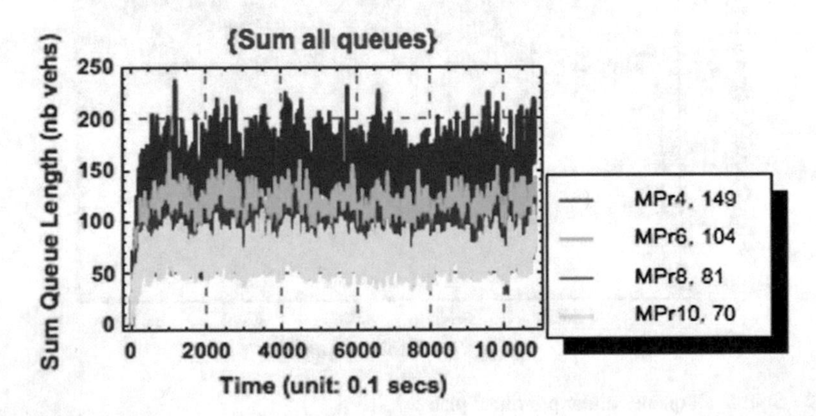

Fig. 4 Sum of all network queues under MP-pract, $\eta = 0.2$: 4–6–8–10 cycle revisions

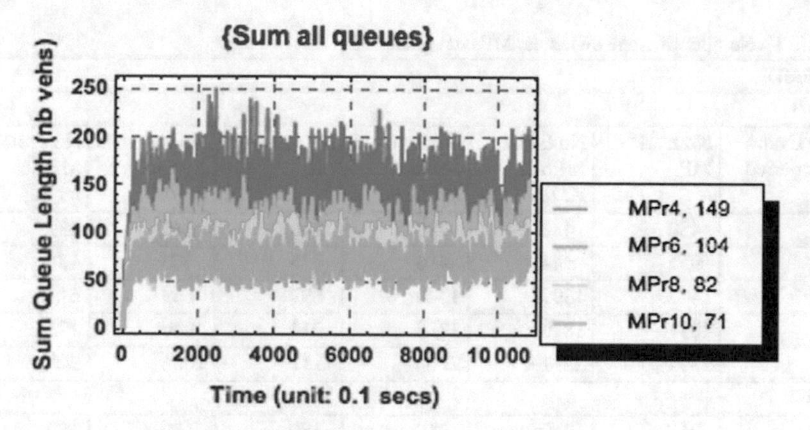

Fig. 5 Sum of all network queues under MP-pract, $\eta = 0.4$: 4–6–8–10 cycle revisions

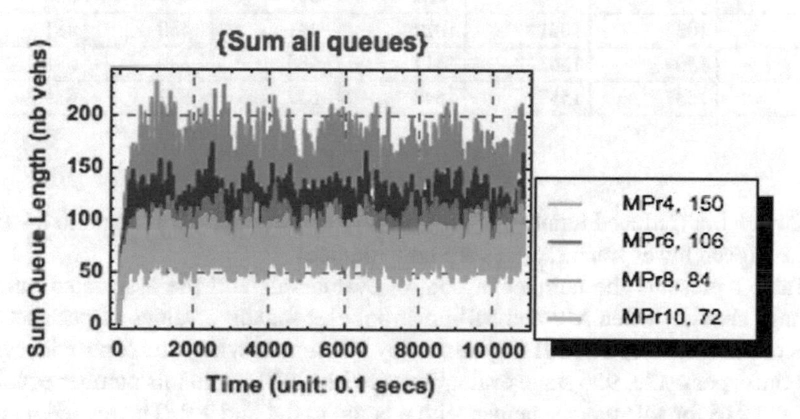

Fig. 6 Sum of all network queues under MP-pract, $\eta = 1.2$: 4–6–8–10 cycle revisions

Figure 5 depicts the network evolution when traffic signals are obtained by the previously presented MP-pract schemes when parameter η values 0.4. The average length of the sum of the queues values 149, 104, 82 and 71 when four, six, eight and ten control revisions.

Similarly, Fig. 6 represents the sum of all the queues when MP-pract signals rule the network and parameter $\eta = 1.2$. The mean measured size of the sum of all the network queues now becomes equal to 150, 106, 84 and 72 under four, six, eight and ten stage revisions per cycle.

Obviously, for each version of MP-pract timing plans, regarding the number of signal revisions and the value of parameter η, the network remains always stable, although the number of signal switches is now reduced. Additionally, the average size of the sum of all the network queues remains practically unchanged for the different values of parameter η. Thus, when the MP-pract stage is revised eight times per cycle, the average queue length values 81 when $\eta = 0.2$ while this value becomes

Table 1 Number of stage switches: MP-MP-pract

No Eval. per period	No Eval. MP	No Eval. MPract $\eta : 0.4$	No switches MPract $\eta : 1.2$	No switches MP	No switches MPract $\eta : 0.4$	No switches MPract $\eta : 1.2$
Node ID						
10001						
2	480	480	480	240	240	239
4	956	944	916	475	463	434
6	1422	1395	1346	699	669	619
8	1877	1810	1738	914	846	772
10	2314	2211	2108	1111	1006	901
10002						
2	360	360	360	180	180	180
4	708	711	701	347	350	340
6	1033	1021	1022	491	480	481
8	1309	1303	1311	586	580	588
10	1537	1537	1544	632	632	639

82 for $\eta = 0.4$ (reduced number of stage switches) and finally it is equal to 84 when $\eta = 1.2$ (even fewer stage changes are now implied).

Table 1 presents the number of control evaluations and the associated number of stage changes when MP and MP-pract policies for three values of parameter η. Hence, when intersection 101 is governed by MP signals where the control is revised four times per cycle, 956 stage evaluations are involved while this number equals to 944 and 916 for MP-pract schemes with η equal to 0.4 and 1.2. The related number of stage switches is 475 under MP plans and it is reduced to 463 and 434 when MP-pract and $\eta = 0.4,\ 1.2$ respectively. However, when ten control revisions per cycle are considered, 2314 and 2108 evaluations are required under MP and MP-pract with $\eta = 1.2$, respectively. In this case, the number of stage changes under MP is 1111 while it is equal to 901 with MP-pract which is strictly inferior to 956 control evaluations which are required when MP with four stage revisions per cycle.

7 Appraising FT Policy Versus MP and MPract Plans

7.1 Probability of the Queue Size

The sum of all the network queues is considered for which the most probable values are going to be compared according to different timing schemes. Figure 7 gives the probability of each potential queue length when Pre-timed and MP plans with four, six, eight and ten control revisions per cycle are considered. Although all

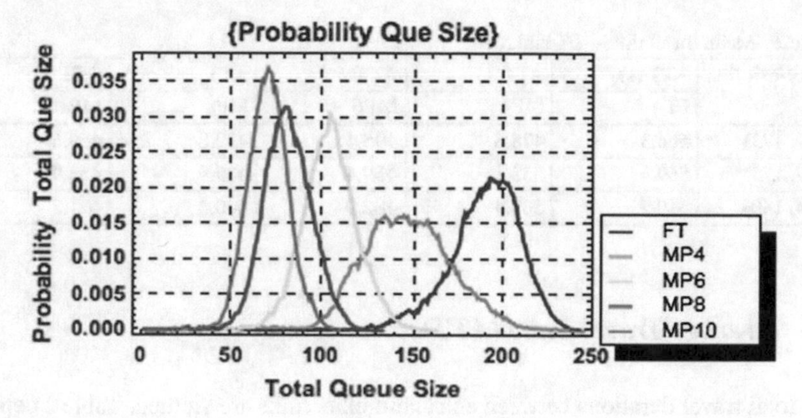

Fig. 7 Probability queue length: FT and MP 4, 6, 8, 10

signal schemes accommodate the demand ensuring the network stability, MP policies (sorted increasing order: MP10, MP8, MP6, MP4) involve the smallest values for the sum of all the network queues with the higher probability. Thus, under MP10 signals, a total queue length value equal to 70 is the most probable one while this value becomes 190 for the FT policy. These results arc in accord with Figs. 2 and 3 representing the evolution of the sum of all the network queues under FT and MP controls.

Figure 8 illustrates the probability of each possible value for the sum of all the network queues under FT and MPract plans ($\eta = 0.4$). One may observe similar results as in the case of MP control. Frequent revisions of the feedback policy imply smaller queue sizes all associated with increased probability values, even though the new decided plan is not applied within every control evaluation.

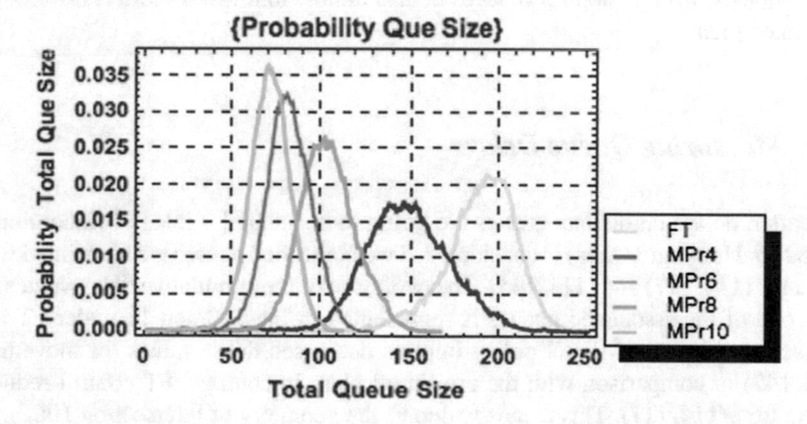

Fig. 8 Probability total queue length: FT and MPract 4, 6, 8, 10

Table 2 Mean travel time—FT and MP

Entry-exit link	ATT (s)	ATT	ATT	ATT	ATT
	FT	MP4	MP6	MP8	MP10
(137, 173)	566.3	478.6	495.4	459.8	459.6
(137, 175)	569.4	558.1	459.4	439.9	433.9
(174, 140)	510.5	587.9	482.9	440.4	421.5

7.2 *Average Travel Time (ATT)*

The total travel durations between entry and exits links are studied. Table 2 depicts the average value of the mean travel time between three entry and exit network links. These durations are computed for the FT and MP policies. For most movements, the feedback plans imply reduced travel durations. For the case of MP4 and phase (174, 140), the average travel time 587.9 s exceeds the related one under FT mostly because travel times are stochastic for the associated implementation.

7.3 *Total Delay Measurement*

Considering a vehicle trajectory, that is a sequence of internal links crossed by a vehicle between the selected entry and exit link, delays are measured. In this model a *trajectory delay* expresses how long cars which followed the corresponding trajectory, were facing red lights and consequently were unable to move. The associated delay distribution is computed and the CDF function is considered. Figure 9 depicts the CDF delay function, for the MPract (purple curve) and FT (green plot) policies.

Obviously, MPract control reduces delays almost four times more regarding the pre-timed plan.

7.4 *Measuring Queue Delays*

Hereafter, delays on distinct queues are going to be studied. Link 114 (incoming at node 106, Huntington area) is considered. Two controlled phases are associated with link 114: (114, 117) and (114, 145). The evolution of the cumulative delay values for each one of the associated queues is represented in Figs. 10 and 11 under FT and MPract signal plans. MPract policy implies decreased delay values for movement (114, 145) in comparison with the pre-timed plan. In contrast, FT control reduces delays for $q(114, 117)$. This is mostly due to the geometry of intersection 106.

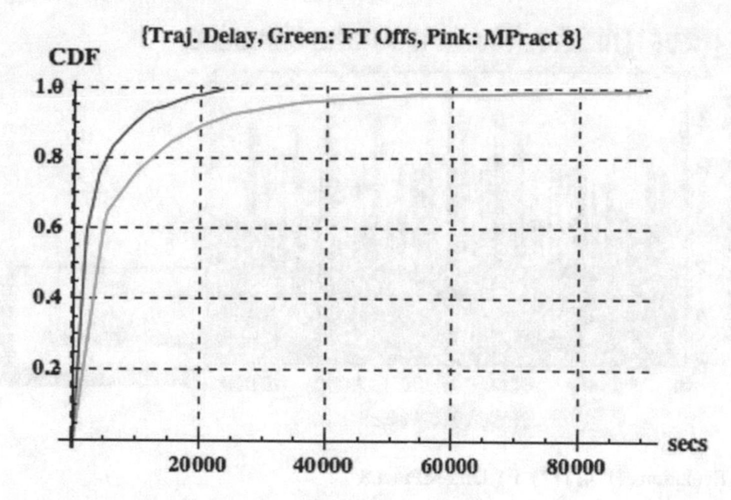

Fig. 9 CDF trajectory delays: MPract (*purple*), FT-Offset (*green*)

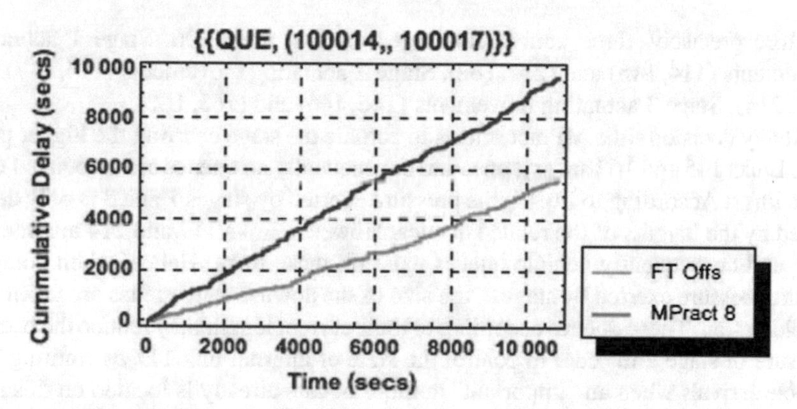

Fig. 10 Cumulative delay: phase (114, 117)

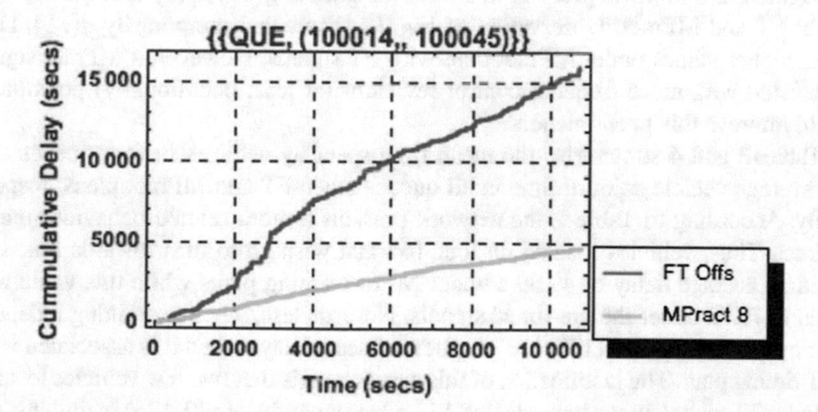

Fig. 11 Cumulative delay: phase (114, 145)

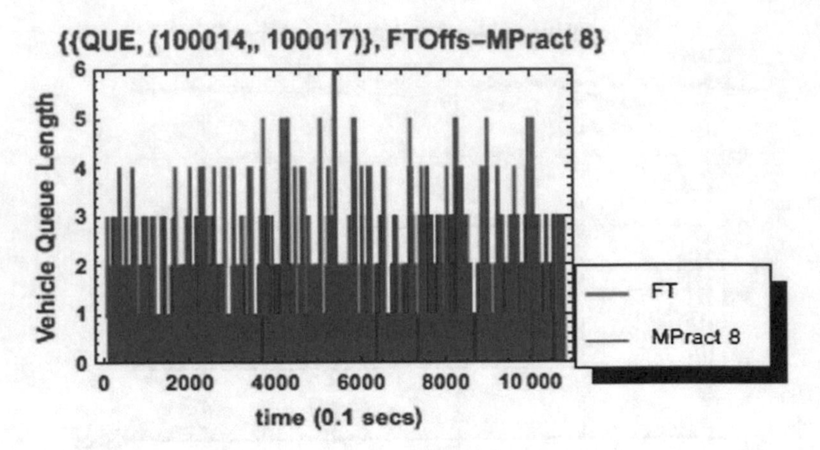

Fig. 12 Evolution $q(114, 117)$, FT Offs-MPract 8

More precisely, three concurrent stages exist at node 106. Stage 1 actuating movements (114, 145) and (217, 162). Stage 2 actuating movements (114, 117) and (217, 214). Stage 3 actuating movements (146, 145) and (163, 162).

At any decision time MPract selects to actuate the stage exerting the higher pressure. Links 145 and 162 are exit ones and consequently no queues are associated with these links. According to Eq. (5) the pressure exerted by stages 1 and 3 is only determined by the lengths of the related queues. However, links 117 and 214 are internal ones and consequently vehicle queues exist on these links. Hence, when computing the pressure exerted by stage 2, the size of the downstream queues are taken into consideration. These queues according to their current length may reduce the exerted pressure of stage 2 in order to control the state of internal link 117 by limiting new vehicle arrivals when an "important" number of cars already is located on this link. Consequently, stage 1 (as well stage 3) receive more green time regarding stage 2.

Figures 12 and 13 depict the evolution of queues $q(114, 117)$ and $q(114, 145)$ under FT and MPract. Observation of Fig. 12 shows that temporarily $q(114, 117)$ takes higher values under MPract than with FT signals. However, a MPract signal associated with more frequent control revisions (at least occasionally) potentially could improve this phenomenon.

Tables 3 and 4 summarise the mean time spent by vehicles in four queues and the average vehicle sojourn time in all queues when FT and MPract plans, respectively. According to Table 4, the network presents a more refined behaviour under MPract. Thus, vehicles located on link 164 and wishing to turn towards link 120 faced an average delay of 49.89 s under MPract timing plans while this value was equal to 79.72 under the pre-timed signals. Nevertheless, when examining independent movements, phase (170, 173) implies reduced delays when it is associated with a FT timing plan. The justification of this assessment is that two few vehicles located on link 170 wish to move towards link 173. Consequently, $q(170, 173)$ maintains low values. Similarly for phase (174, 233) which is simultaneously actuated with phase

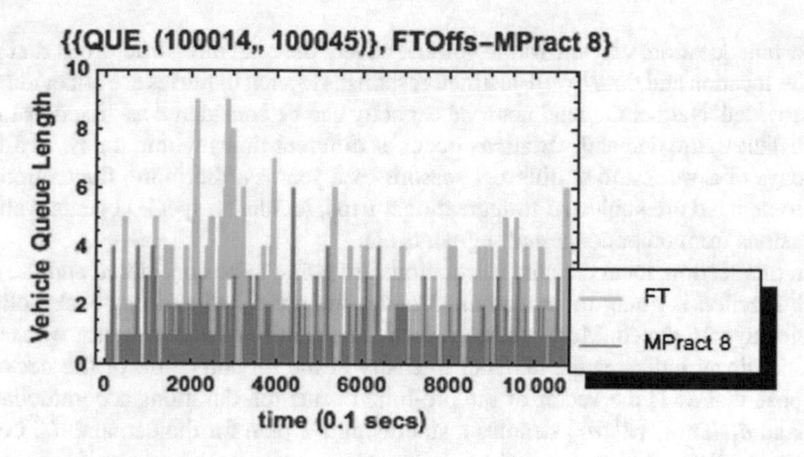

Fig. 13 Evolution $q(114, 145)$, FT Offs-MPract 8

Table 3 Mean time spent by vehicles in distinct queues

Incoming link	Outgoing link	Mean veh. sojourn time (s)	Mean veh. sojourn time (s)
		MPract	FT Offset
164	120	49.96	79.72
167	228	12.16	51.32
117	120	10.72	33.35
170	173	78.46	49.61

Table 4 Average mean time spent by vehicles in all queues

Average mean vehicle sojourn time in queues	MPract (s)	FT Offset (s)
	12.05	22.32

(170, 173). Thus, the only stage actuating movements (174, 233) and (170, 173) is not frequently selected. Associating a different weight to these phases would lead to reduced delay values.

8 Network Behaviour Under Demand Variation

Regarding congestion, various more or less sophisticated definitions can be formulated incorporating traffic engineering constraints, policy maker decisions or economics aspects etc. but what remains common is that congestion is caused by traffic but also impacts that same traffic. Many definitions focusing on the proximate congestion causes, that is increased demand related to the road capacity, imply another important question: why the traffic volume covers the road infrastructure at that particular location and time? Well-justified responses to such inquiries are not evident to be provided. Neither demand not road capacity can be considered as "fixed" values. Important traffic demand variations occur at different times within a day, at different days of a week and at different seasons of a year. Furthermore, fluctuations in traffic demand are subjected to recreational trips, incidents, special events, vehicle diversions from other congested segments etc.

In this section, local demand fluctuations are going to be considered and the network reaction is going to be examined under the pre-timed and feedback policies previously introduced. More precisely, let d_1 be the initial demand vector whose the ith coordinate indicates the demand intensity at the ith entry link of the network. Suppose that λ^1 is the vector of the pre-timed actuation durations accommodating demand d_1. Thus, $\{\lambda^1, d_1\}$ denotes a stabilising FT plan for the demand d_1. Let d_2 denote the demand vector such that

$$d_2(k) = \begin{cases} d_1(k), & \text{if link } k \text{ is not an entry link} \\ & \quad \text{at node 101} \\ \neq d_1(k), & \text{otherwise.} \end{cases}$$

In other words, d_2 represents the new demand after the variation of the demand intensity at node 101. Suppose that for 3 h a demand intensity equal to d_1 is applied to the network, (period [0, 10800)), which is followed by a demand level equal to d_2, for the next 3 h, (period [10800, 21600)) while the employed pre-calculated values of the FT plan remain unchanged during the entire period [0, 21600). Figure 14 illustrates

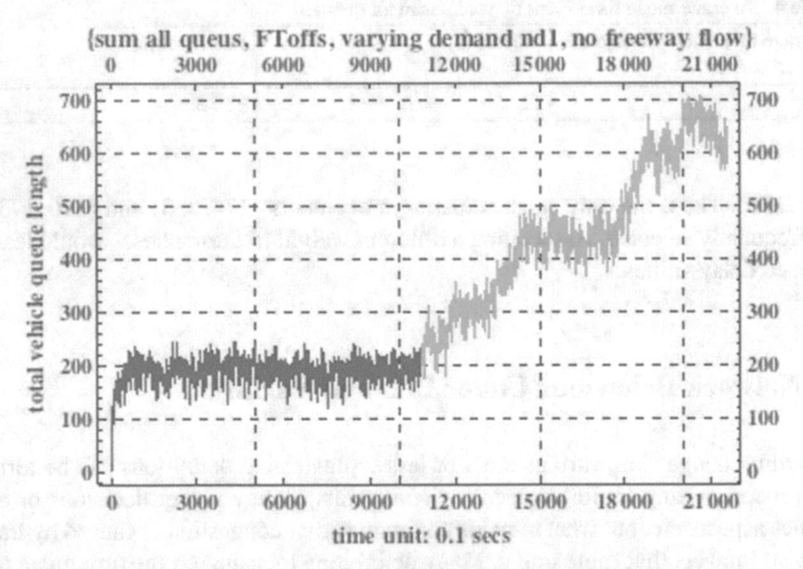

Fig. 14 Network evolution under varying demand: Sum of network queues: $\{\lambda^1, d_1\} \rightarrow \{\lambda^1, d_2\}$

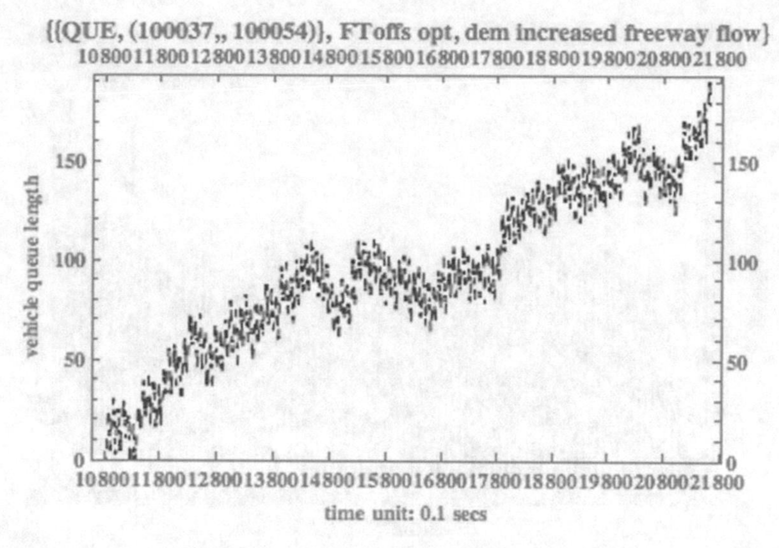

Fig. 15 Evolution of $q(137, 154)$ under demand d_2: applied policy $\{\lambda^1, d_2\}$

the implied network behaviour during the 6 h. Obviously, while the FT signals are associated with demand d_1, that is during the first 3 h, the network remains stable as the orange plot shows. Nevertheless, when demand becomes d_2 the pre-timed plan λ^1 cannot maintain the network stability (cyan curve).

Suppose now, that $\lambda^{'}(101)$ is a new stabilising pre-timed signal plan computed for the intersection 101 where the demand variation occurs and denote the new FT control vector as λ_2. Thus,

$$\lambda_2(i) = \begin{cases} \lambda^1(i), & \text{if node } i \neq 101 \\ \lambda^{'}(101), & \text{otherwise.} \end{cases} \tag{11}$$

One of the most unstable movements at node 101 during the demand variation and under policy λ^1 is the one corresponding to queue $q(137, 154)$. Figure 15 represents the evolution of phase $(137, 154)$ under $\{\lambda^1, d_2\}$. However, when FT policy $\{\lambda_2, d_2\}$ is considered, queue $q(137, 154)$ becomes stable (as all other movements at intersection 101). Figure 16 reveals the stability of $q(137, 154)$ when pre-timed actuations defined by $\lambda^{'}(101)$ are applied to node 101. Similar results hold true for all the other queues associated with this node. However, the network behaviour remains unstable under control $\{\lambda_2, d_2\}$ as Fig. 17 illustrates (purple curve).

More precisely, Fig. 17 depicts the evolution of the sum of all the network queues before and during the demand variation when the initial and λ_2 FT plans are employed respectively on each one of the two periods. Hence, during period $[0, 10800)$ demand d_1 is considered and the stabilising pre-timed plan $\{\lambda^1, d_1\}$ is applied. The network behaviour during this period is depicted by the orange plot. As expected the plan λ^1 accommodates demand d_1. However, when the demand becomes d_2 and the new

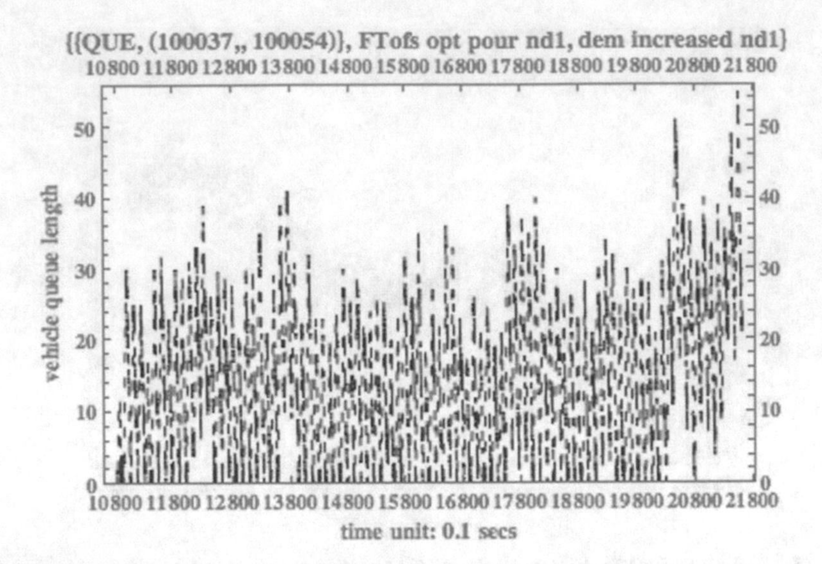

Fig. 16 Evolution of $q(137, 154)$ under demand d_2: applied policy $\{\lambda_2, d_2\}$

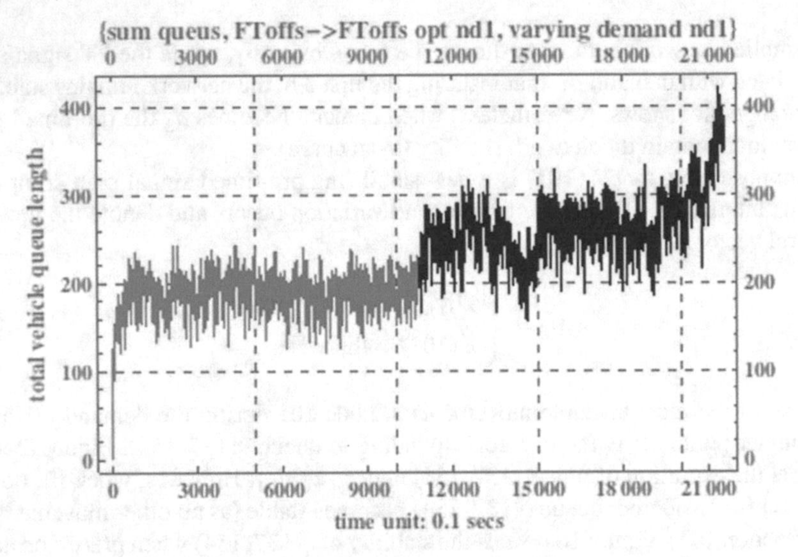

Fig. 17 Network evolution under varying demand: Sum of network queues: $\{\lambda^1, d_1\} \rightarrow \{\lambda_2, d_2\}$

actuation durations as defined by vector λ_2 are applied the network becomes unstable, (dark purple curve), although this control stabilises the new demand level at intersection 101. Indeed, this is an expected result. Since demand on entry links related to node 101 are modified, the outgoing vehicle flows from this intersection will also be different now. Consequently, the vehicle arrivals at the internal network

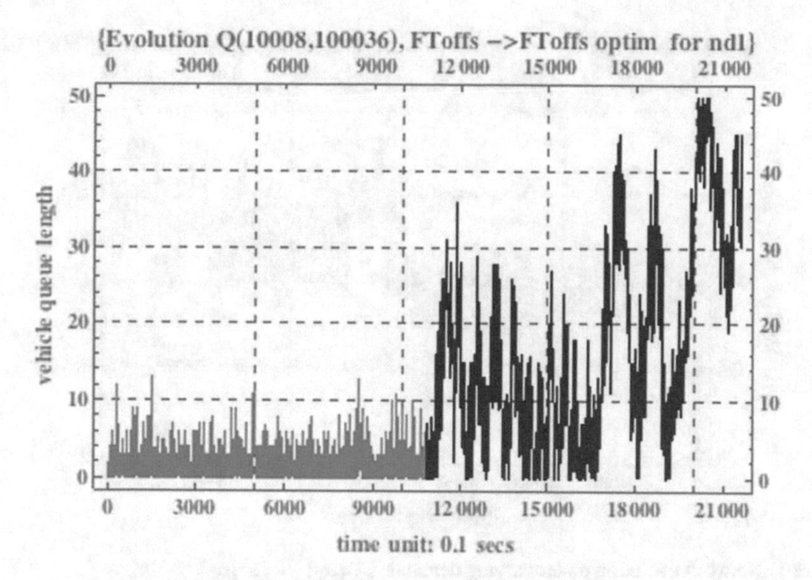

Fig. 18 Evolution $q(108, 136)$ under varying demand: $\{\lambda^1, d_1\} \rightarrow \{\lambda_2, d_2\}$

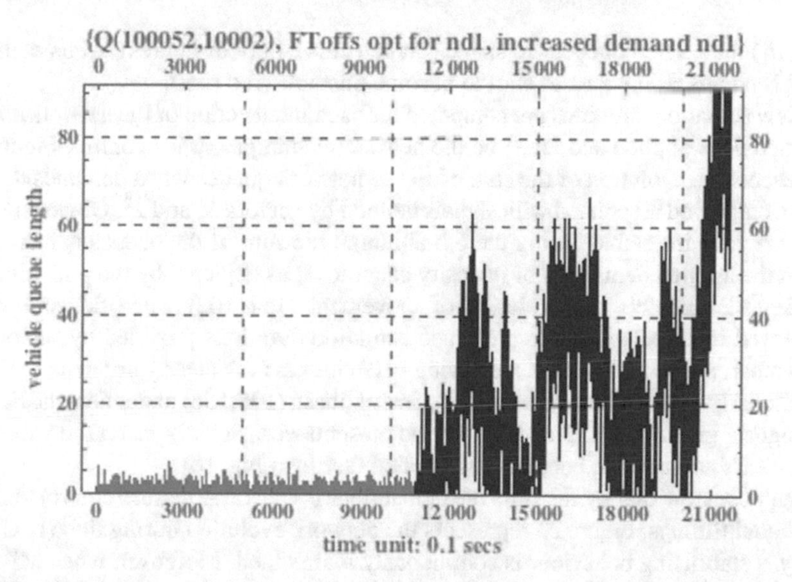

Fig. 19 Evolution $q(154, 102)$ under varying demand: $\{\lambda^1, d_1\} \rightarrow \{\lambda_2, d_2\}$

links will also vary. Obviously, the actuation durations defined by vector λ^1 for any node different to node 101 cannot accommodate the new demand (see also Eq. (11)).

In fact, Figs. 18 and 19 depict the behaviour of phases (108, 136) (movement associated with intersection 103) and (154, 102) (movement associated with intersection 116) under the demand variation. Hence, while during the first 3 h (demand intensity

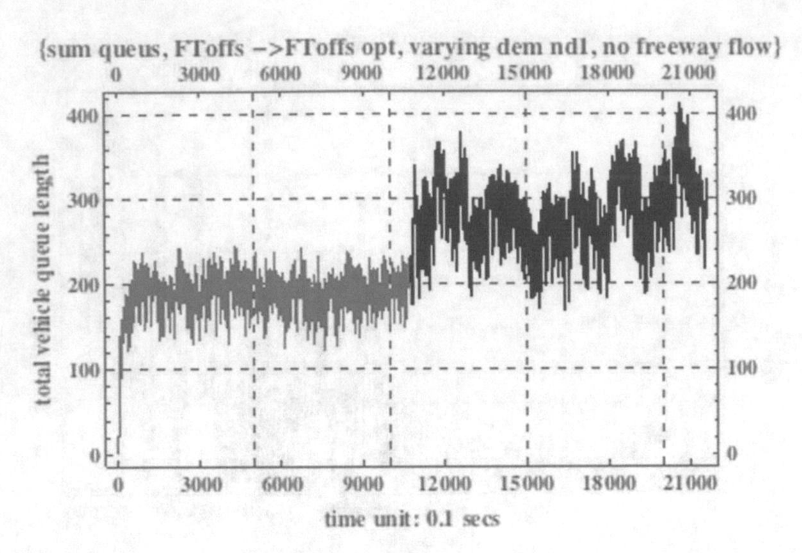

Fig. 20 Network evolution under varying demand: $\{\lambda^1, d_1\} \rightarrow \{\lambda^2, d_2\}$

is of d_1) the related queues are stable (orange curve in both figures), when demand level becomes d_2 the queues start to become unstable (red plot).

New actuation durations are computed for each intersection of the network, when demand d_2 is applied and let λ^2 be the new stabilising pre-timed control. Figure 20 represents the evolution of the sum of all the network queues when demands d_1 and d_2 are employed associated with signals defined by vectors λ^1 and λ^2. Obviously, the network remains stable during the 6 h although the sum of the queues is increased when the applied demand is of intensity equal to d_2 as depicted by the pink plot.

Figure 21 depicts the evolution of movement (154, 102) when demand d_2 is employed in association with pre-timed actuation durations provided by vector λ^2. In comparison with Fig. 19, a stabilising behaviour is now clearly implied.

Similarly Fig. 22 illustrates the evolution of phase (108,136) under demand d_2 and FT signals given by λ^2. Even if the queue presents occasionally increased values, it then clears and a stable behaviour is implied (see also Fig. 18)

Suppose now that by the time the demand varies a Max-Pressure control defines the signal timings. Figure 23 represents the network evolution during the 6 h. Obviously, a stabilising behaviour is continuously maintained. Moreover, when MP signals rule the network (brown plot) the sum of the network queues is lower than when a stabilising pre-timed plan is applied (see purple plot of Fig. 20).

Figures 24 and 25 depict the evolution of queues $q(154, 102)$ and $q(108, 136)$ under demand d_2 and MP policy. Clearly, these movements are not only stable but in addition to that, the related queues take smaller values under MP signals in comparison with the queue values under the stabilising FT plans (see Figs. 21 and 22).

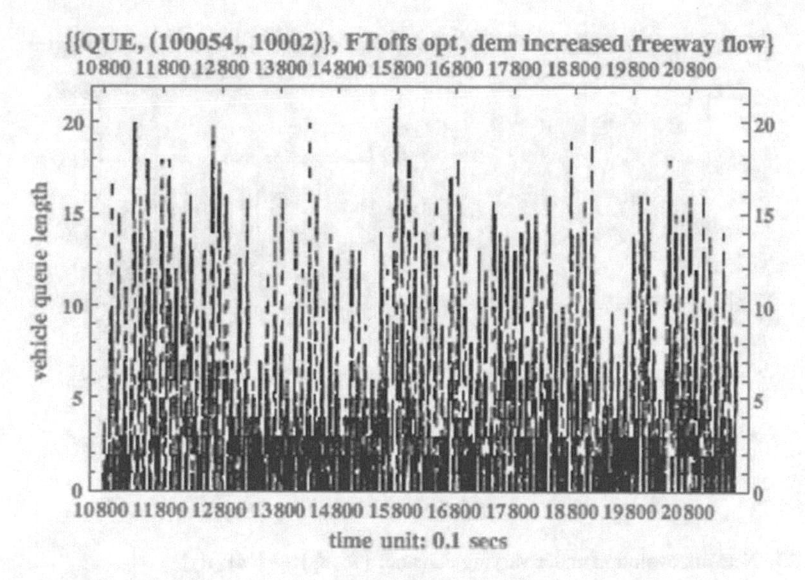

Fig. 21 Evolution $q(154, 102)$ under varying demand: $\{\lambda^2, d_2\}$

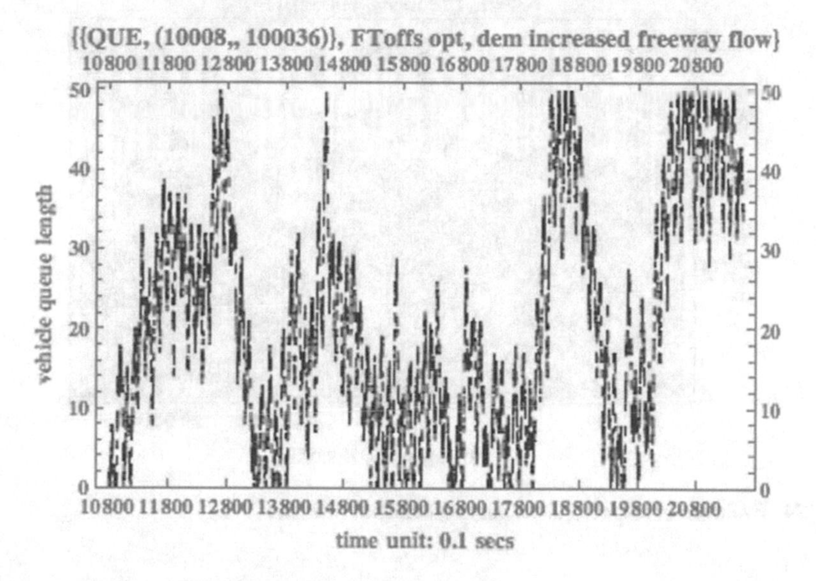

Fig. 22 Evolution $q(108, 136)$ under varying demand: $\{\lambda^2, d_2\}$

Figure 26 proposes a macroscopic point of view in order to capture the aggregate behaviour of the arterial network for approximatively 30 min, when demand d_2 is associated with MP plans. Hence, cumulative external arrivals are depicted in green,

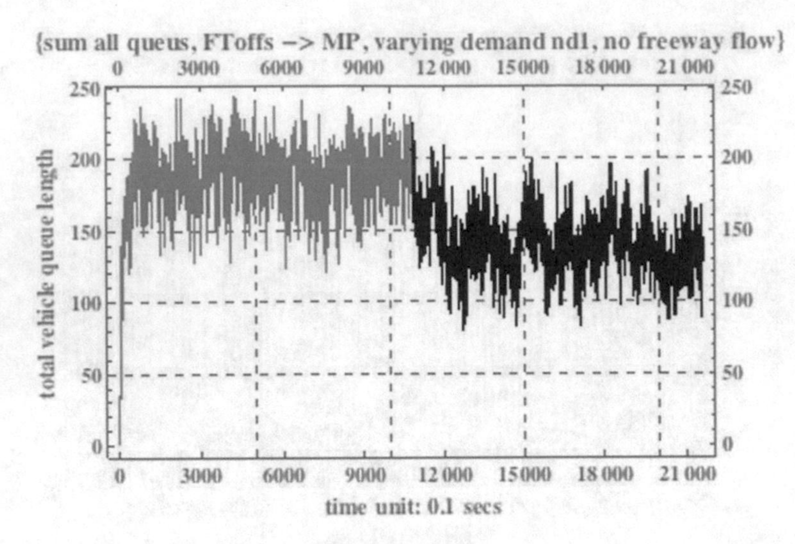

Fig. 23 Network evolution under varying demand: $\{\lambda^1, d_1\} \to \{\text{MP}, d_2\}$

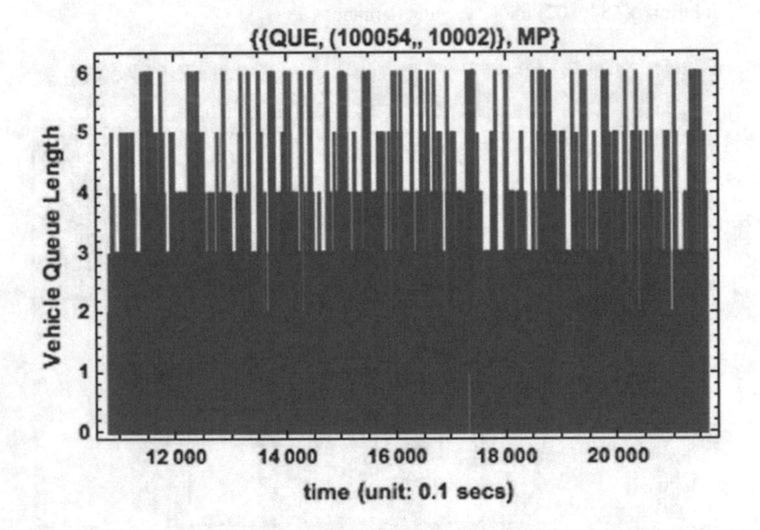

Fig. 24 Evolution $q(154, 102)$ under varying demand: $\{MP, d_2\}$

departures in blue and internal arrivals in red. The three curves almost coincide, indicating that the number of vehicles entering the network is almost the same as the number of exiting vehicles, presenting the network stability.

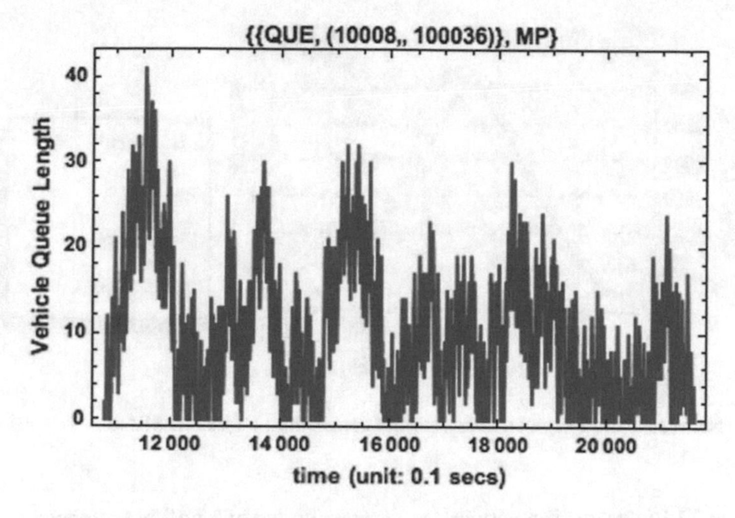

Fig. 25 Evolution $q(108, 136)$ under varying demand: $\{MP, d_2\}$

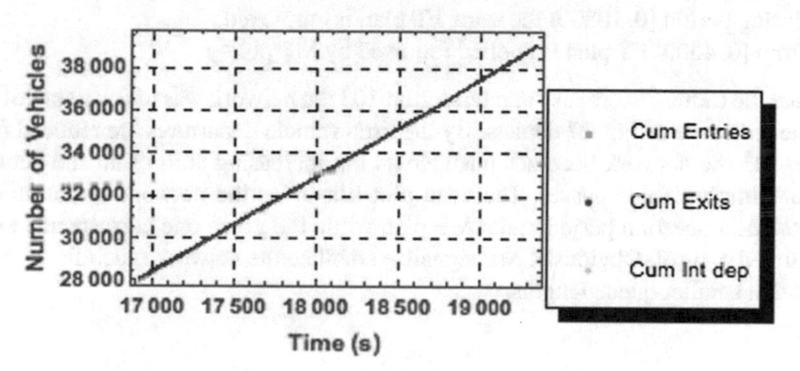

Fig. 26 Macroscopic network behaviour: cumulative entries-exits, internal arrivals

9 Managing Non-recurrent Congestion

9.1 Arterial Management Under Traffic Disturbance

This section considers the network behaviour when the accessibility of an intersection becomes prohibited or partially inaccessible for a limited duration. In particular, a partial disturbance of 1 h (3600 s) is going to be considered prohibiting the majority of vehicles to cross node 103. The manifested network reaction is going to be studied when FT and MP signal plans control the intersections. More precisely, time period [0, 10800) is considered where from $t = 1000$ s to $t = 4600$ s most vehicle departures are dramatically reduced from node 103.

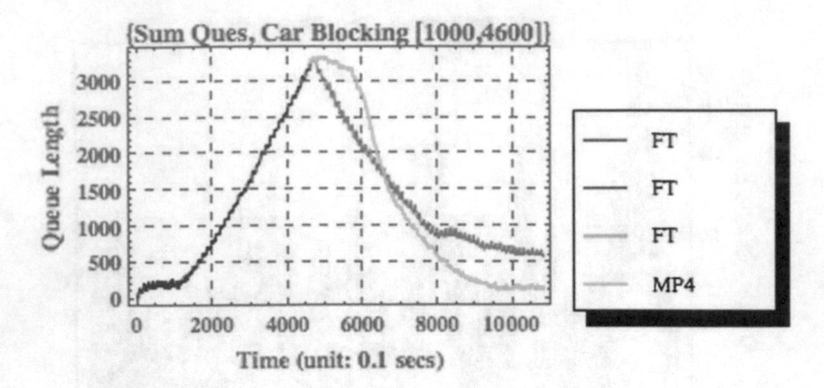

Fig. 27 Network evolution when non-recurrent congestion: FT versus MP

Figure 27 illustrates the network behaviour under two policy patterns:

1. during period [0, 10800) the same FT plan is employed
2. from [0, 4600) FT plan is applied followed by MP policy

Before the traffic disturbance at intersection 103 the network was under a stable state as the red curve of Fig. 27 depicts. By the time vehicle departures are reduced from node 103 the network becomes unstable as the increasing sum of all the network queues implies (blue curve). The cyan plot illustrates the network evolution during the decongestion period under MP plan while the green one corresponds to the pre-timed control. Obviously, MP signalise stabilise the network much faster while maintain smaller queue lengths.

9.2 Congestion: Location and Extend

The impact of the traffic perturbation at intersection 103 to the neighbour nodes is examined. In particular, distinct queues at nodes 103 and 102 are considered and their behaviour during the perturbation time is analysed. The selected phases of which the evolution is studied are: (103, 108), (138, 140) and (105, 103).

Movements (103, 108) and (138, 140) require ability to cross intersection 103 in association with available space on link 108 (for phase (103, 108)). Movement (105, 103) discharges vehicles towards node 103 and for that the currently available vehicle capacity of link 103 interferes. Entry links 138 and 105 are associated with infinite vehicle storage capacity and consequently all queues on these links may reach any value. However, on internal link 103 at most 55 vehicles can be simultaneously located.

Examination of $q(138, 140)$ during the disturbance period as depicted in Fig. 28 shows an unstable behaviour under FT plan. This phase is prioritised under MP

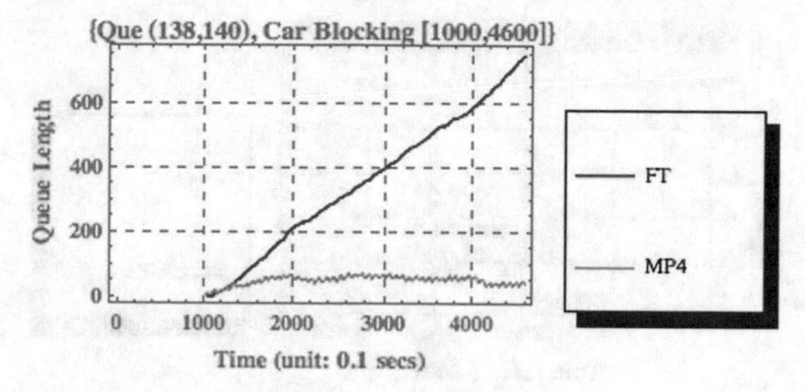

Fig. 28 Evolution $q(138, 140)$ during traffic perturbation: FT versus MP

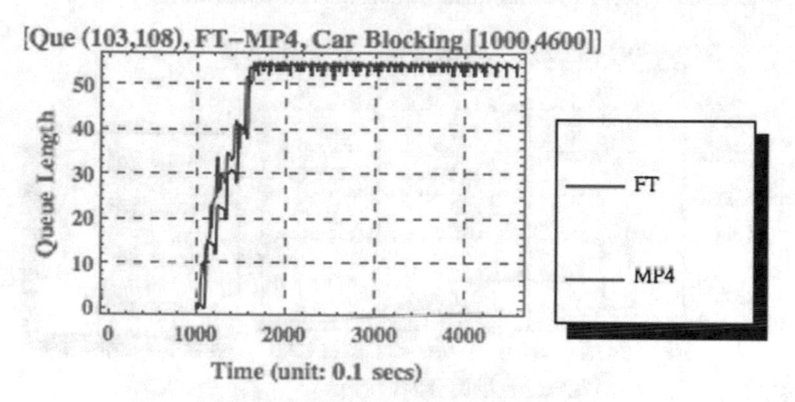

Fig. 29 Evolution $q(103, 108)$ during traffic perturbation: FT versus MP

signals regarding other concurrent phases, since link 140 is an exit link and no queues are considered on this link (see also Eqs. (4) and (5) in MP definition).

Let us now examine phase (103, 108). Two different turns are possible from link 103, one towards link 104 and another towards link 108 (phase (103, 104) forms a right turn and consequently is an uncontrolled movement). At any time the total number of vehicles on link 103 cannot exceed the value of 55. Figure 29 implies a saturated state of link 103 during the perturbation period under both plans. The size of $q(103, 108)$ oscillates at around 52. Due to the perturbation, the number of vehicle departures from this link is decreased during the considered time period. Moreover, within the current context MP actuates less often this phase since the concurrent turn (138, 140) is prioritised (entry link has an important demand and sends vehicles to no congested exit links). By the time some space becomes available on link 103 vehicles from upstream queues are joining this link.

Vehicles appearing at entry link 105 can either turn towards link 103 (lower demand) or select to head towards link 254 (most probable selection). As depicted in Fig. 30 in spite the reduced demand associated with turn (105, 103) this phase is

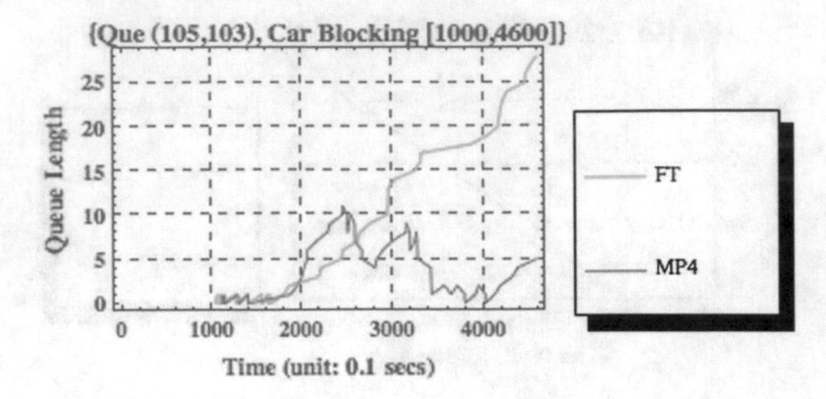

Fig. 30 Evolution $q(105, 103)$ during traffic perturbation: FT versus MP

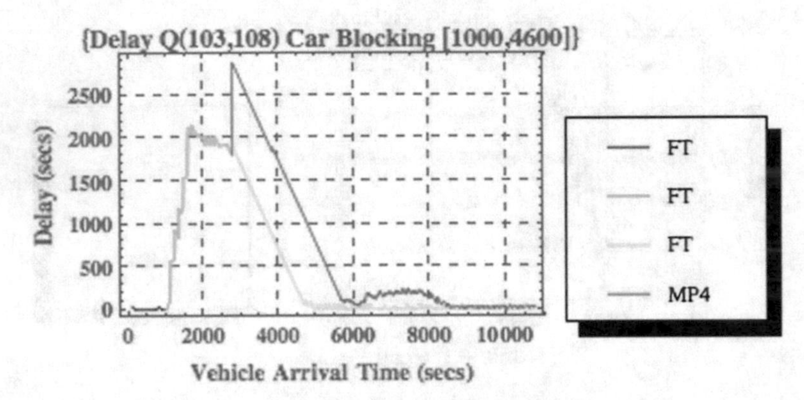

Fig. 31 Delay evolution $q(103, 108)$: FT versus MP

unstable under the pre-timed plan (cyan curve). However, this movement remains stable when MP actuation signals are applied, maintaining low queue values (green plot). To this effect, important role is also played by the fact that since MP policy takes into consideration the current state of both the ingoing and outgoing links an optimised coordination of the movements is implied while that is necessarily the case of FT policy.

Delays are studied for movements at node 103. In particular, the time during which vehicles were queued as a function of their arrival time will be measured for the contradictory phases (138, 140) and (103, 108). Figure 31 illustrates the delay values associated with queue $q(103, 108)$ under two scenarios. For both cases, a pre-timed plan is applied from the beginning of the simulation to the end of the perturbation time, that is during period [0, 4000). The sequence of red-cyan curve illustrates this situation. The difference between the two scenarios concern the control applied just by the end of the disturbance. The yellow curve indicates the delays met by vehicles when they joined ($q103, 108$) during period [4000, 10800) when FT plans and the green plot illustrates the same metric but under MP signals. Obviously, MP

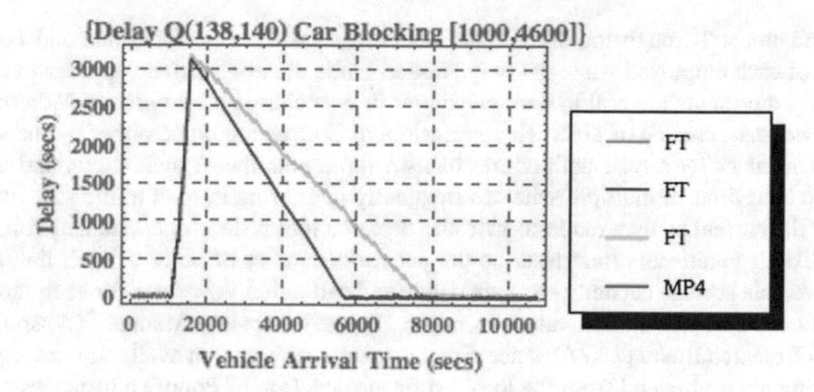

Fig. 32 Delay evolution $q(138, 140)$: FT versus MP

increases delays for $q(103, 108)$. One would wonder if that observation is reasonable and what the justification for such a behaviour would be. Two stages are associated with intersection 103. Stage 1 simultaneously actuates movements (103, 108) and (208, 203). Similarly stage 2 actuates phases (138,140) and (139, 104). Internal links 103 and 208 are of finite vehicle storage capacity equal to 55 and 45 respectively while entry links 138 and 139 may reach any value. As previously discussed, $q(138, 140)$ becomes unstable during the disturbance period and under FT control, reaching values greater than 700 (see Fig. 28). The MP policy applied just after the end of the disturbance, will meet the queue length indicated by the blue plot of Fig. 28. The exerted pressure by stage 2 actuating this movement is much higher regarding the one of stage 1 since $q(138, 140)$ had higher values at $t = 4000$ s and heads vehicles towards exit link 140. Consequently stage 2 will be frequently actuated by MP policy until a stable situation will be attained. Thus, movement (103, 108) belonging to stage 1 will receive less green for t greater to 4000 s and thus delays associated with this phase will be increased regarding the ones of the FT plan which periodically actuates $q(103, 108)$. Furthermore, Fig. 32 represents the delays associated with phase (138, 140) for both FT and MP timing plans. Clearly, delays are lower under MP plans and the queue stability is now obtained much faster regarding the case of pre-timed plans (cyan curve corresponding to FT remains higher than the purple one associated with MP).

10 Simulation Technique: Addressing Scenario Reconstruction

10.1 Analysing Observations

As argued in Sect. 3 a DES approach is considered for the current study, aiming at optimal traffic management, in association with intense discrete event simulations. Through the simulation model, the involved performance and benefits of each employed strategy are quantified while the *event-driven* approach accurately represents the real system behaviour in a stochastic environment. Moreover, the *nondeterminism* of DES where transitional choices are made either by the system itself or by a well-defined mechanism represents the asynchronous and random behaviour of multiple schemes frequently interfering in road traffic structures. For the current study a made-to-measure decision tool is developed, named PointQ, involving parameters that describe the network (number of lanes in each link and its vehicle storage capacity etc.), demands and saturation flow rates for each movement (obtained as default values from the Highway Capacity Manual (Transportation Research Board (2000)), since direct measurements are not available) and signal timing plans obtained from the local traffic agency. One of PointQ advantages consists in its accuracy while minimal input information is required in order to run any strategy. Each observation is recorded within a database and hence the entire simulated scenario can be accurately reproduced. One of the major advantages of this aspect consists in examining whether an observation result is justified or is due to simulation errors and so on. Hereafter, a representative example is discussed.

Let us consider the realised trips of vehicles entering the network from entry link 174 and leaving it from exit link 237 and measure the average journey duration when pre-timed and MP signal plans are employed.

Table 5 illustrates the average realised travel time under FT and MP signals with 4, 6, 8 and 10 control revisions per cycle, while no demand variation or other traffic incident causing non-recurrent congestion occurs. One observes that the average travel time under MP controls with 4 (MP4) control revisions per cycle is of 678.6 s while the corresponding time with MP8 plans equals to 693 s. These findings contradict previous conclusions on the length of the sum of all the network queues and the related probabilities, discussed in Sects. 7 and 6.

All experiments of Table 5 employ the same demand (level and geometry), saturation flow rates and phase actuation plans as the ones utilised for obtaining the results of Table 2 (indicating average travel times of other vehicle journeys) as well plots of Figs. 3 and 8 representing the evolution of the sum of all the network queues under MP versions and the associated queue probability. In particular Figs. 3 and 8 verify theoretical results implying that the frequent the control decision is updated the smaller values the total network queue length reaches. Since the demand and saturation flow rates are unchanged the average trip duration should also decrease by the time the number of MP control revisions is increased.

Table 5 Mean travel time—MP versions

Entry-exit link	ATT (s)	ATT	ATT	ATT
	MP4	MP6	MP8	MP10
(174, 234)	678.6	499.4	693	459.8

Consequently, one would wonder of the coherency of these results and the reliability of the decision tool.

10.2 *MP8 Increases ATT. Reasonable Conclusion?*

Would it be reasonable to state that MP signals associated with eight stage revisions per cycle involve larger trip durations?

- Observation of the evolution of the sum of all network queues under MP8 policy, as depicted in Fig. 3, implies smaller queue lengths than the related ones under MP4 and MP8 plans.
- Similarly, the probability of the queue size decreases as the number of MP control revisions increases, as illustrated in Fig. 8.

Hence, could one conclude that a computation error or a PointQ bug has occurred and consequently that is the reason for which the average travel time under MP8 is increased regarding the one of MP4, for trips between entry link 174 and exit link 234?

Before proceeding to any assessment let's make use of all the simulator advantages and proceed to the computation of the delays associated with these trips. Table 6 resumes the related total trip delays for MP4, MP6, MP8 and MP10. Obviously delays are reduced as the number of control revisions increases. This results ensures that no computational or programming error is involved.

But then what reasonable justification could be given for the increased travel times under MP8?

10.3 *Microscopic Analysis*

A microscopic analysis seems necessary in order to examine in detail the simulation context involved with trips from entry link 174 towards exit link 234.

During the employed implementation, stochastic travel times are associated with each link (based on the mean free flow speed and depending upon the current link state) following shifted Log-Normal distributions (the shift ensures a non zero travel duration). A detailed analysis of the travel time of vehicles which entered the network from link 174 and exited at link 237 is proceeded for the MP6 and MP8 signal plans.

Table 6 Average of the mean sojourn time of vehicles in queues for MP versions

Average mean sojourn				
Time of vehicles in queues	MP4	MP6	MP8	MP10
(s)	**4276.2**	**3006.1**	**2345.6**	**2009.3**

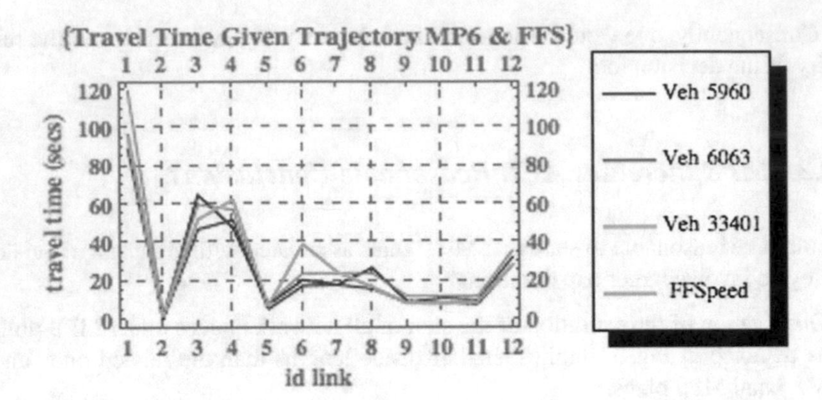

Fig. 33 Travel time—MP6: three trips from entry link 174 towards exit link 234

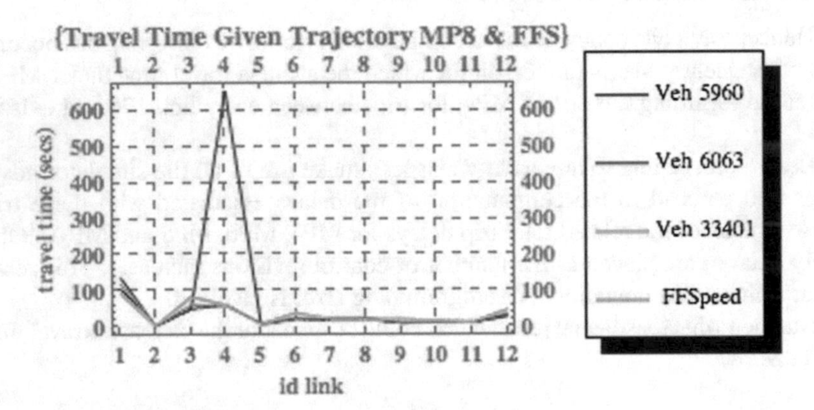

Fig. 34 Travel time—MP6: three trips from entry link 174 towards exit link 234

The micro statistical analysis implies that three vehicles entered the network from link 174 and exited it at link 237.

Figure 33 represents the realised travel time, for each one of the three vehicles and for each crossed link, for the implementation corresponding to MP6 traffic plans, (red, blue and green curves). Moreover, in the same figure is also depicted the related mean travel time for each one of the considered links, based on the free flow speed (orange plot). Clearly, for all vehicles and for each travelled link, the realised travel time is close to the one indicated by the considered mean travel value.

Figure 34 illustrates similar results as Fig. 33 but when MP8 signal plans govern each intersection node. Manifestly, the vehicle of id 6063 traveled too slowly link 4 (blue plot). That vehicle consumed approximatively 600 more seconds when traveling link 4, than the mean travel time indicated. A deeper examination implied that a higher value of travel time was provided by the probability law defining link travel times but still this value could be accepted according to the corresponding variance and standard deviation values. As a conclusion, one should consider large simulation durations associated with series of simulations for stochastic models involving a multitude of possibilities for each considered approach. Thus, in the case of road traffic networks where fore example various trip possibilities exist, when a particular demand is low, concluding from a series of simulations will limit similar effects. However, a detailed analysis will show whether a particular system behaviour is due to different sorts of errors or it can be reasonably justified.

11 Discussion

Various approaches tackling optimal road management exist, each one based on different concepts and assumptions. Moreover, when addressing congestion and its impacts there is rarely a uniform conceptual framework appraising the suggested management, due to the variety and scope of the goals involved. Nevertheless, according to the manner in which congestion is evaluated and analysed, including the employed methodological tools and the related assumptions, important policy consequences are involved. Furthermore, the *peak-spreading* describing how urban congestion is spread, has become a fact, in the sense that the period during which traffic is very dense is largely extended. Likewise, many urban regions experience degraded travel conditions and consequently travel times predictability are reliability are also reduced.

This study suggests a framework related to a next traffic generation management, where signal plans decisions are determined in real depending upon the present state of the related intersection. No demand knowledge is required, the involved decentralised aspect of the control easies decision making while the computational and implementation cost are drastically reduced. The so called Max-Pressure algorithm, provides theoretical stability guarantees for the network, based on queue measurements and turning ratios. The Max-Pressure Adaptive version estimates the turning probabilities (since this information cannot always be available) while the network equilibrium remains ensured. In order to reduce frequent signal switches, the Max-Pressure Practical algorithm is presented and appraised regarding Max-Pressure, illustrating how similar network performance can be obtained within reduced costs (time lost, signal changes etc.). Various versions of the feedback policies are considered depending upon the frequency of the control revision and compared with the pre-timed control for the same network, demand and vehicle routing under regular traffic conditions. Higher network performance is clearly obtained when intersections are governed by the closed loop timing plans. Performance metrics based on queue lengths, average travel times, trip durations and delays are mostly measured.

Different traffic patterns are modelled within the purpose to derive the benefits of each signal policy plans.

In particular, the stabilising pre-timed actuation durations λ^1 corresponding to a given demand level d_1 cannot accommodate the new demand intensity d_2 occurred after a slight fluctuation on the demand intensity associated with a single intersection. Likewise, a saturated node can quickly give rise to queues whose upstream propagation can swamp local roads and intersections. As a consequence, multiple intersections become congested and a new computation of the actuation durations associated with each network node, λ^2 where $\lambda^2 \neq \lambda^1$, is revealed necessary in order to maintain the network stability under pre-timed signal timings. Nevertheless, the Max-Pressure actuation stages are automatically adapted to the new demand scheme and accordingly all network queues are stable avoiding any intersection saturation.

A non-recurrent congestion pattern is modelled occurring during a limited period of time. Thus, during half an hour most vehicle departures are prohibited from a particular intersection. The network effects are studied and the congestion impacts are measured according to both policies: Fixed Time versus Max-Pressure plans. Analysis assessing congestion, reveals how the feedback signals mitigate congestion expansion while allow for a faster network decongestion when the traffic conditions return back to the normal state. Max-Pressure control, prioritises movements requiring more actuation duration at the current decision time and thus re-establishes the network state within the best possible conditions.

A discrete event approach is considered for the study of the related traffic control problem associated with discrete event simulations in order to obtain knowledge of the system behaviour under the employed strategy. Deep statistical analysis of the recorded observations, allow a precise and detailed reconstruction of the simulated scenario revealing all the schemes through which the considered implementation passed. A representative example suggesting a methodology for examining whether the induced metrics can be justified or should considered as absurd ones, is developed. As a result, all the obtained statistical indicators can be examined, verified and validated since well-founded justifications can be provided any time that is necessary. Since the reflected results are enhanced reliable conclusions can be provided to any system enquiries.

Estimation of the queue size is aimed to be measured and employed within a short future in order to contribute to a realistic application able to be employed within a real time traffic control. In the USA more than 90% of the controllers are pre-timed ones, consequently the benefits of MP and MP-Pract policies worth to be explored.

References

1. Varaiya, P. (2013). Max pressure control of a network of signalized intersections. *Transportation Research Part C*, *36*, 177–195.
2. Aboudolas, E. K., & Papageorgiou, M. (2009). Store-and-forward based methods for the signal control problem in large-scale congested urban road networks. *Transportation Research Part C-Emerging Technologies*, *17*, 163–174.
3. Mirchandani, L. H. P. (2001). A real-time traffic signal control system: Architecture, algorithms, and analysis. *Transportation Research Part C-Emerging Technologies*, *9*(6), 415–432.
4. Heydecker, B. (2004). Objectives, stimulus and feedback in signal control of road traffics. *Journal of Intelligent Transportation Systems*, *8*, 63–76.
5. Allsop, R. E. (1971). Delay-minimizing settings for fixed-time traffic signals at a single road junction. *IMA Journal of Applied Mathematics*, *2*(8), 164–185.

6. Lo, H. K. (2001). A cell-based traffic control formulation: Strategies and benefits of dynamic timing plans. *Transportation Science*, *35*(2).
7. Baccelli, F., Cohen, G., Olsder, J., & Quadrat, G. (1992). *Synchronization and linearity—An algebra for discrete event systems*. New York: Wiley.
8. Muralidharan, A., Pedarsani, R., & Varaiya, P. (2015). Analysis of fixed-time control. *Transportation Research Part B: Methodological*, *73*, 81–90.
9. Varaiya, P., Kurzhanski, A. B. (1988). *Discrete event systems: Models and applications*. Springer.
10. Glasserman, P., & Yao, D. (1991). Monotone structure in discrete-event systems (p. 297). New York: Wiley.
11. Ho, Y.-C. (1991). *Discrete-event dynamic systems: Analyzing complexity and performance inn the modern world* (p. 291). New York.
12. Peterson. (1981). Petri net theory and the modeling of systems. New York: Wiley.
13. Chao, C., Kwon, J., Varaiya, P. (2005). *An empirical assessment of traffic operations*. Elsevier.
14. Daganzo, C. F., Cassidy, M. J., Bertini, R. L. (1999). *Possible explanations of phase transitions in highway traffic* (Vol. 33a, pp. 365–3791). Elsevier.
15. CALTRANS. (1999). *California life-cycle benefit/cost analysis model*. Inc.
16. Ho, Y. C., & Cao, X. (1991). *Perturbation analysis of discrete event dynamic systems*. Boston: Kluwer
17. Glasserman, P., & Gong, W. B. (1990). Smoothed perturbation analysis for a class of discrete event systems. *IEEE Transaction on Automatic Control*, *AC-35*, 1218–1230.
18. Reisig, W. (1985). *Petri nets*. Springer.
19. Cieslak, R., Desclaux, C., Fawaz, A., & Varaiya, P. (1988). Supervisory control of discrete-event processes with partial observations. *IEEE Transaction on Automatic Control*, *33*(3), 249–260.
20. Baccelli, F., Cohen, G., & Gaujal, B. (1992). Recursive equations and basic properties of timed petri nets. *Journal of Discrete Event Dynamic Systems*.
21. Chen, C. J., & Varaiya, P. (2001). Causes and cures of traffic congestion. *IEEE Control Systems Magazine*.
22. Gomes, G., Horowitz, R., Kurzhanskiy, A., Kwon, J., & Varaiya, P. (2008). Behaviour of the cell transmission model and effectiveness of ramp metering. *Transportation Research Part C: Methodological*, *16*(4), 485–513.

Simulation-Optimization of the Mexico City Public Transportation Network: A Complex Network Analysis Framework

Idalia Flores De La Mota and Aída Huerta-Barrientos

Abstract The urban transport mobility is one of the most important problems for the cities, and involves many aspects that concern to citizens, governments and the economical growth of the countries. Mobility in Mexico City is also a huge problem since the city size makes it insoluble and citizens prefer to use private transportation instead of the public transport network because it offers a poor coverage and a lack of modal transfer centers. With the purpose of analyzing the mobility problems in Mexico City as well as detecting areas of opportunity, the objective of this chapter is to model and simulate the public transportation network from the complex network perspective to asses network structural vulnerability and resilience, considering mobility and accessibility aspects. Firstly, we analyze the urban transport infrastructure in Mexico City taking into account the planning process and sustainability criteria. Secondly, we model and simulate the Mexico City's public transportation network as a complex network. Thirdly, we characterize the complex network topology of the Mexico City's public transportation network, and finally we present the main results.

Keywords Urban mobility · Optimization-simulation · Synthetic microanalysis · Mexico City

I. Flores De La Mota (✉) · A. Huerta-Barrientos
Facultad de Ingeniería, Deparatmento de Ingeniería de Sistemas, Universidad Nacional Autónoma de México, Mexico City, Mexico
e-mail: idalia@unam.mx

A. Huerta-Barrientos
e-mail: aida.huerta@comunidad.unam.mx

M. Mujica Mota and I. Flores De La Mota (eds.), *Applied Simulation and Optimization 2*, DOI 10.1007/978-3-319-55810-3_2

1 Introduction

"Adding highway lanes to deal with traffic congestion is like loosening your belt to cure obesity."

Lewis Mumford. The Roaring Traffic's Boom.

Mexico City is divided by 16 geo-political sectors (see Fig. 1) where each sector has its own government authority. The majority of the Mexican population is urban (78% of total population lives in cities) as in the United States and Brazil (see Table 1). Like many countries around the globe, urban population in Mexico is

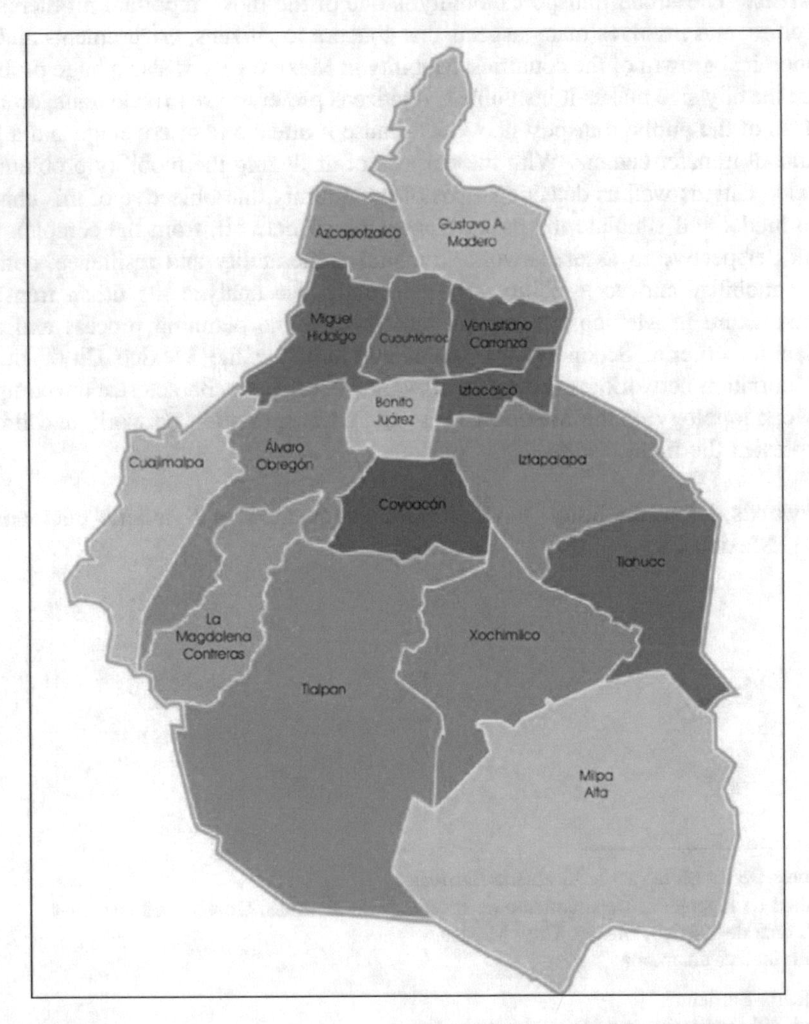

Fig. 1 Mexico City sectors, reproduced from http://mapamexicodf360.com.mx/carte/image/es/mapa-delegaciones-mexico.jpg

growing at higher rates compared to the total population, making Mexican cities local engines for national growth [60].

Varela [60] emphasizes that as competitiveness and growth in Mexican cities are increasingly compromised by congestion, air quality problems, and increased travel times; city officials not only face the challenge of accommodating a growing urban population but also sustaining a constant provision of basic urban services (e.g. clean water, health, job opportunities, transportation, and education). Varela [60] adds that unfortunately, periods of high growth without effective planning and increasing motorization, have pushed Mexican cities towards a "3D" urban growth model: distant, disperse, and disconnected. It is important to note that the 3D model is a direct result of national policies subsidizing housing projects in the outskirts of urban agglomerations, managing urban and rural land poorly, and prioritizing car-oriented solutions for transportation. In consequence, over the past 30 years Mexico City's population has doubled and its size has increased seven-fold and nowadays it is considered the most populated metropolitan area in the western hemisphere. Table 2 shows some socio-economic KPI's of Mexico City from 2008

Table 1 Urbanization and economic growth, adapted from [58]

	Brazil	China	India	Mexico	United States	Global
Population (billions)	0.5	1.3	1.2	0.1	0.3	7.1
Annual growth rate of population	0.83%	0.46%	1.28%	1.07%	0.9%	1.01%
Urban population	87%	47%	30%	78%	82%	50%
Change on annual level of urbanization (2010–2015)	1.10%	2.30%	2.40%	1.20%	1.20%	1.85%
GDP per capita (in U.S. Dollars)	12,000	9,100	3,900	15,300	49,800	12,400
GDP growth rate per capita (annual percent in 2011)	1.80%	8.00%	4.90%	2.70%	1.00%	–

Table 2 Socio-economic KPI's of Mexico City, adapted from [14]

Administrative organization	The metropolitan area of Mexico is composed of 16 Delegations in the Federal District, 58 municipalities in State of Mexico, and 1 Municipality in State of Hidalgo
Population (2008)	Federal District: 8.8 million Metropolitan area (Federal District and State of Mexico): 19.2 million
Area (2010)	Federal District: 1,487 km^2 Metropolitan area: 7,180 km^2 (40.1% of which is urbanized)
Population density (2010)	Federal District: 5,958 people/km^2 Metropolitan area: 6,671 people/km^2
Annual population growth rate (2005–2010)	Federal District: 1.49% Metropolitan area: 3.96%
GDP and growth (2011)	163.6 billion USD (17% of the national GDP, Federal District only) Annual GDP growth (2008–2011): 4%
Unemployment rate (2011)	6.5%

to 2011. Floater et al. [26] Believe that one alternative to the 3D model is the 3C urban growth model: compact, connected and coordinated. In this direction, it is mandatory that a study about urban mobility needs to consider a variety of aspects such as the urban development, the land use, the environmental conditions, the weather, the security, and the social welfare.

Mobility in Mexico City is also a huge problem since the city size makes it insoluble. Mexico City presents the highest congestion level on the road network, causing more than 90% extra travel time for citizens during busy hours. The traffic congestion affects directly on the quality of life, however citizens prefer to use private transportation instead of the public transport network because it offers a poor coverage and a lack of modal transfer centers. Table 3 (at the end of the chapter) shows the mobility KPI' for Mexico City during 2001, 2007 and 2010.

In the last years, an increasing amount of literature has been devoted to modeling public transportation networks as complex networks [8, 9, 20, 63]. Interesting contributions are found in the literature. For instance, In [61] authors used complex network concepts to analyze statistical properties of urban public transport networks in several major cities of the world. Cheung et al. [20] analyzed the air transportation network in the U.S. Recently Háznagy et al. [33] analyzed the urban public transportation systems of five Hungarian cities performing a comprehensive network analysis of the systems with the main goal of identifying significant similarities and differences of the transportation networks of these cities. Háznagy et al. [33] considered directed and weighted links, where the weights represented the capacities of the vehicles (bus, tram, trolleybus) in the morning peak hours. Reggiani et al. [51] Establishes that the following questions need to be answered with respect to transport networks as complex networks:

(a) Is a complex network a necessary condition for the emergence or presence of transport resilience and vulnerability?
(b) Several indicators of resilience and vulnerability co-exist; are these differences related to specific fields of transportation research?
(c) Can connectivity or accessibility be considered as a unifying framework for understanding and interpreting—in the transport literature—the concepts of resilience and vulnerability?

In this direction, connectivity as the ability to create and maintain a connection between two or more points in a spatial system is one of the essential elements that characterize complex networks. Given the relevance of the connectivity pattern in complex networks, it may seem plausible that complex networks—and connectivity—are a *sine qua non* for the development of resilience and reduce vulnerability in transportation systems. More recent studies show how the topological properties of a network can offer useful insights into the way a transport network is structured and into the question of which are the most critical nodes (hubs). In this case, resilience and vulnerability conditions associated with such hubs can then affect the resilience/vulnerability of the whole network.

Table 3 Mobility KPI's for Mexico City during 2001, 2007 and 2010, adapted from [15]

Total trips per day (2007)	48.8 million (Metropolitan area) and 32.0 million (Federal District)			
Daily trips per person (2007)	2.5 (Metropolitan area) and 3.6 (Federal District)			
Trips and modal share in the Federal District (2007)	Mode	Trips	% Total	%Public transport
	Non motorized	8,600,000	26.9%	
	Private vehicles	4,800,000	15.0%	
	Microbuses	9,448,800	29.5%	50.8%
	Metro	4,984,800	15.6%	26.8%
	Autobuses	1,878,600	5.9%	10.1%
	Taxis	1,041,600	3.3%	5.6%
	Metrobus	762,600	2.4%	4.1%
	Trolley (RTP)	204,600	0.6%	1.1%
	Suburban train	167,400	0.5%	0.9%
	Light train	111,600	0.3%	0.6%
	Total	32,000,000	100.0%	58.1%
Road network (2007)	10,200 km (91% local roads)			
Total vehicles (Federal District, 2001)	Cars			4,460,386
	Taxis			225,302
	Motorcycle			11,920
	Microbuses			20,459
	Buses			8,240
	Combis			3,519
	Metrobus–articulated buses			322
	Metrobus–regular buses			54
	Metrobus–biarticulated buses			27
	Totals			4,730,228
Road safety (2010)	Total number of accidents			14,729
	Number of deaths			1,026
	Involved vehicle in deaths			3.5% Microbus 81.0% Car 5.6% Truck
	Involved victim in deaths			14.0% Motorcycle driver 52.0% Pedestrian 20.0% Car driver

Additionally, Lin and Ban [42] presented the current state of topological research on transportation systems under a complex network framework, as well as the efforts and challenges that have been made in the last decade.

In this chapter, we propose to model and simulate the public transportation network in Mexico City from the complex network perspective to asses' network structural vulnerability and resilience, considering mobility and accessibility aspects. We consider that a research about the public transportation network in Mexico City should be conducted at different levels. The first one can be done considering the networks as a whole, while the second one should take into account the relationship between geo-political sectors and the third one should analyze each geo-political sector individually. For the purpose of this study, we consider the public transportation network in Mexico City as a whole. In addition, we take into account the lack of connections in the multimodal public transportation network to make some tests based on networks algorithms.

This chapter is divided into five main sections. In Sect. 2, the urban transport infrastructure is analysed considering the planning process and sustainability criteria. In Sect. 3, the complex network modeling and simulation of the Mexico City's public transportation network is carried out. The complex network topology of the Mexico City's public transportation network is characterized in Sect. 4. The concluding remarks are drawn in Sect. 5.

Note: Due to the use of the nomenclature of both network theory and graph theory, some authors cited in this chapter use terms such as nodes and vertices to refer to the same, as well as arcs and edges.

2 Urban Transport Infrastructure

Nowadays one of the biggest problems in cities is the transportation system and its infrastructure. There has been a lost of studies and research in recent decades trying to find solutions. In general, there is an economic impact when countries make an investment in this sector. Most of the studies on transportation infrastructure, in particular, focus on its impact on economic growth. In the past two decades, the analytical literature has grown enormously with studies carried out using different theoretical approaches, such as a production function (or cost) and growth regressions, as well as different variants of these models (using different data, methods and methodologies). The majority of these studies have found that transportation infrastructure has a positive effect on output, productivity or economic growth rate [16]. For instance, Aschauer [3] in his empirical study provided substantial evidence that public transport is an important determinant of economic performance. Another example is the study of Alminas et al. [1], who found that transport in general has contributed to growth in the Baltic region.

Another study on the Spanish plan to extend roads and railways that connect Spain with other countries concludes that these have a positive impact in terms of Gross Domestic Product [2]. In a study of the railroad in the United States, it was mentioned that many economists believe that the project costs exceed the benefits [6]. However, the traditional model of cost-benefit assessment does not include the impact of development projects [23].

In these studies focused on growth, we see there is a bias towards economic rather than social goals. That is why it is important to emphasize the impact of transport infrastructure on development and not just growth. Transport infrastructure has to deal with accessibility, mobility and traffic mainly, but if we want to establish a sustainable public transport, is important to consider factors as, economy, land use, trips, environment and social welfare. According to The City of Calgary [57] we divide the urban transport infrastructure as follows:

- Transportation Planning
- Transportation Optimization
- Transportation Simulation

Transportation planning covers many different aspects and is an essential part of the socio-economic system. According to Levy [41], "Most regional transport planners employ what is called the rational model of planning. The model considers planning as a logical and technical process that uses the analysis of quantitative data to decide how to best invest resources in new and existing transport infrastructure."

Phases for Transportation Planning

There are three phases: The first, preanalysis, considers what problems and issues the region faces and what goals and objectives it can set to help address those issues. The second phase is technical analysis. The process involves the development of the models that are going to be used later. The post-analysis phase involves plan evaluation, program, implementation and monitoring of the results, [35].

Transportation planning involves the following steps:

- Monitoring existing conditions;
- Forecasting future population and employment growth, including assessing projected land uses in the region and identifying major growth corridors;
- Identifying current and projected future transportation problems and needs and analyzing, through detailed planning studies, various transportation improvement strategies to address those needs;
- Developing long-range plans and short-range programs of alternative capital improvement and operational strategies for moving people and goods;
- Estimating the impact of recommended future improvements to the transportation system on environmental issues, including air quality; and
- The development of a financial plan to ensure sufficient income to cover the costs of implementing strategies.

In order to consider these aspects is important to study them into an urban infrastructure scope [25].

Urban Infrastructure

Urban infrastructure, a human creation, is designed and directed by architects, civil engineers, urban planners among others. These professionals design, develop and implement projects (involved with the structural organization of cities and companies) for the proper operation of important sectors of society. When governments are responsible for construction, maintenance, operation and costs, the term "urban infrastructure" is a synonym for public works. Road infrastructure is the set of facilities and equipment used for roads, including road networks, parking spaces, traffic lights, stop signs laybys, drainage systems, bridges and sidewalks. Urban infrastructure includes transportation infrastructure, which in turn, can be divided into three categories: land, sea, and air, they can be found in the following modalities:

The problem in the case of Mexico City is the fragmented government that makes more difficult to implement strategies for plans. This is shown in next Fig. 2.

"Such institutional and operational fragmentation has significant implications especially for users. In Buenavista—an area of Mexico City where three modes of transport converge—travelers must walk up to 1.5 km to transfer from one mode to another. About 150,000 people use this disconnected transport hub everyday"

2.1 Transportation Analysis

Manage and plan the services of cities entails a lot of work and participation of experts in different areas. Such is the case of transport that currently represents a challenge for

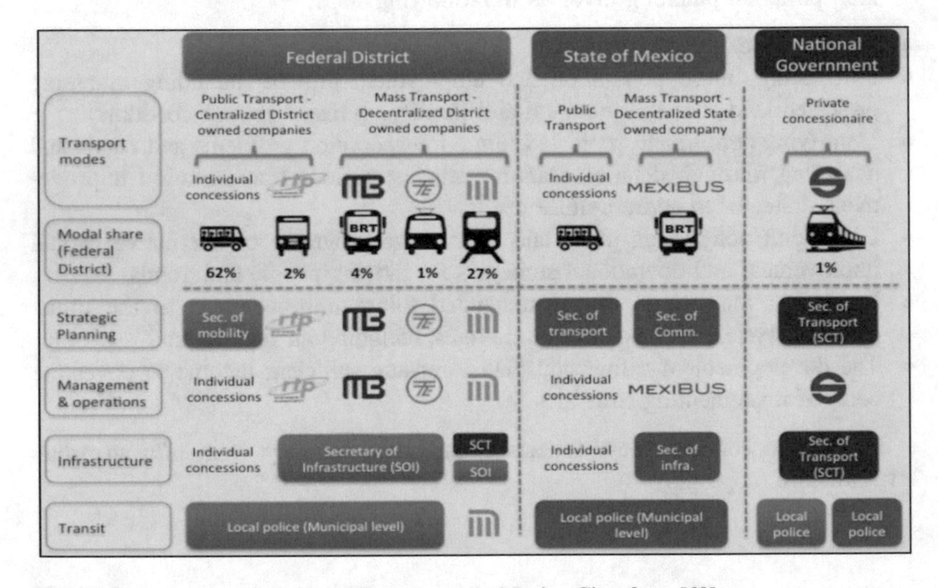

Fig. 2 Governance system for public transport in Mexico City, from [60]

researchers from different areas. There are three measures used for transportation analysis: traffic, mobility and accessibility [43]. As is observed in Fig. 3, the aspects taken into account to compare the three measures are definition of transportation, unit of measure, modes considered, assumptions concerning what benefits consumers, consideration of land use and favored transport improvement strategies (Fig. 4).

Fig. 3 Modal connection in Buenavista, adapted from [60]

	Traffic	Mobility	Access
Definition of Transportation	Vehicle travel.	Person and goods movement.	Ability to obtain goods, services and activities.
Unit of measure	Vehicle-miles and vehicle-trips	Person-miles, person-trips and ton-miles.	Trips.
Modes considered	Automobile and truck.	Automobile, truck and public transit.	All modes, including mobility substitutes such as telecommuting.
Common performance indicators	Vehicle traffic volumes and speeds, roadway Level of Service, costs per vehicle-mile, parking convenience.	Person-trip volumes and speeds, road and transit Level of Service, cost per person-trip, travel convenience.	Multi-modal Level of Service, land use accessibility, generalized cost to reach activities.
Assumptions concerning what benefits consumers.	Maximum vehicle mileage and speed, convenient parking, low vehicle costs.	Maximum personal travel and goods movement.	Maximum transport options, convenience, land use accessibility, cost efficiency.
Consideration of land use.	Favors low-density, urban fringe development patterns.	Favors some land use clustering, to accommodate transit.	Favors land use clustering, mix and connectivity.
Favored transport improvement strategies	Increased road and parking capacity, speed and safety.	Increased transport system capacity, speeds and safety.	Improved mobility, mobility substitutes and land use accessibility.

Fig. 4 Comparing transportation measurements, reproduced from [43]

Litman [43] defines these three measures as follows:

Traffic Definition

Traffic refers to vehicle movement. This perspective assumes that "travel" means vehicle travel and "trip" means vehicle-trip. It assumes that the primary way to improve transportation system quality is to increased vehicle mileage and speed.

Mobility Definition

Mobility refers to the movement of people or goods. It assumes that "travel" means person- or ton-miles, "trip" means person- or freight-vehicle trip. It assumes that any increase in travel mileage or speed benefits society.

Accessibility Definition

Accessibility (or just access) refers to the ability to reach desired goods, services, activities and destinations (collectively called opportunities). Access is the ultimate goal of most transportation, except a small portion of travel in which movement is an end in itself (jogging, horseback riding, pleasure drives), with no destination. This perspective assumes that there may be many ways of improving transportation, including improved mobility, improved land use accessibility (which reduce the distance between destinations), or improved mobility substitutes such as telecommunications or delivery services."

For transportation analysis it is important to consider diverse measures that are used for it, and according to the selected method, different results are obtained. In this chapter we use three different measures in order of importance according to the level of analysis in three levels; macro, mezzo and micro as it will be explained below. It is important to note that sustainability and quality of life of the inhabitants are priority for any proposal or alternative arises.

2.2 Sustainable Urban Transport Infrastructure

According to HABITAT [31] mean by sustainable mobility the following:

Sustainable Urban Mobility: The goal of all transportation is to create universal access to safe, clean and affordable transport for all that in turn may provide access to opportunities, services, goods and amenities. Accessibility and sustainable mobility is to do with the quality and efficiency of reaching destinations whose distances are reduced rather than the hardware associated with transport. Accordingly, sustainable urban mobility is determined by the degree to which the city as a whole is accessible to all its residents, including the poor, the elderly, the young, people with disabilities, women and children. Moreover quality of life and sustainability corresponds to [30]:

In its original definition, sustainable development focuses on "meeting the needs of the present without compromising the ability of future generations to meet their own needs" [45]. The fulfillment of needs is not only a precondition for sustainable development but also for individual well-being and thus for a high quality of life.

Quality of life is most commonly defined as consisting of two parts, the objective (the resources and capabilities that are given for a person) and the subjective (the well-being of a person).

There are some sustainable and environmental friendly transport indicators recommended [46] as reported by the European Environmental Agency (EEA) in Copenhagen suggests that appropriate environmental indicators should be able to respond to the following simple questions: what is actually happening of environmental change? is it related to (significant) policy goals? is progress possibly measurable? moreover, how does overarching welfare development influenced? important criteria to select suitable indicators that are both descriptive, able to measure performance as well as progress, are thus that they are:

- Policy relevant, consisting of parameters that actually might be influenced by policy and administration;
- Accessible for measuring and comparison—over time or in space; in goals versus results;
- Representative and valid, covering a broad scope of the environmental problems at stake;
- Reliable and, based on accessible data, of high quality with regular updating;
- Simplified, able to manage and reduce complex relationships;
- Informative in order to promote an improved policy performance and broader understanding of the environment transport relationships

Drawing on well-established international indicator sets on environment and transport, ideal and possible (accessible) indicators are discussed, and an indicator for environmentally friendly urban transport is suggested, divided in five main areas: driving forces, transport factors, environmental factors, urban and societal impacts from transport, urban planning, policies and measures (Fig. 5).

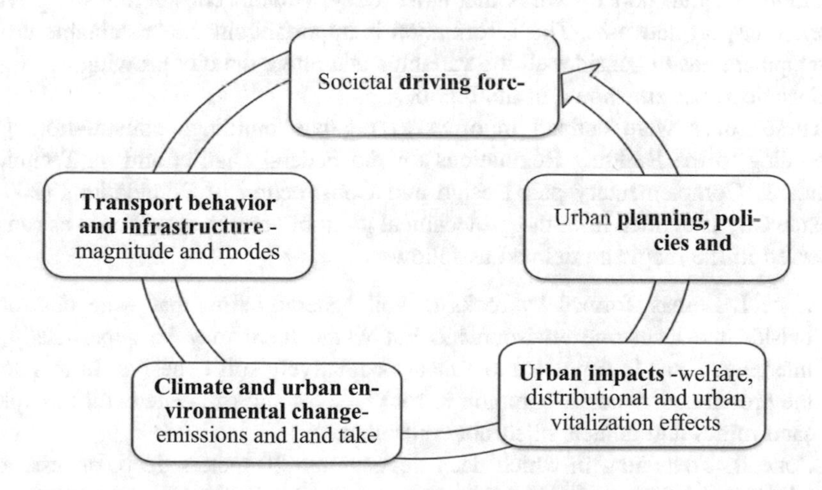

Fig. 5 Indicators for urban transport, environment and climate. From [46]

Other authors offer a slight different view about indicators as Paz et al. [49] that points out: Numerous studies have established different measures to quantify sustainability [64]. According to Bell and Morse [7], sustainability primarily is measured by means of three components: (i) time scale, (ii) spatial scale, and (iii) system quality. The time and spatial scale corresponds to the analysis period and the geographical region of interest, respectively. On the other hand, system quality corresponds to the quantification of the overall system performance or state. In order to quantify system quality, Sustainability Indicators (SIs) have been developed in a diverse range of fields, including biology and the life sciences, hydrology, and transportation.

It is clear that a truly sustainable state for a system requires all the relevant interdependent subsystems/sectors and components, at levels so that the consumption of and the impact on the natural and economic resources do not deplete nor destroy those resources. Hence, the assessment of a system state requires a holistic analysis in order to consider all the relevant sectors and impacts. [49].

As Paz et al. [49] say the analysis should be holistic, and we agree with it, just the approach is different since they propose a study of a system of systems and use fuzzy logic for qualitative indicators.

2.3 The Public Transport Network in Mexico City Context

Mexico City like all other cities has very specific features as the subsoil conditions, and the geographical location; as it is a seismic zone and is filmed by mountains, has two nearby volcanoes and was a lake 1500 years ago. So everything with regard to infrastructure, urban development and air as quality water have to be considered in a study on mobility. The following maps show aspects such as subsoil, environmental pollution and transport networks that exist today, without considering the private public transport networks. This information is important since a sustainable urban development has to consider all the variables that affect the city growing.

Seismic zones are shown in the Fig. 6.

These zones were defined in order to regulate buildings construction, [4]. According to the Building Regulations for the Federal District and its Technical Standards Complementary pair Design and Construction of Foundations (2004), Mexico City is divided from the geotechnical point of view in three zones as can be observed in the map, and defined as follows:

(a) Zone I. Lomas, formed by rocks or soil generally firm that were deposited outside the lacustrine environment, but where there may be superficially or interleaved, sandy deposits loose state or relatively soft cohesive. In this area, the presence of voids is common in rocks, caves and excavated soil to exploit sand mines and tunnels filled not controlled;
(b) Zone II. Transition, in which deep deposits are 20 meters deep, or less, and which it consists predominantly sandy and sandy silt layers interspersed with

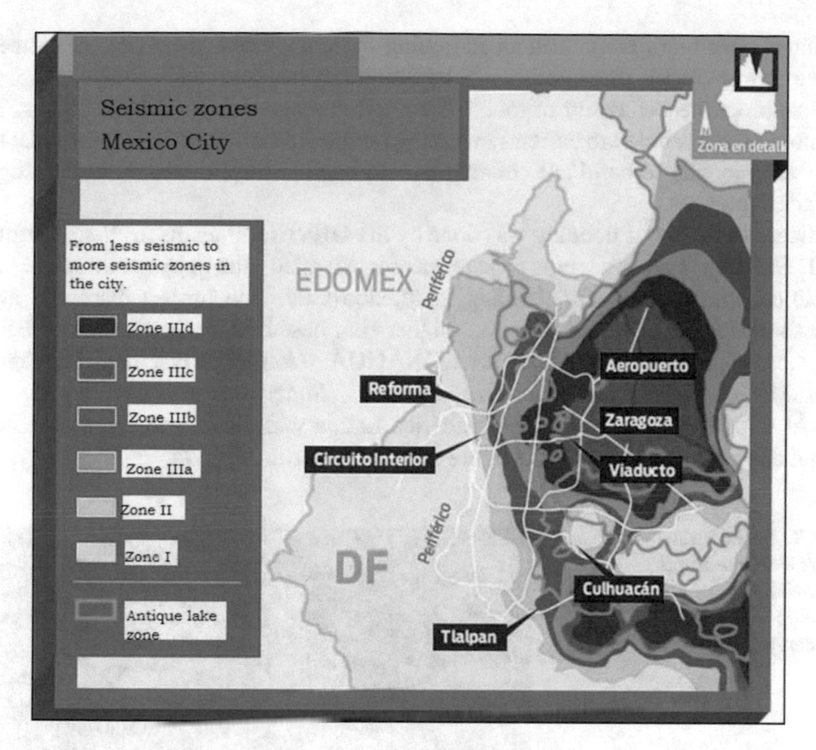

Fig. 6 Seismic zones in Mexico City. *Source* http://www.eluniversaldf.mx/home/especial-en-que-zonas-se-sienten-mas-los-sismos.html

layers lacustrine clay; the thickness thereof varies between a few tens of centimeters and meters;

(c) Zone III. Lacustrine, composed of powerful deposits of highly compressible clay, separated by layers with different sandy silt or clay content. These layers are generally fairly sandy compact to very compact and variable thickness from centimeters to several meters.

Lacustrine deposits usually they covered superficially by alluvial soils, dried materials and artificial fillers; the thickness of this set can be greater than 50 m.

Geotechnical anomalies within the lake area. Auvinet [4].

The lake area is far from having uniform characteristics. In this area there are sites easily where the subsoil has identifiable characteristics. It stresses in particular the existence in the historic center of prehispanic thick fillings. Many farms have on the other hand a complex loading history under colonial buildings; some of them have now disappeared, amending substantially the behaviour of the subsoil under the weight of buildings and seismic conditions.

A similar situation occurs along traces of old roads or albarradones, in areas of channels that were filled and places of ancient human settlements established in all islands or partially artificial lakes within the former, known as tlateles (Tlatelolco,

Tlahuac, Iztacalco, etc.), without forgetting the chinampas areas. The presence of these abnormalities, often undetected by designers, has been the source of problems of inappropriate behaviour of foundations and damage structural in buildings. The authors of this article are currently working on a micro zoning to bring the risks that may arise locally to build in a certain place and define recommendations to mitigate its consequences.

Other study about flooding was done by the DEVELOP teams in Wise, Virginia, and Saltillo, Mexico, and researches investigated the physical, social and socio-economic aspects of flooding in Mexico City. The project discerned areas most susceptible to flooding and of higher risk based on socio-economic characteristics. The team partnered with CONAGUA (Comisión Nacional del Agua), ITESM (Instituto Tecnológico de Estudios Superiores de Monterrey), and CAALCA (Centro del Agua para América Latina y el Caribe) to assist with decisions and policy making. Next figure shows the result (Fig. 7).

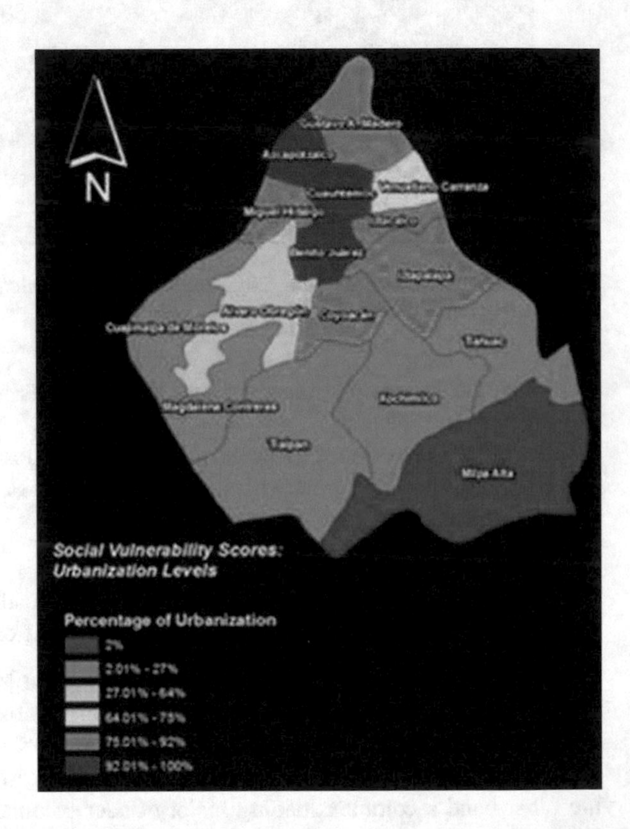

Fig. 7 Social vulnerability scores. *Source* http://earthzine.org/category/develop-virtual-poster-session/page/29/

Puente [50] states that, a methodology for assessing urban vulnerability rests on two premises: (1) the material conditions of a city are good indicators of vulnerability; (2) the main components of vulnerability can be mapped at the scale of urban neighborhoods. Based on them last step consists on creating a matrix that displays the appropriate indicators (factors) on one axis and the areal units of analysis on the other axe.

For the purposes of this chapter, we just mention some of them.

Another important factor is air pollution in Mexico City, "Environmental pollution is an increasingly serious problem in third world cities. Pollution arises from both fixed and mobile sources. Industrial facilities in the mega-cities of developing countries have rarely been subject to policies of pollution control. Equally important is pollution generated by urban transportation systems, especially those that

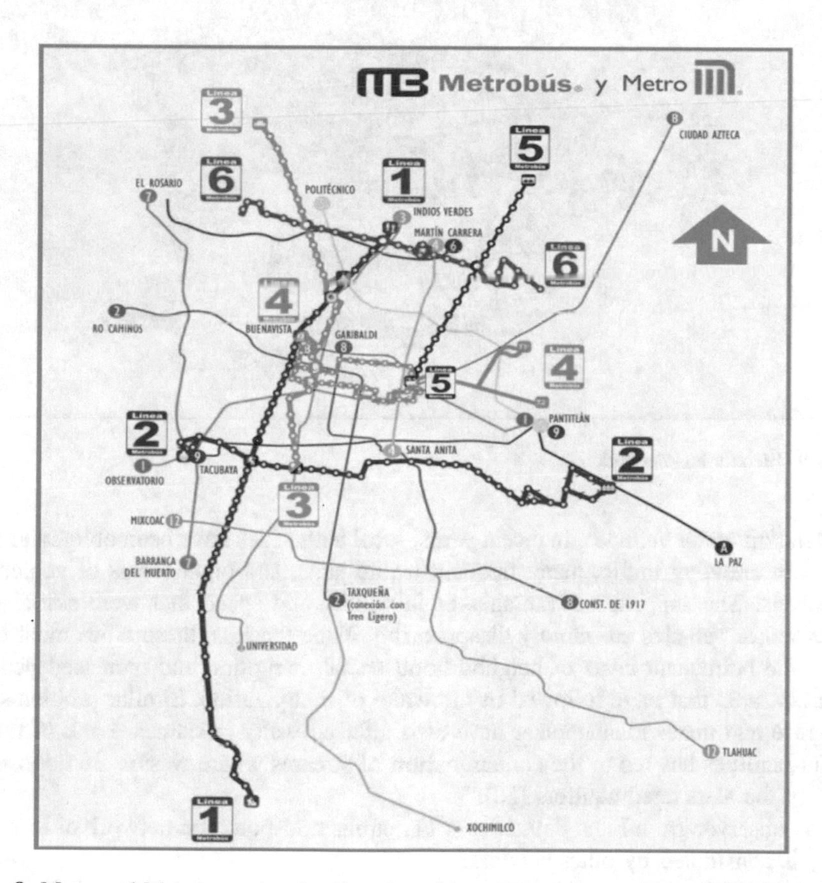

Fig. 8 Metro and Metrobus networks. From http://www.juliotoledo.com/mapas%20juliotoledo

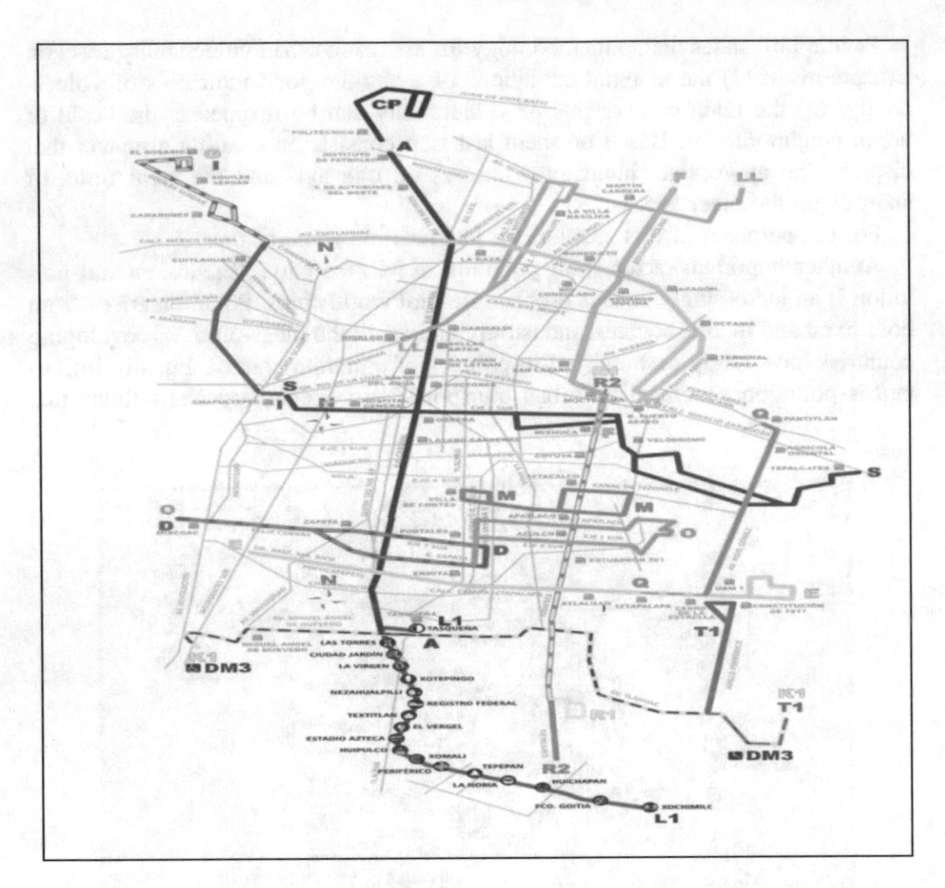

Fig. 9 Electric bus network

depend on motor vehicles. In recent years, local authorities have been obliged to put up with crawling traffic, many frequent traffic jams, and other forms of vehicular paralysis. The supposed advantages of flexibility and speed that were associated with motor vehicles are rapidly disappearing. None the less, these cities must live with the permanent costs of neighborhood social disruption and increased pedestrian hazards that have followed in the wake of motorization. Similar problems of overuse and under management have also affected water resources. Lack of treatment facilities has led to the contamination of streams where wastes are deposited and of the associated aquifers [50]."

As observed from Figs. 8, 9, 10 and 11, public transportation network of Mexico City is constituted by other networks.

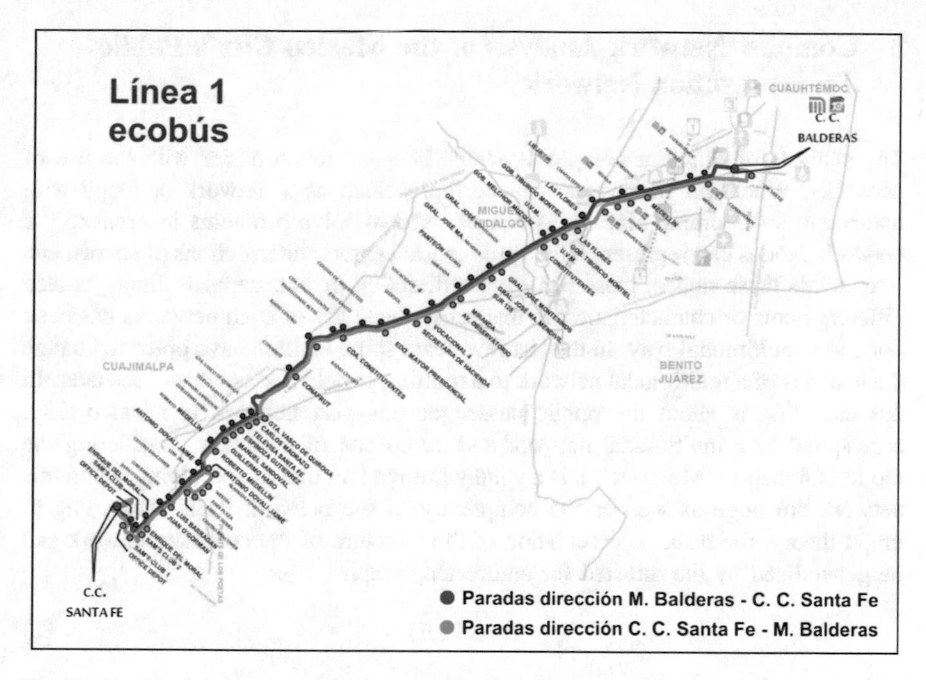

Fig. 10 Eco bus line 1. *Source* http://modulom1.blogspot.mx/2014/12/servicio-ecobus.html

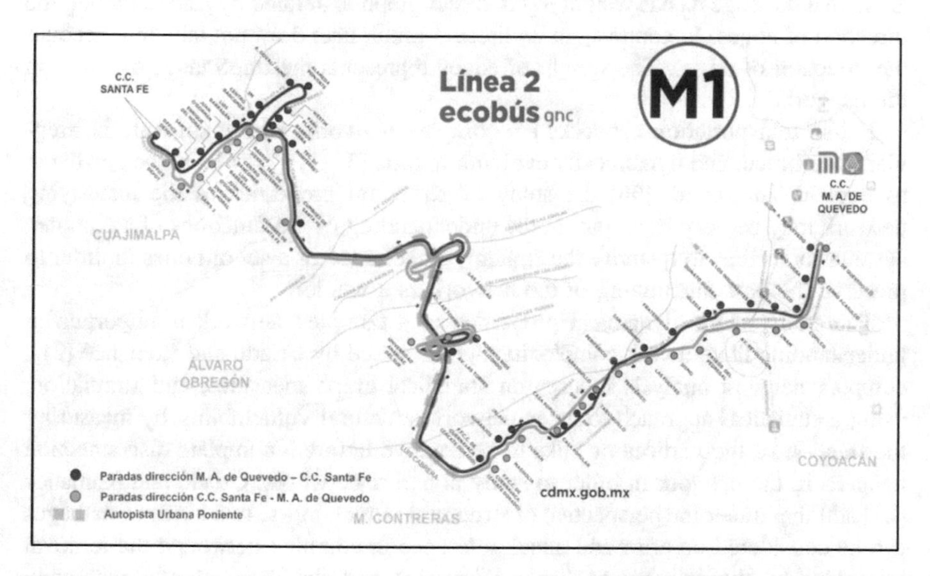

Fig. 11 Eco bus line 2. *Source* http://modulom1.blogspot.mx/2014/12/servicio-ecobus.html

3 Complex Network Analysis of the Mexico City's Public Transportation Network

The analysis of different complex systems is made much easier with the use of networks. Whenever the system can be represented as a network or graph with nodes and arcs, simple algorithms can be used to solve problems inherent to the network. Nodes can represent cities, production centers, intersections of streets, etc. Arcs relate these nodes, these can have a direction or not, capacity limits or also different items or characteristics, in that sense the study of such networks has been done in a multimodal way. In the last few years some authors have opted to change the analysis of a multimodal network to a multilayer network as we will see later. In our case that is about the public passenger transport network in Mexico City, correspond to a multimodal network and composed of networks considering the mode of transport. Moreover, it is a widely known fact that the problems facing this network are huge as well as the complexity of the network itself. According to graph theory, the basic representation of the structure of the complex network can be generalized by the directed (or undirected) graph

$$G = \langle V, E \rangle \tag{1}$$

where V describes set of nodes (vertices) and E describes set of arcs (edges) that compose the network. Let $W = (w_{ij})$ be the adjacency matrix associated to the graph G, so that the edge e_{ij} has weight w_{ij}. A direct graph is defined by differentiating the direction of edges. In contrast, an undirected graph take does not take into account the direction of edges. The weight of edges represents the importance of edges in the network.

Public transportation networks are complex networks whose structure is irregular, distributed, and dynamically evolving in time [5, 13, 17, 54]. On the one hand, as explain Thai et al. [56] the study of structural properties of the underlying network may be very important in the understanding of the functions of a complex systems as well as to quantify the strategic importance of a set of nodes in order to preserve the best functioning of the network as a whole.

The study of the dynamical properties of a complex network is important in understanding the network complexity. As discussed by Criado and Romance [21], complex network analysis focuses on statistical graph measures, and simulation, using a statistical approach to asses network structural vulnerability by measuring the fraction of the vertices or links to be remove before a complete disconnection happens in the network in order to study complex networks. Criado and Romance [21] add that under the perspective of structural vulnerability, two kinds of damages can be considered on error and attack tolerance in complex networks: the removal of randomly chosen vertices (error tolerance) and the removal of deliberately chosen vertices (attack tolerance).

In this section, we model and simulate the public transportation network in Mexico City from the complex network perspective to asses network structural vulnerability and resilience, considering mobility and accessibility aspects.

To model the Mexico City's public transportation network as a complex network and evaluate their empirical characteristics we used Gephi, and open source software for the visual exploration of complex networks developed since 2008.

Gephi was created by Mathieu Bastian, Sebastien Heymann, and Mathieu Jacomy, and extended by Eduardo Ramos Ibañez, Cezary Bartosiak, Julian Bilcke, Patrick McSweeney, André Panisson, Jeremy Subtil, Helder Suzuki, Martin Skurla, and Antonio Patriarca from Web Atlas. It is suitable for the analysis of all kind of complex networks. For the purpose of this study, we consider the public transportation network in Mexico City as a whole taking into account the *trolebus, metro, metro-bus, ecobus, tren ligero and suburbano transportation systems.*

It is important to note that the original configuration of public transport networks from the network analysis was L-space, also referred as the space of stops or space of stations, in which stops or stations are vertices. In this way, two vertices are connected on an arbitrary route [42, 53]. In this study, we built a complex network where a node represents a station from the public transportation networks in Mexico whereas a directed arc represents the physical connection between two stations. The weight of an arc represents the physical linear distance between two stations. The complex network consists of 923 nodes and 1203 arcs. The layouts included in Gephi are algorithms that position the nodes in the 2-D or 3-D graphic space. The patterns created, based on the different layouts, emphasis the properties of the structure of networks. For instance, using the force-algorithms the connected nodes tend to be closer, while disconnected nodes tend to be further [38].

The force directed layout optimizes Martin et al. [44]:

$$\min_{x, \ldots, x_{n1}} \sum_i \left(\sum_j (w_{ij} d(x_i, x_j)^2) + D_{x_i} \right), \tag{2}$$

where x_i are positions of nodes, w_{ij} are arcs weights and D_{x_i} is the density of edges near x_i.Where D_{xi} denotes the density of the points $x_1, \ldots, x_n$ near x_i. The sum in (2) contains both an attractive and a repulsive term. The attractive term $\sum_j (w_{ij} d(x_i, x_j)^2)$ attempts to draw together nodes, which have strong relations via w_{ij}. The repulsive term D_{xi} attempts to push nodes into areas of the plane that are sparsely populated. The minimization in (2) is a difficult nonlinear problem. For that reason, we use a greedy optimization procedure based on simulated annealing Martin et al. [44]. The procedure is greedy in that we update the position of each vertex by optimizing the inner sum $\sum_j (w_{ij} d(x_i, x_j)^2) + D_{x_i}$ while fixing the positions of the other nodes.

In order to select the pertinent layout in Gephi software (that means random, force atlas, Fruchterman and Reingold [28], Noverlap, OpenOrd, Hu [34]); it is important to take into account the capability of the algorithm to handle the given

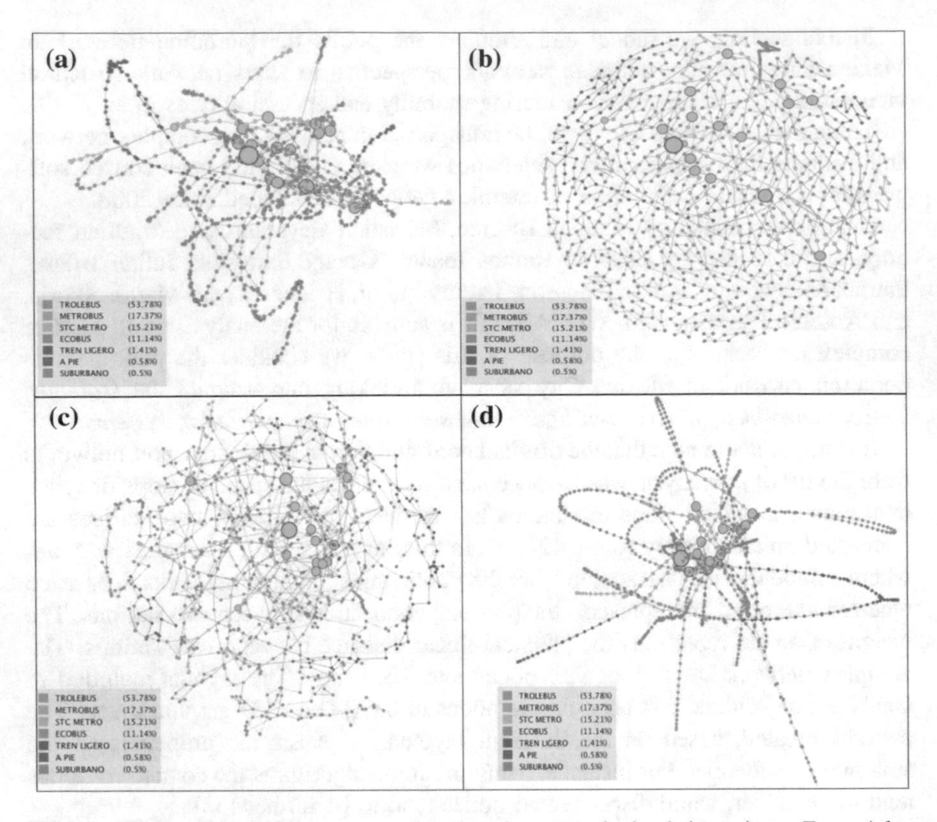

Fig. 12 Mexico City's public transportation complex network simulation using **a** Force Atlas, **b** Fruchterman Reingold, **c** OpenOrd, and **d** Yifan Hu algorithms

data (nodes and arcs), the user time constraint, and the structural network properties to analyze.

For our simulation, we have used Force Atlas, Fruchterman and Reingold [28], and Hu [34] algorithms. As Dey and Roy [24] due to Force Atlas algorithm uses different techniques such as degree-dependent repulsive force, Barnes Hut simulation, and adaptive temperatures for their simulation process.

In this direction, Dey and Roy [24] add that the main idea of simulation is that the nodes repulse and the arcs attract. The network layout using Force Atlas algorithm is shown in Fig. 12a.

Fruchterman and Reingold [28] propose to model a continuous network depending on even distribution of the nodes, making arc lengths uniform and reflects inherent symmetry. The network layout using [28] algorithm is shown in Fig. 12b.

Cherven [19] notes that the OpenOrd algorithm helps to generate network graphs very fast, and is best suited to very large networks that operate at a very high rate of speed while providing a medium degree of accuracy.

The importance of using OpenOrd algorithm is because it uses edge-cutting, average-link clustering, multilevel graph coarsening, and a parallel implementation of a force directed method based on simulated annealing Martin et al. [44]. An advantage of this algorithm over Fruchterman and Reingold [28] one is that for large graphs is the running time, Fruchterman and Reingold [28] is $O(n^2)$ in the number of nodes n, The running time can be improved using a grid based density calculation, and by employing a multilevel approach Martin et al. [44].

The goal of OpenOrd is to draw G in two dimensions. Let $x_i = (x_{i,1}, x_{i,2})$ denote the position of v_i in the plane. OpenOrd draws G by attempting to solve Eq. (2).

All nodes are initially placed at the origin, and the update is repeated for each node in the graph to complete one iteration of the optimization. The iterations are controlled via a simulated annealing type schedule, which consists of five different phases: liquid, expansion, cool-down, crunch, and simmer Martin et al. [44].

During each stage of the annealing schedule, authors vary several parameters of the optimization: temperature, attraction, and damping. These parameters control how far nodes are allowed to move. At each step of the algorithm, they compute two possible node moves. The first possible move is always a random jump, whose distance is determined by the temperature. The second possible move is analytically calculated (known as a barrier jump22). This move is computed as the weighted centroid of the neighbors of the vertex. The damping multiplier determines how far towards this centroid the vertex is allowed to move and the attraction factor weights the resulting energy to determine the desirability of such a move. Of these two possible moves, we choose the move which results in the lowest inner sum energy $\sum_j (w_{ij}d(x_i, x_j)^2) + D_{x_i}$ (Part of Eq. 2).

OpenOrd uses simulated annealing to solve the problem of Eq. (2). The network layout using OpenOrd algorithm is shown in Fig. 12c.

As Cherven [19] states, Fruchterman and Reingold [28] algorithm produces faster results compared to other force-directed methods by focusing on attraction and repulsion at the neighborhood (rather than the entire network) level. The network layout using Yifan Hu algorithm is shown in Fig. 12d.

3.1 Statistical Graph Measures of the Mexico City's Public Transportation Complex Network

According to complex networks framework is necessary to have some measures as centrality ones, in order to answer the question "What is the most important or central node in a given network?" Centrality measures (defined below) are the most basic and frequently used methods for analysis of complex networks Tarapata [56].

Based on this, here is a list of some statistical graph measures from Eq. (3) to Eq. (9) to evaluate the empirical characteristics of the Mexico City's public transportation complex network mostly based on Dey and Roy [24] and Tarapata [55].

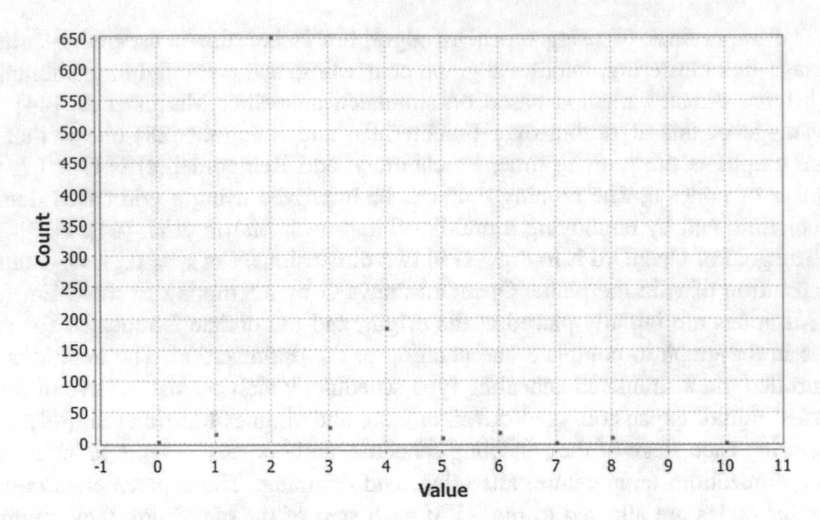

Fig. 13 Mexico City's distribution degree for complex public transport network

Mean Degree

Degree k_i is defined as the number of links connected to the node. The mean degree represents the average degree of all nodes in a network.

$$\langle k \rangle = \sum_{i=1}^{N} k_i / N \tag{3}$$

where $|V| = N$ the average degree calculated was 2.607 (Fig. 13).

Connectivity and Accessibility

According to Rodrigue et al. [52], accessibility is defined as the measure of the capacity of a location to be reached by, or to reach different locations therefore, the capacity and the structure of transport infrastructure are key elements in the determination of accessibility. Following Rodrigue et al. [52], two spatial categories are applicable to accessibility problems: topological accessibility and contiguous accessibility. In the first case, it is related to measuring accessibility in a system of nodes and paths, for instance a transportation network, assuming that accessibility is a measurable attribute significant only to specific elements of the transportation system. In the second case, the measure of accessibility is carried out over a surface, being a measurable attribute of every location, as space is considered in a contiguous manner.

Rodrigue et al. [52] adds that the most basic measure of accessibility involves network connectivity through the degree node. As shown in Table 4, the nodes Bellas Artes and Aquiles Serdan of the Mexico City's public transportation complex network are the most connected. These nodes are subway stations from line 2 and 7. Based on the average degree calculated, and considering the Mexico City's

Table 4 Most connected nodes of the Mexico City's public complex transportation network

Station	Transport type	Transport line	Degree	Weighted degree
Bellas Artes	STC metro	2	10	5892
Aquiles Serdan	STC metro	7	10	6721
Balderas	STC metro	1	8	4437
Miguel Angel de Quevedo	STC metro	3	8	5573
Oceania	STC metro	5	8	6017
Av. Copilco	Trolebus	K1	8	4876
Centro Bancomer	Ecobus	34B	8	5068
Luis Barragan	Ecobus	34B	8	5068
Juan Ogorman	Ecobus	34B	8	5068
Enrique del Moral	Ecobus	34B	8	5068
Sams 1	Ecobus	34B	8	5068
Office Depot	Ecobus	34B	8	5068
Tacubaya	STC metro	1	7	6724
La Raza	STC metro	3	7	5752
Salto del Agua	STC metro	1	6	3043
Cuahutemoc	STC metro	1	6	3426
Sevilla	STC metro	1	6	3370
Chapultepec	STC metro	1	6	3698
Hidalgo	STC metro	2	6	3387
Chabacano	STC metro	2	6	5433
Jamaica	STC metro	4	6	4602
Juarez	STC metro	3	6	3444
Centro medico	STC metro	3	6	5117
Mixcoac	STC metro	7	6	4106
San Juan de Letran	STC metro	8	6	3316
Doctores	STC metro	8	6	3893
Lazaro Cardenas	STC metro	9	6	4627
Centro Scop	Metro bus	2	6	3650
San Lazaro	Metro bus	4	6	4088
Dr. Aceves	Trolebus	A	6	2960
Calz. De los Misterios	Trolebus	G	6	2582
Salonica	Trolebus	G	6	2728
Deportivo 18 de marzo	STC metro	3	5	3927
Zapata	STC metro	3	5	2967
Politecnico	STC metro	5	5	2124
Garibaldi	STC metro	8	5	2495
Insurgentes Sur	STC metro	12	5	2717
Deportivo 18 de marzo	Metro bus	1	5	2586

(continued)

Table 4 (continued)

Station	Transport type	Transport line	Degree	Weighted degree
Patriotismo	Trolebus	D	5	1985
Insurgentes	Trolebus	D	5	1595
20 de Noviembre	Trolebus	D	5	1595
Division del Norte	Trolebus	D	5	1595

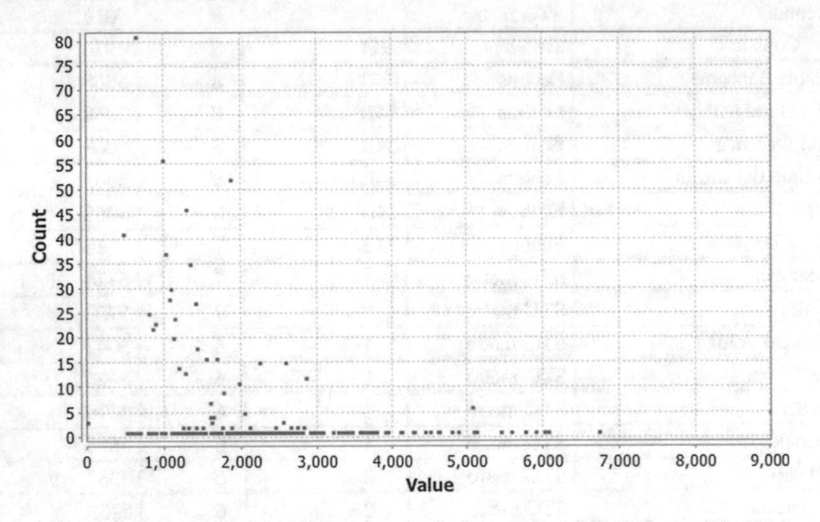

Fig. 14 Mexico City's public complex transportation network weighted degree distribution

public transportation complex network as a whole, this degree was 2.607, that means this kind of network has a low accessibility.

Weighted Degree Distribution

Considering that the weight of an edge represents the physical linear distance between two stations, we calculate the average weighted degree.

The weighted degree of a node is like the degree. It's based on the number of edge for a node, but ponderated by the weight of each edge. It's doing the sum of the weight of the edges.

For example, a node with 4 edges that weight 1 ($1 + 1+1 + 1 = 4$) is equivalent to:

a node with 2 edges that weight 2 ($2 + 2 = 4$) or
a node with 2 edges that weight 1 and 1 edge that weight 2 ($1 + 1+2 = 4$) or
a node with 1 edge that weight 4 etc.…

In the Mexico City case and based on the Table 4, the weighted degree is 1539.835 (see Fig. 14).

Betweenness Centrality
A node is central if it structurally lies between many other nodes, in the sense that it is transversed by many of the shortest paths connecting pairs of nodes. The betweenness centrality is defined as follows.

$$bc_i = \sum_{l \in V} \sum_{k \neq l \in V} \frac{p_{l,i,k}}{p_{l,k}} \tag{4}$$

where $p_{l,i,k}$ count of the shortest paths in G between l and k nodes visiting the i-th node, $p_{l,k}$ count of the shortest path in G between l and k nodes. The higher bc_i value, the better (the i-th node is more important or more central). In order to calculate the betweenness centrality, the Gephi software uses A Faster Algorithm for Betweenness Centrality Brandes [12]. The betweenness centrality distribution is shown in Fig. 15.

Eccentricity Distribution
As Hage and Harary [32] states, the eccentricity ec_i of the i-th node is calculated using Eq. (5).

$$ec_i = max_{j \in V} d_{ij} \tag{5}$$

where, d_{ij} represents the length of the shortest path in G between the i-th, and the j-th node (number of edges on the shortest path from l to j). The lower ec_i value, the better (the i-th node is more important or more central). In order to calculate the eccentricity, the Gephi software uses A Faster Algorithm for Betweenness Centrality Brandes [12]. The eccentricity distribution calculated is shown in Fig. 16.

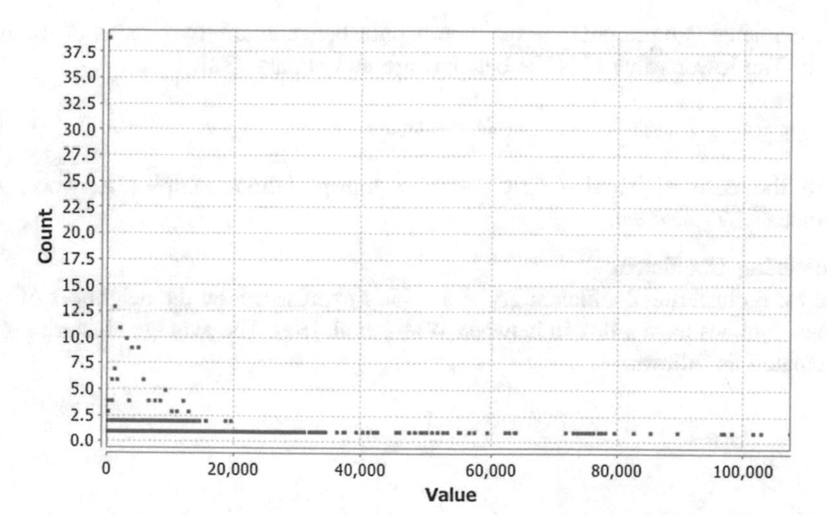

Fig. 15 Betweenness centrality distribution of Mexico City's public complex transport network

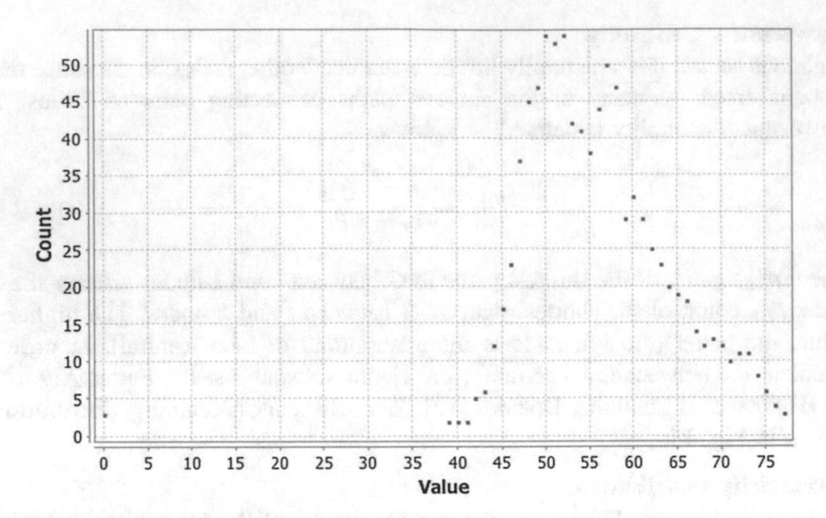

Fig. 16 Eccentricity distribution of Mexico City's complex public transport network

Average Shortest Paths Length
The average shortest paths length L denotes the average minimum distance between any two nodes. The lower L value is the better Watts et al. [62]. In this case, the average path length of the network was 23.611 km.

$$L = \frac{1}{N(N-1)} \sum_{i \neq j \in V} d_{ij} \tag{6}$$

Diameter
The diameter D represents the maximum path between any two nodes of the network. The lower value D is the better Hage and Harary [32].

$$D = max_{i \in V} ec_i \tag{7}$$

In the case of Mexico City's public transportation complex network, the diameter is 77 segments.

Clustering Coefficient
The local clustering coefficient gc_i of a node i expresses how the neighbors of two adjacent nodes have a link in between Watts et al. [62]. The average clustering C is calculated as follows.

$$C = \frac{1}{N} \sum_{i \in V} gc_i \tag{8}$$

$$gc_i = \frac{2E_i}{k_i(k_i - 1)}, \; k_i > 1 \tag{9}$$

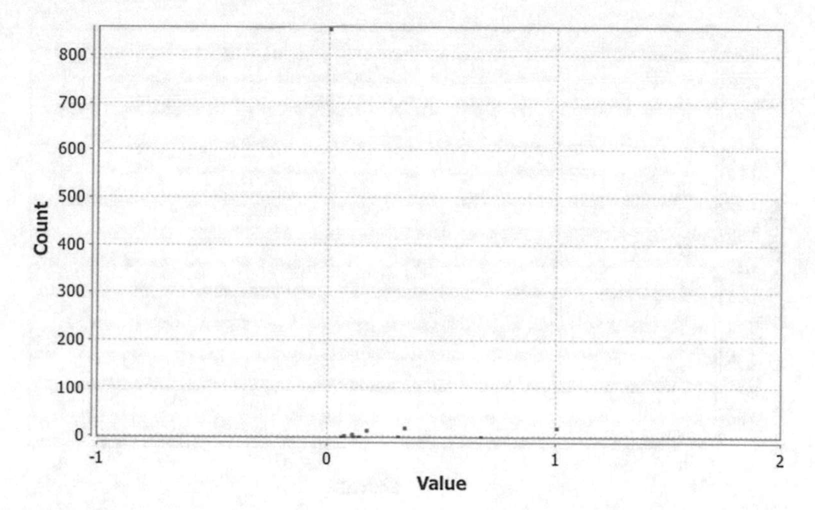

Fig. 17 Clustering coefficient distribution of the Mexico City's public complex transport network

Here, E_i count the edges between first-neighbours of the i-th node. The higher gc_i value, the better (the i-th node is more central). The average clustering coefficient was calculated using Gephi software, based on the algorithm proposed by Latapy [40], and is equal to 0.033. Figure 17 shows the clustering coefficient distribution of the Mexico City's public transportation complex network.

4 Complex Network Topology of the Mexico City's Public Transportation Network

4.1 Topology of the Mexico City's Public Transportation Network

Hubs Distribution

The hubs are nodes with much higher degrees than the average node degree. The occurrence of hubs tends to form clusters in the network. It is important to note that the hubs distribution is assessed in Gephi software based on the algorithm of Kleinberg [37]. The hubs distribution of Mexico City's public transportation complex network is shown in Fig. 18.

Authority Distribution

The authority is defined as nodes with the smaller degrees than the average node degree. The authority distribution is also assessed in Gephi software based on the algorithm of Kleinberg [37]. The authority distribution of Mexico City's public transportation complex network is shown in Fig. 19.

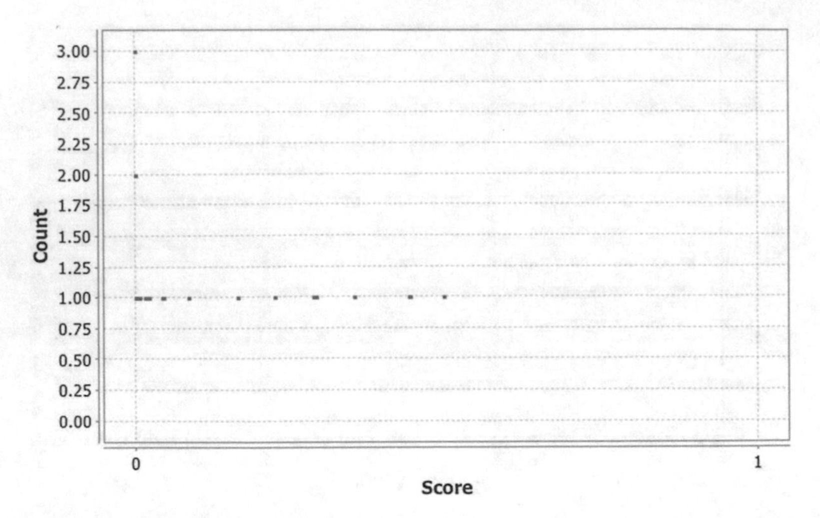

Fig. 18 Hubs distribution of Mexico City's public complex transport network

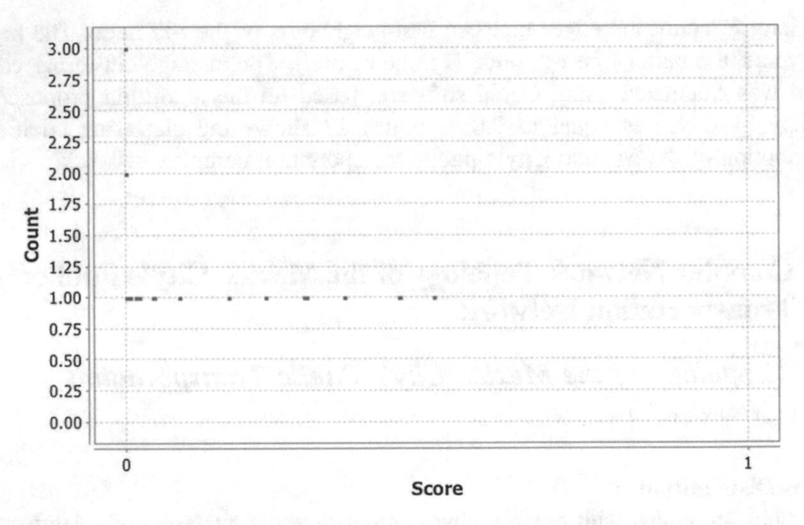

Fig. 19 Authority distribution for Mexico City's public complex transportation network

Modularity

Fortunato and Castellano [27] describe the modularity as the decomposition of the networks into sub-units or communities, which are sets of highly inter-connected nodes. Following Blondel et al. [10], the identification of such communities is of crucial importance as they help to uncover a priori unknown functional modules. As Fortunato and Castellano [27] explain: identify modules and their boundaries allow a classification of vertices, according to their topological position in the modules. In this direction, vertices with a central position in their cluster may have an important

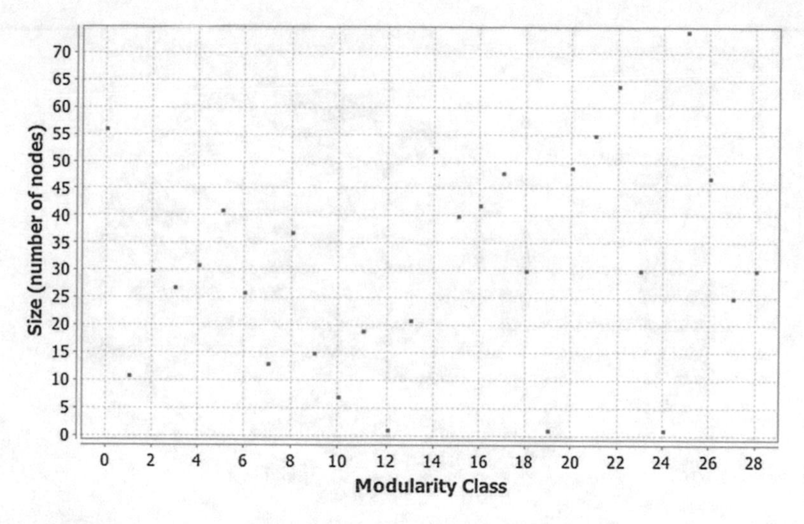

Fig. 20 Communities distribution size detected in the Mexico City's public complex transportation network

function, for instance, control and stability within the group, while vertices at the boundaries between modules play the role of mediation between different communities. Modularity analysis using Gephi software is based on the algorithm proposed by Blondel et al. [10]. It is important to note that the communities detection in graphs is based only on the topology. In the case of Mexico City's public transportation complex network, the modularity is 0.895 and 29 communities were detected. Figure 20 shows the size distribution of the communities detected in the Mexico City's public transportation complex network.

4.2 Assessment of Structural Vulnerability and Resilience of Mexico City's Public Transportation Complex Network Based on Simulation

According to Criado and Romance [21], under the perspective of structural vulnerability, two types of damage can be considered on error and attack tolerance in complex networks: the removal of randomly chosen nodes (error tolerance) and the removal of deliberately chosen nodes (attack tolerance). To analyze the resilience of the Mexico City's public transportation complex network, we remove nodes, which correspond to the stations of the *trolebus, metro, metro-bus, ecobus, trenligero and suburbano transportation systems*, and edges, which correspond to the physical distance between stations. We chose them both randomly and deliberately. In the Mexico City's public transportation complex network, eliminating 20% of stations

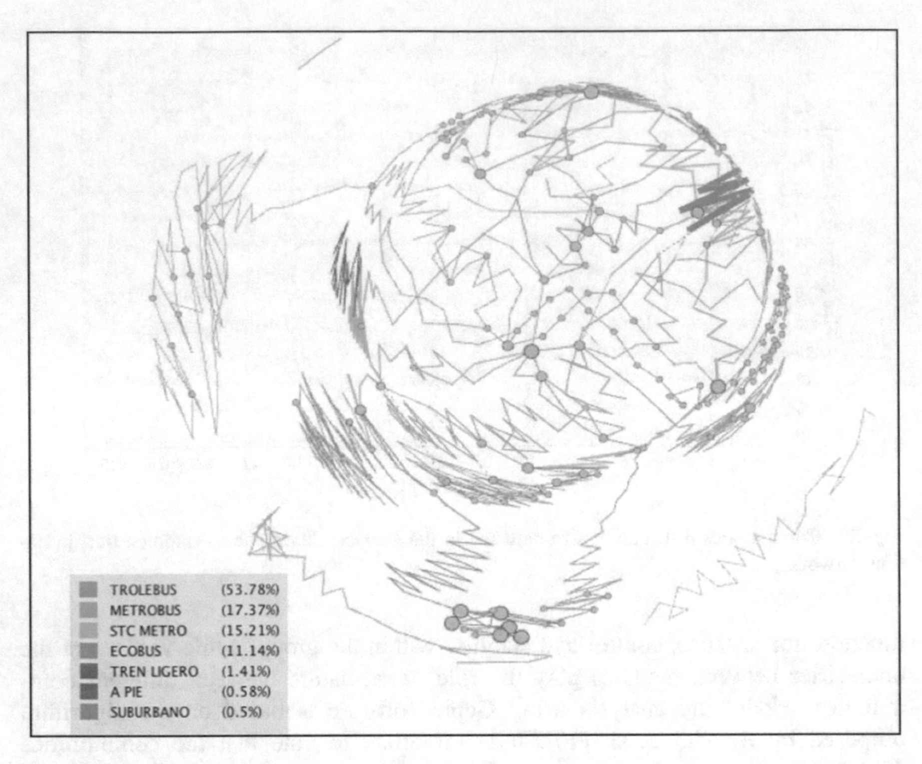

Fig. 21 Simulation of Mexico City public complex transportation network using Fruchterman Reingold algorithm, by eliminating 20% of edges randomly

randomly from the network (see Fig. 21), the average degree calculated reduces from 2.607 to 2.488 and the average weighted degree from 1539.835 to 1450.403.

Eliminating 30% of the highest degree nodes from the network (see Fig. 22), the average degree calculated reduces from 2.607 to 1.79 and the average weighted degree from 1539.835 to 1068.313. It is important to note that resilience and vulnerability conditions associated with the hubs can then affect the resilience/vulnerability of the whole network.

4.3 Multimodal Networks and Multilayer Networks

Krygsman et al. [39] observe that much of the effort associated with public transport trips is performed to simply reach the system and the final destination. In this sense, access and exit stations (together with wait and transfer times) are the weakest part of a multimodal public transport chain and their contribution to the total travel disutility is often substantial [11].

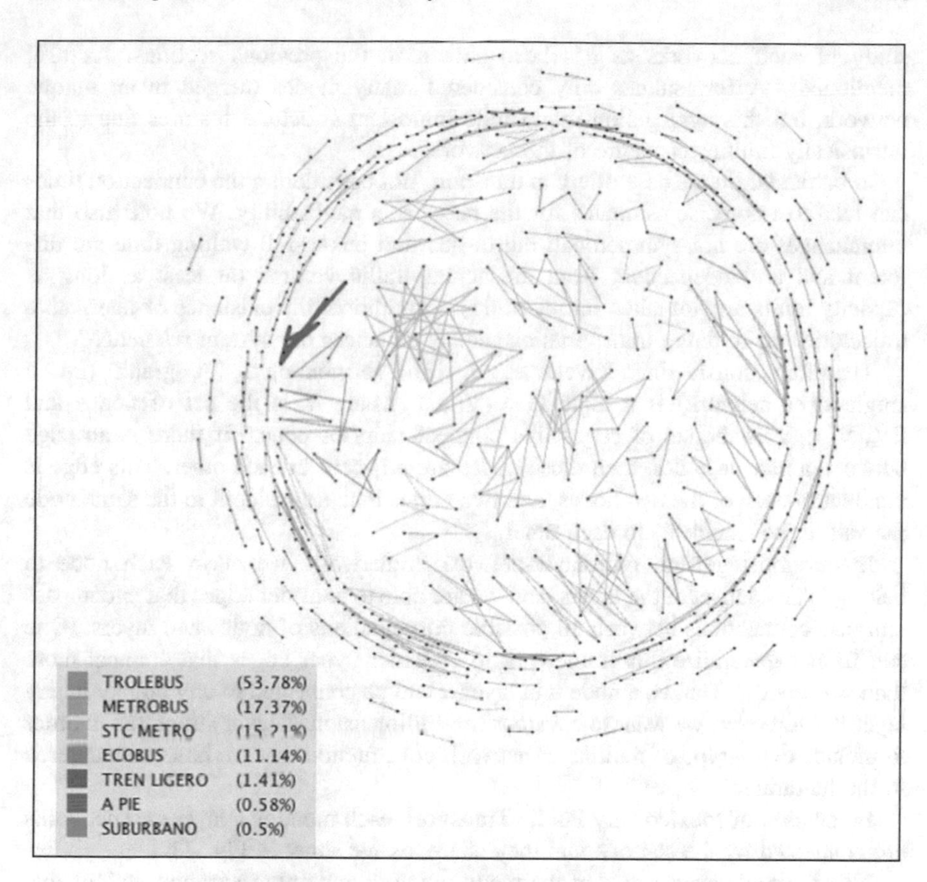

Fig. 22 Simulation of Mexico City public complex transportation network using Fruchterman Reingold algorithm, by eliminating 30% of the highest degree nodes

Access and exit determine, importantly, the availability (or the catchment area) of public transport [11, 45, 47]. Generally, an increase in access and egress (time and/or distance) is associated with a decrease in the use of public transport [18, 48].

In this direction, two scenarios are observed: in the first one, if access and egress exceed an absolute maximum threshold; users will not use the public transport system, while in the second one, if the access and egress trip components are acceptable, users may use the system, however; much will depend on the convenience of the system. Therefore, we consider that making public transport attractive, safe, self-sustaining and efficient to users is a task that must consider several aspects that are often over looked in studies of this type. Some of the factors that have not been considered are the connection between modes of transportation, which have to do with cycling, walking or using some short-route transport. This has to do with land use, climate and distance.

Due to the complexity of the system and considering the different transport modes and networks involved, it is important to take into account the complete study of such networks as has been shown in the previous sections. As [29] mentioned: “A few studies only considered many modes merged in an unique network, but this aggregation might hide important structural features due to the intrinsically multilayer nature of the network”.

In particular, in the case of urban transport, not considering the connection times can lead to imprecise estimates for the network’s navigability. We note also that interchanges are not symmetrical: rail-to-bus and bus-to-rail waiting time are different and are independent from the actual traffic volume (at least as long as capacity limits are not taken into account). In addition, the existence of alternative trajectories on different transportation modes enhances the system resilience”.

Therefore considering Kivela et al. [36] terminology: “A graph (i.e. a single-layer network) is a tuple G = (V, E), where V is the set of nodes and $E \subseteq V \times V$ is the set of edges that connect pairs of nodes. If there is an edge between a pair of nodes, then those nodes are adjacent to each other. This edge is incident to each of the two nodes, and two edges that are incident to the same node are said to be ‘incident’ to each other.

In our most general multilayer-network framework, we allow each node to belong to any subset of the layers, and we are able to consider edges that encompass pairwise connections between all possible combinations of nodes and layers. (One can further generalize this framework to consider hyper edges that connect more than two nodes.) That is, a node u in layer α can be connected to any node v in any layer β. Moreover, we want to consider ‘multidimensional’ layer structures in order to include every type of multilayer network construction that we have encountered in the literature.”

In the case of Mexico City Public Transport, each mode is a layer and networks are connected by the stations that they share, as we show in Fig. 23.

This figure displays a part of the metro network and only metrobus stations that have connection with it, however these connections are mainly in the central area of Mexico City. This figure shows more clearly the need to analyze the problem as a Multi-layer network, not all layers are considered since there is more means of public transport.

Layer α represents Metro stations, while layer β represents Metrobus stations, and they are connected with other modes of public transport.

In a multilayer network, we need to define connections between pairs of node-layer tuples. As with monoplex networks, we will use the term *adjacency* to describe a direct connection via an edge between a pair of node-layers and the term *incidence* to describe the connection between a node-layer and an edge.

Two edges that are incident to the same node-layer are also ‘incident’ to each other. We want to allow all of the possible types of edges that can occur between any pair of node-layers—including ones in which a node is adjacent to a copy of itself in some other layer as well as ones in which a node is adjacent to some other node from another layer. In normal networks (i.e. graphs), the adjacencies are defined by an edge set $E \subseteq V \times V$, in which the first element in each edge is the starting node and the second element is the ending node. In multilayer networks, we also need to specify the starting and ending layers for each edge. We thus define an

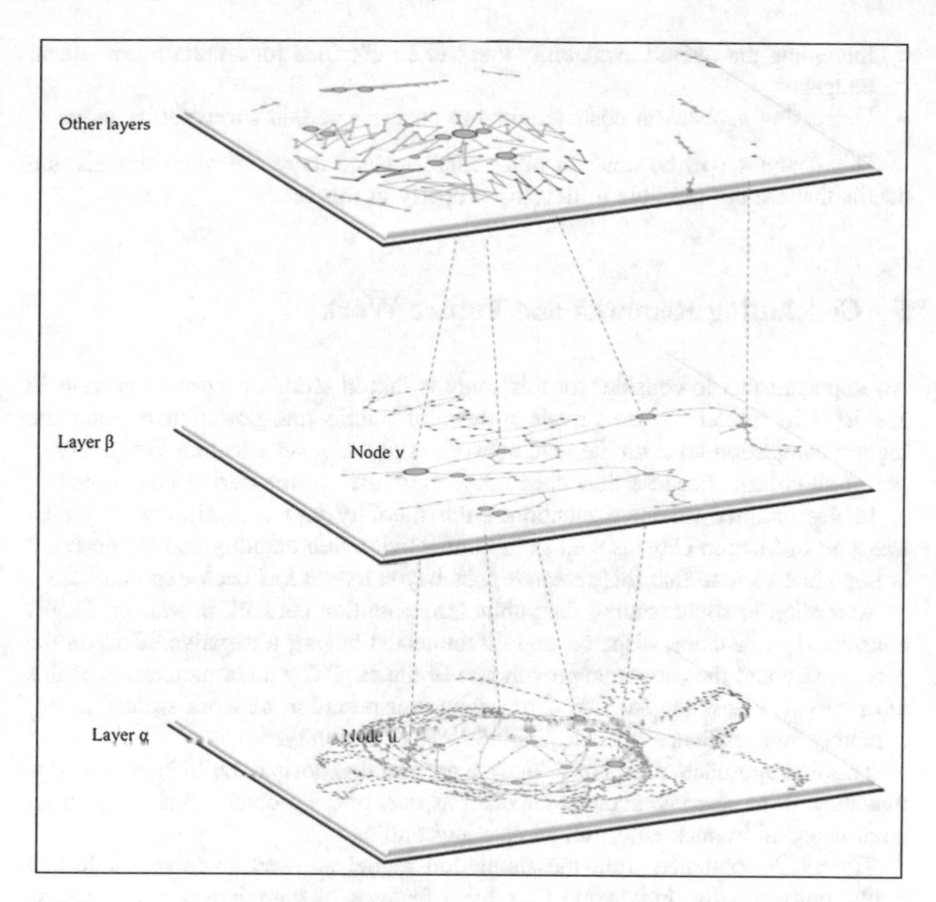

Fig. 23 Mexico City multilayer network. Layers represent public transport networks

edge set *EM* of a multilayer network as a set of pairs of possible combinations of nodes and elementary layers. That is, $EM \subseteq VM \times VM$.

Using the components that we set up above, we define a *multilayer network* as a quadruplet $M = (VM, EM, V, \mathbf{L})$. [36].

For the general analysis is important to take into account the connectivity not only by layer but intra layers, and how to consider strategies that create a resilient network.

In this way, Demeester et al. [22] set up some objectives for an integrated approach to multilayer survivability that includes:

- Avoiding contention between the different single-layer recovery schemes
- Promoting cooperation and sharing of spare capacity

- Increasing the overall availability that can be obtained for a certain investment budget
- Decreasing investment costs required to ensure a certain survivability target

This analysis will be done in other chapter since there are more models and details that are not possible to develop properly in this one.

5 Concluding Remarks and Future Work

An important fact to consider for this study is that in terms of mobility citizens in Mexico City prefer to use private instead of public transportation causing the highest congestion level on the road network at global level affecting the quality of life of all citizens because they spend 90% extra travel time during busy hours.

In this chapter we have mentioned the mobility and accessibility of public transport in Mexico City, as well as its connectivity, vulnerability and resilience. It is important to note that this research goes beyond what has been exposed here.

According to some studies, the public transportation network in Mexico City is considered as distance, disperse, and disconnected having a negative effect on the productivity and the economic growth rate of the city. The main motivation of this work was to assess the Mexico City public transportation network structural vulnerability and resilience for detecting areas of opportunity.

This first approach allows us to make a general diagnosis to build later scenarios that allow us to take into account the other aspects of the problem, such as security, environmental impact, land use, climate and traffic.

The results obtained from the simulation model allowed us to conclude that public transportation in Mexico City have features of complex networks whose structure is irregular, distributed and dynamically evolving in time.

The study of structural properties of Mexico City public transportation network allowed us to quantify the strategic importance of a set of nodes (stations) to preserve the functioning of the network as a whole. In order to carry out the assessment we modeled and simulated the network using Gephi software. Our simulations were executed using Force Atlas, Fruchterman Reingold, OpenOrd, and Yifan Hu algorithms.

On the one hand, we observed that the network had a low accessibility because the average degree is low, 2.607. It means that it has a low capacity to be reached by different locations. On the other hand, when the 20% of the total nodes were randomly eliminated to test the resilience of the network, the average degree reduces from 2.607 to 2.488. While eliminating the 30% of the highest degree nodes, the average degree reduces to 1.79. In conclusion, Mexico City public transportation network also presents high vulnerability.

The importance of having this research is that measures to take make the public transport an attractive option against the private one.

Acknowledgements This research was supported by UNAM-PAPIIT grant IT102117.

References

1. Alminas, M., Vasiliauskas, A. V., & Jakubauskas, G. (2009). The impact of transport on the competitiveness of national economy. *Department of Transport Management, 24*(2), 93–99.
2. Álvarez-Herranz, A., & Martínez-Ruíz, M. P. (2012). Evaluating the economic and regional impact on national transport and infrastructure policies with accessibility variables. *Transport, 27*(4), 414–427.
3. Aschauer, D. A. (1991). *Transportation spending and economic growth: The effects of transit and highway expenditures*. Report. Washington, D.C.: American Transit Association.
4. Auvinet, G., Méndez, E., Juárez, M., & Rodríguez, J. F. (2013). *Geotechnical risks affecting housing projects in Mexico valley*. México: Sociedad Mexicana de Mecánica de Suelos.
5. Bagler, G. (2008). Analysis of the airport network of India as a complex weighted network. *Physica A: Statistical Mechanics and its Applications, 387*(12), 2972–2980.
6. Balaker, T. (2006). Do economists reach a conclusion on rail transit? *Econ Journal Watch, 3* (3), 551.
7. Bell, S., & Morse, S. (2008). *Sustainability indicators: Measuring the immeasurable* (2nd ed.). London, UK: Earthscan.
8. Barabási, A., & Abert, R. (1999). Emergence of scaling in random networks. *Science, 286* (5439), 509–512.
9. Berche, B., Ferber, C. V., Holovatch, T., & Holovatch, Y. (2012). Transportation network stability: A case study of city transit. *Advances in Complex Systems, 15*(supp01), 1–18.
10. Blondel, V. D., Guillaume, J. L., Lambiotte, R., & ELefebvre, E. (2008). Fast unfolding of communities in large networks. *Journal of Statistical Mechanics: Theory and Experiment* (10).
11. Bovy, P. H. L., Jansen, G. R. M. (1979). Travel times for disaggregate travel demand modelling: a discussion and a new travel time model. In G. R. M. Jansen, et al. (Eds.), New *Developments in modelling travel demand and urban systems* (pp. 129–158). Saxon House, England.
12. Brandes, U. (2001). A faster algorithm for betweenness centrality. *Journal of Mathematical Sociology, 25*(2), 163–177.
13. Buckwalter, D. W. (2001). Complex topology in the highway network of Hungary, 1990 and 1998. *Journal of Transport Geography, 9*(2), 125–135.
14. CAF. (2011a). Desarrollo urbano y movilidad en América Latina and INEGI 2013.
15. CAF. (2011b). Desarrollo urbano y movilidad en América Latina. CTSEMBARQ México, Based on 2007 mobility survey. Estadísticas SETRAVI. Fideicomiso para el mejoramiento de lasvías de DF. 2001. Secretaría del MedioAmbiente. 2008. Inventario de emisiones de la ZMVM, 2006.
16. Calderón, C., & Servén, L. (2008). Infrastructure and economic development in Sub-Saharan Africa. *The World Bank Policy Research Working Paper*, 4712.
17. Caschili, S., & De Montis, A. (2013). Accessibility and complex network analysis of the US commuting system. *Cities, 30,* 4–17.
18. Cervero, R. (2001). Walk-and-ride: Factors influencing pedestrian access to transit. *Journal of Public Transportation, 3*(4), 1–23.
19. Cherven, K. (2015). *Mastering Gephi network visualization*. Birmingham: Packt Publishing.
20. Cheung, D. P., & Gunes, M. H. (2012). A complex network analysis of the United States air transportation. In *Proceedings of the 2012 IEEE/ACM International Conference on Advances in Social Networks Analysis and Mining*, 26–29 August 2012 (pp. 699–701). Istanbul, Turkey: Kadir Has University 2012.

21. Criado, R., & Romance, M. (2012). Optimization and its applications. In M. T. Thai & P. M. Pardalos (eds.), *Handbook of optimization in complex networks: Communication and social networks* (Vol. 58). Springer. doi:10.1007/978-1-4614-0857-41.
22. Demeester, P., & Michael, G. (1999). Resilience in multilayer networks. *IEEE Communications Magazine*, 0163-6804/99.
23. De Rus, G. (2008). *The economic effects of high speed rail investment.* University of Las Palmas, Spain, Discussion Paper, 2008-16.
24. Dey, P., & Roy, S. (2016). A comparative analysis of different social network parameters derived from Facebook profiles. In S. C. Satapathy, et al. (Eds.), *Proceedings of the Second International conference on Computer and Communication Technologies*. Advances in Intelligent Systems and Computing (Vol. 379).
25. Flores I., Mújica, M., & Hernández, S. (2015). Urban transport infrastructure: A survey. In *EMSS 2015 Proceedings*, Bergeggi, Italy.
26. Floater, G., & Rode, P. (2014). Cities and the new climate economy: the transformative role of global urban growth. In *The new climate economy: The transformative role of global urban growth*, November 2014. http://newclimateeconomy.net/.
27. Fortunato, S., & Castellano, C. (2007). Community structure in graphs. arXiv:0712.2716.
28. Fruchterman, T. M. J., & Reingold, E. M. (1991). Graph drawing by force-directed placement. *Software—Practice & Experience*, *21*(11), 1129–1164.
29. Galloti, R., & Barthelemy, M. (2014). Anatomy and efficiency of urban multimodal mobility. *Scientific Reports*, *4*, 6911. doi:10.1038/srep06911.
30. Grünberger, S., & Omann, I. (2011). Quality of life and sustainability. Links between sustainable behaviour, social capital and well-being. In *9th Biennial Conference of the European Society for Ecological Economics (ESEE): "Advancing Sustainability in a Time of Crisis"* from 14th to 17th June 2011, Istanbul, Turkey.
31. HABITAT. (2015). http://www.gob.mx/sedatu/documentos/reglas-de-operacion-del-programa-habitat-2015. Reporte Nacional de la Movilidad Urbana en México 2014–2015.
32. Hage, P., & Harary, F. (1995). Eccentricity and centrality in networks. *Social Networks, 17* (1), 57–63.
33. Háznagy, A., Fi, I., London, A., & Tamas, N. (2015) Complex network analysis of public transportation networks: A comprehensive study. In *2015 Models and Technologies for Intelligent Transportation Systems (MT-ITS)*, 3–5 June 2015. Budapest, Hungary.
34. Hu, Y. F. (2005). Efficient and high quality force-directed graph drawing. *The Mathematica Journal, 10,* 37–71.
35. Johnston, R. A. (2004). The urban transportation planning process. In S. Hansen, & G. Guliano (Eds.), *The Geography of Urban Transportation. Multi-modal transportation planning* (pp. 115–138). The Guilford Press.
36. Kivelä, M., et al. (2014). Multilayer networks. *Journal of Complex Networks*, *2*, 203–271.
37. Kleinberg, J. M. (1999). Authoritative sources in a hyperlinked environment. *Journal of the ACM, 46*(5), 604–632.
38. Kobourov, S. G. (2012). Force-directed drawing algorithms. In *Handbook of graph drawing and visualization*. CRC Press.
39. Krygsman, S., Dijsta, M., & Arentze, T. (2004). Multimodal public transport: An analysis of travel time elements and the interconnectivity ratio. *Transport Policy, 11*(2004), 265–275.
40. Latapy, M. (2008). Main-memory triangle computations for very large (Sparse (Power-Law)) graphs. *Theoretical Computer Science (TCS), 407*(1–3), 458–473.
41. Levy, J. M. (2011). *Contemporary urban planning*. Boston: Longman.
42. Lin, J., & Ban, Y. (2013). Complex network topology of transportation systems. *Transport Reviews, 33*(6), 658–685. doi:10.1080/01441647.2013.848955.
43. Litman, T. (2011). *Measuring transportation traffic, mobility and accessibility*. Victoria Transport Policy Institute. http://www.vtpi.org, Info@vtpi.org.
44. Martin, F., & Boyack, K. (2011). OpenOrd: An open-source toolbox for large graph layout. In *Proceedings of SPIE—The International Society for Optical Engineering*, January 2011.

45. Murray, A. T. (2001). Strategic analysis of public transport coverage. *Socio-Economic Planning Sciences, 35,* 175–188.
46. Nenseth, V., & Nielsen, G. (2009). Indicators for sustainable urban transport—state of the art. TØI report 1029/2009.
47. Ortúzar, J.d.D., Willumsen, L.G., 2002. Modelling Transport, third ed, Wiley, West Sussex, England.
48. O'Sullivan, S., & Morrall, J. (1996). Walking distances to and from light-rail transit stations. *Transportation Research Record*, 1538.
49. Paz, A., Maheshwari, P., Kachroo, P., & Ahmad, S. (2013). Estimation of performance indices for the planning of sustainable transportation systems. Hindawi Publishing Corporation. *Advances in Fuzzy Systems, 2013*, Article ID 601468, 13 pp.
50. Puente, S. (1999). Social vulnerability to disasters in Mexico City: An assessment method. In J. K. Mitchell (Eds.), *Crucibles of hazard: Mega-cities and disasters in transition* (pp. 296–297). Tokyo New York, Paris: United Nations University Press.
51. Reggiani, A., Nijkamp, P., & Lanzi, D. (2015). Transport resilience and vulnerability: The role of connectivity. *Transportation Research Part A, 81,* 4–15.
52. Rodrigue, J.-P., Comtois, C., & Slack, B. (2009). *The geography of transport systems*. New York, NY: Routledge.
53. Sienkiewicz, J., & Holyst, J. A. (2005). Statistical analyses of 22 public transport networks in Poland. *Physical Review E, 72,* 046127.
54. Shen, B., & Gao, Z.-Y. (2008). Dynamical properties of transportation on complex networks. *Physica A: Statistical Mechanics and its Applications, 387*(5), 1352–1360.
55. Tarapata, Z. (2015). Modelling and analysis of transportation networks using complex networks: Poland case study. *The Archives of Transport, 4*(36), 55–65.
56. Thai, My. T., & Pardalos, P. M. (2012). *Handbook of optimization in complex networks*. Springer.
57. The City of Calgary, Transportation Department. Retrieved 8 March, 2015, from http://www.calgary.ca/Transportation/Pages/Transportation-Department.aspx.
58. Tsay, S., & Herrmann, V. (2013). *Rethinking urban mobility: Sustainable policies for the century of the city (68 pp.)*. Washington, DC: Carnegie Endowment for International Peace.
59. UN Documents Gathering a Body of Global Agreements. (1987). Our common future, Chapter 2. In *Towards sustainable development*. United Nations Decade of Education for Sustainable Development in a Creative Commons, Open Source Climate. Geneva, Switzerland, June 1987.
60. Varela, S. (2015). Urban and suburban transport in Mexico City: Lessons learned implementing BRTs lines and suburban railways for the first time. Integrated Transport Development Experiences of Global City Clusters, International Transport Forum, 2–3 July 2015, Beijing China.
61. Von Ferber, C., Holovatch, T., Holovatch, Y., & Palchykov, V. (2009). Public transport networks: Empirical analysis and modeling. *The European Physical Journal B, 68*(2), 261–275.
62. Watts, D. J., & Strogatz, S. H. (1998). Collective dynamics of 'small-world' networks. *Nature, 393*(6684), 440–442.
63. Zanin, M., & Lillo, F. (2013). Modelling the air transport with complex networks: A short review. *The European Physical Journal Special Topics, 215*(1), 5–21.
64. Zheng, J., Atkinson-Palombo, C., McCahill, C., O'Hara, R., & Garrick, N. W. (2011). Quantifying the economic domain of transportation sustainability. In *Proceedings of the Annual Meeting of the Transportation Research Board CDROM*, Washington, DC, USA.

Integrating Data Mining and Simulation Optimization for Decision Making in Manufacturing

Deogratias Kibira and Guodong Shao

Abstract Manufacturers are facing an ever-increasing demand for customized products on the one hand and environmentally friendly products on the other. This situation affects both the product and the process life cycles. To guide decision-making across these life cycles, the performance of today's manufacturing systems is monitored by collecting and analyzing large volumes of data, primarily from the shop floor. A new research field, Data Mining, can uncover insights hidden in that data. However, insights alone may not always result in actionable recommendations. Simulation models are frequently used to test and evaluate the performance impacts of various decisions under different operating conditions. As the number of possible operating conditions increases, so does the complexity and difficulty to understand and assess those impacts. This chapter describes a decision-making methodology that combines data mining and simulation. Data mining develops associations between system and performance to derive scenarios for simulation inputs. Thereafter, simulation is used in conjunction with optimization is to produce actionable recommendations. We demonstrate the methodology with an example of a machine shop where the concern is to optimize energy consumption and production time. Implementing this methodology requires interface standards. As such, this chapter also discusses candidate standards and gaps in those standards for information representation, model composition, and system integration.

D. Kibira (✉)
Department of Industrial and System Engineering, Morgan State University, Baltimore, MD, USA
e-mail: deogratias.kibira@morgan.edu

G. Shao
Engineering Laboratory, National Institute of Standards and Technology (NIST), Gaithersburg, USA
e-mail: gshao@nist.gov

M. Mujica Mota and I. Flores De La Mota (eds.), *Applied Simulation and Optimization 2*, DOI 10.1007/978-3-319-55810-3_3

1 Introduction

In 2014, the United States manufacturing industry produced $2.1 trillion worth of goods and supported 12.3 million jobs [1]. While these figures are impressive, there has been a declining trend in manufacturing's share of the Gross Domestic Product (GDP). The reasons for this decline include increasing global competition, sustainability concerns, and uncertainties in the cost and supply of materials [2].

Traditionally, cost, quality, productivity, and throughput are the major considerations when selecting materials, manufacturing processes or developing production plans. However, environmental sustainability is now considered to be the fourth such consideration. Even though it may negatively impact the other three, sustainability is deemed critical for an organization to succeed in today's markets.

To better understand and predict those impacts, a type of manufacturing systems, called smart manufacturing systems (SMS), is being proposed. SMS are characterized by the wide availability of data that can shed light on those impacts and predictions. This data is expected to improve real-time system planning and operational decision-making. But this can be achieved only if context and meaning can be deduced from it. Data collected by smart sensors, radio frequency identification (RFID), and wireless communications, is described by volume, velocity, variety, veracity, validity, volatility, and value—the so-called 7 Vs of big data [3]. Figure 1 from United Nations Economic Commission for Europe (UNECE) [4] shows the recent past, current, and projected "explosion" of business data.

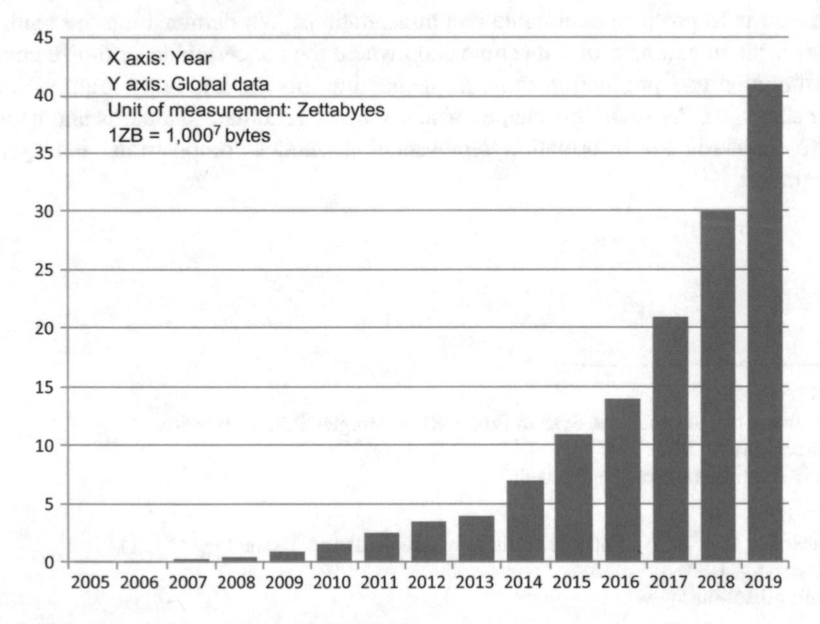

Fig. 1 Growth in big data (*Source* UNECE)

Gröger et al. [5] identifies two types of collected data: process data and operational data. Process data is made up of execution data, which includes flow-oriented machine and production-events data recorded by the Manufacturing Execution System (MES). Operational data mainly encompasses Computer-Aided Design (CAD), Computer-Aided Process Planning (CAPP), and Enterprise Resource Planning (ERP) data. These data are interrelated and influenced by many factors including hidden patterns, correlations, associations, and trends. It is our view that uncovering these factors can contribute to the decision-making process considerably. The conventional approaches have inherent limitations for deriving actionable recommendations based on the process data and operational data [6]. Thus, a new methodology is needed.

In this chapter, we describe such a methodology combining simulation, data mining, and optimization specifically for utilizing the large amount of process and production data. This methodology is demonstrated with a case study of a machining process where environmental impacts and production time are the performance measures. The objective is to choose the process sequence, production plan, manufacturing resources, and parameter settings that optimize both above-mentioned measures during production operation.

Traditionally, simulation has been used to investigate the performance of manufacturing systems under a predefined set of scenarios. Better et al. [7] observe, however, that in a system with a high degree of complexity and uncertainty, (1) it is not always obvious which parameters and variables to focus on to improve performance nor (2) is it evident to what extent these variables should be changed for optimal operation. Brady and Yellig [8] also observe that in complex simulation systems of real-world problems, it is challenging to determine a set of input variables and their values to obtain optimal output. One major reason is that data contains intricate dependencies, which must be determined before generating the input to simulation models. Techniques and methods are needed to discover such dependencies before performing simulation analysis. In summary, the main contribution of this chapter is the novel methodology that integrates data mining, simulation, and optimization techniques for more effective model parameter identification, simulation input preparation, and actionable recommendation derivation.

In our proposed methodology, data-mining is used to develop high-level association rules among various kinds of data, including performance data. The outputs from using those rules are used as inputs to the simulation model. Simulation optimization then determines the best process and operational parameter settings to obtain actionable recommendations for decision makers and operators. We believe that the combined effect of data mining, simulation, and optimization can improve manufacturing decision making in face of big data and system complexity.

We use a case of a small machine shop with two performance objectives: minimize production time and resource—material, energy, and water—usage during the machining processes. Each part design has a different process plan. Some machines can perform more than one process. However, the sequencing of parts through the shop depends on the users' objectives. The choice of a machine for a given process will produce different impacts on both performance objectives.

The chapter is structured as follows: Section 2 provides a background to data mining with a focus on the unsupervised learning techniques: association and clustering. Section 3 overviews simulation modeling for manufacturing applications. Section 4 reviews simulation optimization methods and techniques as they are currently applied to decision-making in manufacturing. Section 5 illustrates the integration of data mining, simulation, and optimization. Section 6 presents the proposed methodology and the strengths of a combined-methods approach. Section 7 presents a case study demonstrating how energy and production time can be optimized in a machine shop based on the methodology. Section 8 presents a summary and discussion of how the methodology can be implemented highlighting integration needs.

2 Background to Data Mining

This section provides a background to data mining techniques in manufacturing particularly association and clustering that are relevant to the work of this chapter.

2.1 Data Mining Techniques

Data mining is the process of discovering knowledge hidden in large amount of data [9]. The data being mined is typically observed data—as opposed to experimental data—so the data mining techniques employed have no influence on the data-collection methods. In Agard and Kusiak [10], for example, the authors show how to mine data stored in ERP, previous schedules, and MES to gain knowledge about the best choice of manufacturing processes based on defined design characteristics.

Data mining techniques draw from several disciplines including statistics, visualization, information retrieval, neural networks, pattern recognition, spatial data analysis, image databases, signal processing, probabilistic graph theory, and inductive logic programming. Data mining can in general be distinguished into two groups: descriptive and predictive. Descriptive techniques describe events from data and factors that are responsible for them. Predictive techniques attempt to predict the behavior of new data sets. Both techniques use the same general approach which is to (1) identify data fields and types and (2) specify the data as discrete or continuous.

Our current focus is on predictive data mining. Predictive data mining types include supervised learning, unsupervised learning, and semi-supervised learning [11]. With supervised learning, output variables are known or predetermined and the purpose of a learning algorithm is to develop a function that maps output variables to the inputs. Output variables corresponding to any given inputs can then be predicted using the learned function. Semi- supervised learning problems have only some of the data associated with output variables.

Unsupervised learning is where none of the input data is associated with prior defined responses, called data labels. The objective for unsupervised learning is to model the underlying structure or distribution in the data in order to learn more about or discover any patterns in the data. In other words, the intent of unsupervised learning is to understand hidden data concepts where the data labels are not known beforehand.

The case study described in this chapter involves predicting the best operational performance of a manufacturing system based on collected data, from which the best parameters are determined. Therefore, we investigate a case of unsupervised learning. Unsupervised learning techniques include clustering and association rule data mining [11]. We further discuss association and clustering as follows.

2.1.1 Association

Association techniques (1) discover relationships among large volumes of data and (2) represent those relationships as rules that "describe" the data. Discovery is based on the probability of co-occurrence of items in a collection of a large data set. The relationships between co-occurring items are expressed as association rules. Conceptually, an association rule indicates that the occurrence of certain items in a transaction would imply the occurrence of other specific items in the same transaction [12]. In other words, there is a supposed phenomenon within the system that makes these two types of items to concurrently occur.

The aim of association data mining is not to try to understand the underlying phenomenon. Rather, the association learning process attempts to determine the relating association rules. The idea of mining association rules originates from the market analysis where rules such as "a customer who buys products A and B also buys product C with probability p." In theory, given enough manufacturing data, such rules could be derived, rules that help explain the relationship between the values of the input data and the values of output data representing system performance. These rules could be of the type "**if** parts are sequenced such that process A is performed before process B, **then** there is an increase in the total energy consumed per part with probability p." Our focus then is on understanding the relationship between input and output variables (performance data)—that is, the rules—as well as the ranges that these variables can take. Algorithms for association rules learning concentrate on obtaining statistically significant patterns, and deriving rules from those patterns [13]. This is done by finding the frequency of concurrence of items from a transaction dataset and generating association rules based on user specified minimum confidence.

For example, one rule might say "that if we pick any product at random and find out that it was processed according to a given processing sequence through the factory floor, we can be confident, quantified by a percentage, that its production time is larger than average."

Association rules derived from data should be reliable. The measures of rules obtained include *support* and *confidence* of a rule. Support is a proportion of items in the data that contain a given set of items that occur together. Confidence is an indicator of how often a rule has been found to be true.

The main challenge of association rule induction is that there are so many possible rules. For example, the large product range of a typical job shop results in several classes of product designs, materials, and processing requirements. The rules cannot be processed by inspecting each one in turn. Therefore, efficient algorithms are needed to restrict the search space and check only a subset of all rules, but if possible, without missing important rules. One such algorithm is the Apriori algorithm [14], which is the algorithm used in the work of this chapter.

2.1.2 Clustering

Clustering is the process of identifying a finite set of categories, called clusters, that "describe" the data. Clustering techniques segment large data sets into smaller homogeneous subsets that can be easily managed, separately modelled, and analyzed [15]. Clusters are formed such that objects in the same cluster are more similar to each other than objects in different clusters.

Clusters correspond to hidden patterns in the data. Clusters can overlap or be at multi-level dimensions such that a data point can belong to more than one cluster. A clustering algorithm creates clusters by identifying points closest to the center of a cluster and expanding outwards up to a certain threshold when a new cluster needs to be formed. The process continues until all data points are assigned to a cluster [11].

In manufacturing applications, Kerdprasop [16] used clustering techniques to determine patterns and relationships in multidimensional data to indicate a potential poor yield in high volume production environments. Another potential application is determining relationships that can help differentiate categories of parts that can be processed by similar machines. The features of such parts are used to form a cluster and are useful in developing cell manufacturing systems through "group technology."

3 Modeling and Simulation for Manufacturing Applications

A simulation is a computerized model of a real, or a proposed, system. Users can conduct experiments with such a model to better understand the likely behavior of that system for a given set of conditions and scenarios [17]. Because of the dynamic nature of manufacturing operations, most simulation models are stochastic.

Table 1 Simulation application in different stages of manufacturing system life cycle

Phase	Objective	Model type
Planning	Production volume, factory requirements	System dynamics, control theory
Basic design	Department layout options, throughput analysis, aggregate analysis	System dynamics, discrete event simulation, agent based simulation
Detailed design	Layout, equipment specification, production management options	Discrete event simulation, agent based simulation
Setup and optimization	Production validation	Discrete event simulation

Law and Kelton [18] summarize the benefits of modeling and simulating manufacturing systems. First, they help identify and quantify the equipment and personnel. Second, they can predict the probability distributions associated with performance. Third, they can be used to evaluate operational procedures. Fourth, they can take into consideration of stochastic behavior of the system.

Objectives and types of simulation models needed may differ within each life cycle phase of a manufacturing system. Table 1 shows typical objectives and simulation types in various phases of a system life cycle. A review of 317 simulation papers by Negahban et al. [19] shows three major research topic areas: simulation language development, manufacturing system design, and manufacturing system operation. For manufacturing system operations, simulation helps users understand, assess, and evaluate the operation so that the 'best' configurations that result in 'optimum' performance can be determined.

Manufacturing simulation has been widely used for determining policies or rules to be employed in specific operational situations [20]. But, for a long time, its application to real-time control was limited by computational capacity, system reconfiguration time, data, and optimization issues. These days, a number of technologies is helping to overcome those limitations. Those technologies include high-speed computation, communication, integration technologies, standards, and automated data collection and processing [21, 22]. Simulation models can be updated with data to provide capacity to foresee the impact of new orders, equipment failures, and changes in operations.

Simulation can also be used for generating and filling gaps in missing data for analysis by other methods. Shao et al. [23] demonstrate how simulation could be used to generate data to help evaluate the performance of data-analytics applications. For this approach to be effective in real applications, however, data-generating models require improved verification and validation methods.

One of the major activities of simulation projects is input-data preparation. Previous research efforts have attempted to address this issue. For example, Skoogh et al. [22] demonstrates a Generic Data Management Tool (GDM—Tool) for data extraction, conversion, cleaning, and distribution fitting. The GDM—Tool enables data reuse, thereby, reducing needed time for carrying out simulation projects.

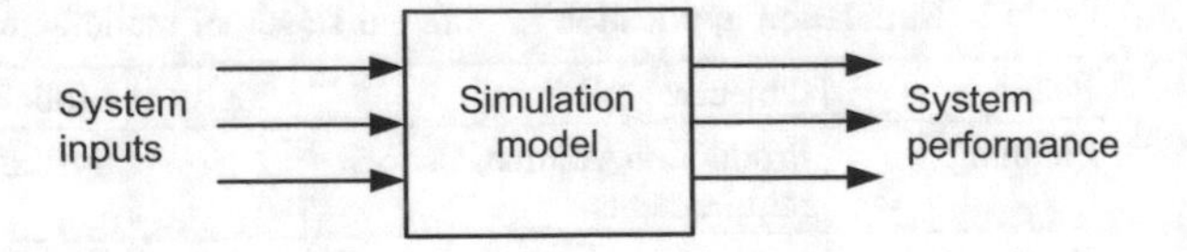

Fig. 2 The conceptual relationship between inputs and outputs of a simulation model

However, finding optimal parameters and settings from a large volume and variety of streaming data, as addressed by this chapter, cannot be carried out using the GDM-Tool.

The common approach for simulation-based decision-making is to prepare and run a number of scenarios and select the one with best outcome, as shown in Fig. 2. However, this approach is very difficult for a complex system with several inputs, particularly if model execution time is long. In addition, the quality of the answer obtained largely depends on the skill of the analyst who selects and defines the scenarios.

Identifying the "best solution" requires an optimization process, which is mostly the maximization or minimization of the expected value of the objective function of a problem [24]. Brady and Bowden [25] proposed two approaches for integrating simulation and optimization. The first is to construct an external optimization framework around the simulation model. The second is the internal approach, to investigate the relationship between input variables based on the dynamics of their interaction within the simulation model. This chapter uses the first approach. The importance of optimization has led simulation vendors to include optimization modules as part of their tools.

4 Simulation Optimization

Simulation optimization is the search for specific values or settings of controllable input parameters to a simulation such that a target objective is achieved [26]. This objective depends on simulation input. The procedure for optimization is to define a set of decision variables and optimize (i.e., maximize or minimize) the designated performance subject to constraints and bounds on range of the decision variables. Azadivar [24] formulated one form of the simulation optimization as:

$$\begin{array}{ll} \text{Maximize (or minimize)} & \mathbf{f}(\boldsymbol{X}) = \mathrm{E}[\boldsymbol{z}(\boldsymbol{X})] \\ \text{subject to} & \mathbf{g}(\boldsymbol{X}) = \mathrm{E}[\boldsymbol{r}(\boldsymbol{X})] \leq 0 \\ \text{and} & \mathbf{h}(\boldsymbol{X}) \leq \mathbf{0}. \end{array}$$

where $\boldsymbol{z}$ and $\boldsymbol{r}$ are random vectors representing several responses of the simulation model for a given $\boldsymbol{X}$, a multi-dimensional vector of decision variables. The functions $\mathbf{f}$ and $\mathbf{g}$ are the unknown expected values of these vectors, which can only be

Table 2 Optimization search strategies for selected simulation tools

Optimization package	Search strategy (optimization method)	Simulation software
SimRunner	Evolutionary, genetic algorithms	ProModel
OptQuest	Scatter search, tabu search, neural networks	Arena, Quest, FlexSim, Micro Saint Sharp, Simio, etc.
AutoStat	Evolutionary, genetic algorithms	AutoMod
Optimiz	Neural networks	Simul8
Optimizer	Simulated annealing, Tabu search	Witness
ExtendSim evolutionary optimizer	Evolutionary	ExtendSim

estimated by observations on $\mathbf{z}$ and $\mathbf{r}$. That means that the objective function (objective functions, in case of a multi-criteria problem) and/or constraints, are responses that can only be evaluated by simulation. The variable $\boldsymbol{h}$ is a vector of deterministic constraints on the decision variables.

The specific optimization algorithms used often depends on the type of simulation method. Because running a simulation model requires significant computations, efficient optimization algorithms are crucial. Some of the optimization methods that are applicable to different simulation types are overviewed by Amaran [26]. Carson and Maria [27] categorize simulation optimization methods into gradient based search methods, stochastic optimization, response surface methodology, heuristic methods, A-teams, and statistical methods. Fu et al. [28] reviews the state of practice for simulation optimization.

Table 2 shows a sample of commonly used simulation-based optimization tools. Researchers also often develop custom-made optimization tools based on simulation software for particular situations. Phatak et al. [29] introduce an example of an in-house optimization tool for manufacturing problems based on the particle swam optimization algorithm.

Unlike mathematical-programming formulations of optimization problems, there is no way of telling whether an optimum has been reached using simulation-based formulations. The optimization packages, such as those shown in Table 2, seek improved system performance by changing settings of system parameters. Consequently, these packages develop a solution incrementally by building upon earlier solutions to obtain a better one. The packages do this by proposing new simulation inputs, executing the simulation, and evaluating the performance iteratively [7]. Figure 3 illustrates this procedure.

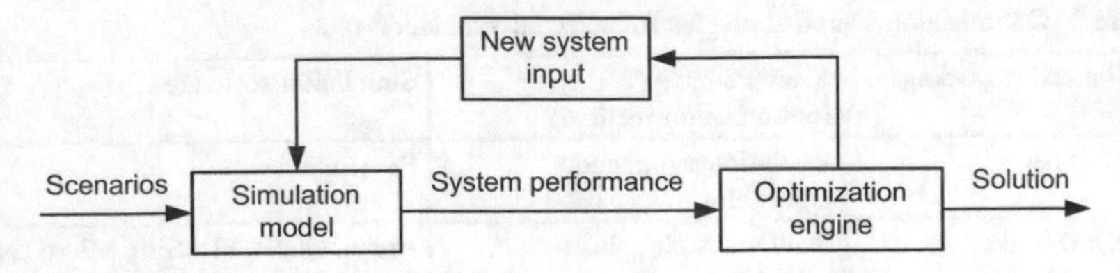

Fig. 3 Process of getting a solution using simulation-based optimization

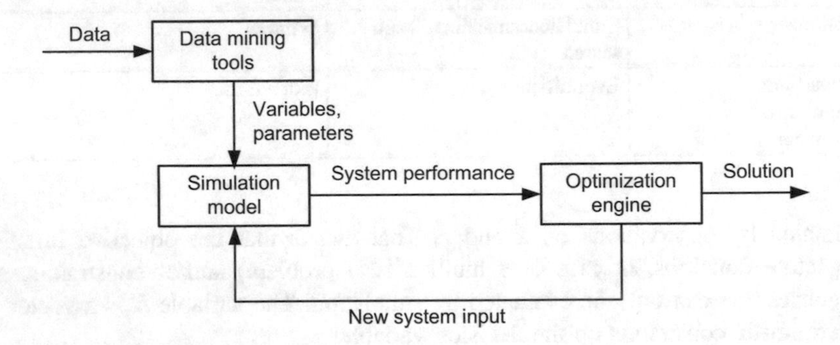

Fig. 4 Data mining integrated approach to simulation optimization

5 Integrating Data Mining with Simulation and Optimization

Embedding an optimization module into simulation tools, as described in Sect. 4, provides actionable solutions. However, determining the set of inputs that optimize system performance is challenging because of the large volume of data, and number of possible input parameters and their interactions. Although tools such as the input data management to simulation have been developed from previous research [22], they do not address the data challenges discussed in the previous section. We propose using data mining as a technique to help obtain simulation scenarios through association of collected data with system performance. Remondino et al. [30] described two ways of combining data mining with simulation. The first, called micro-level modeling, is where data mining is applied on historical data to (1) develop the appropriate scenarios and (2) tune scenario-based simulation input parameters.

The second, called macro-level modeling, is where data mining analyzes simulation output data to (1) reveal patterns describing system behavior and (2) develop ways to use those patterns to aid decision-making [31, 32].

Our proposed methodology is based on micro-level modeling the first approach. Figure 4 shows the high-level components and their interactions. Two features are combined with the classical simulation modeling and analysis: data-mining and optimization. This approach is suitable for both static and dynamic data.

6 A Methodology for Manufacturing System Optimization

Based on the review and discussion of Sects. 1 and 2, we conclude that (1) modeling and simulation tools cannot directly use streaming data, and (2) further analysis is needed to obtain actionable recommendations from the patterns and rules obtained by data mining. Therefore, a methodology combining different methods is needed. Operational steps for this methodology illustrated in Fig. 5 are next described.

In summary, the user first formulates the problem by specifying the scope, high-level performance objectives, indicators, and metrics. This is followed by acquiring domain knowledge and developing a conceptual model to understand model requirements, activities, and processes. The next step is to collect data and apply data mining techniques on the data. The final steps are simulation modeling

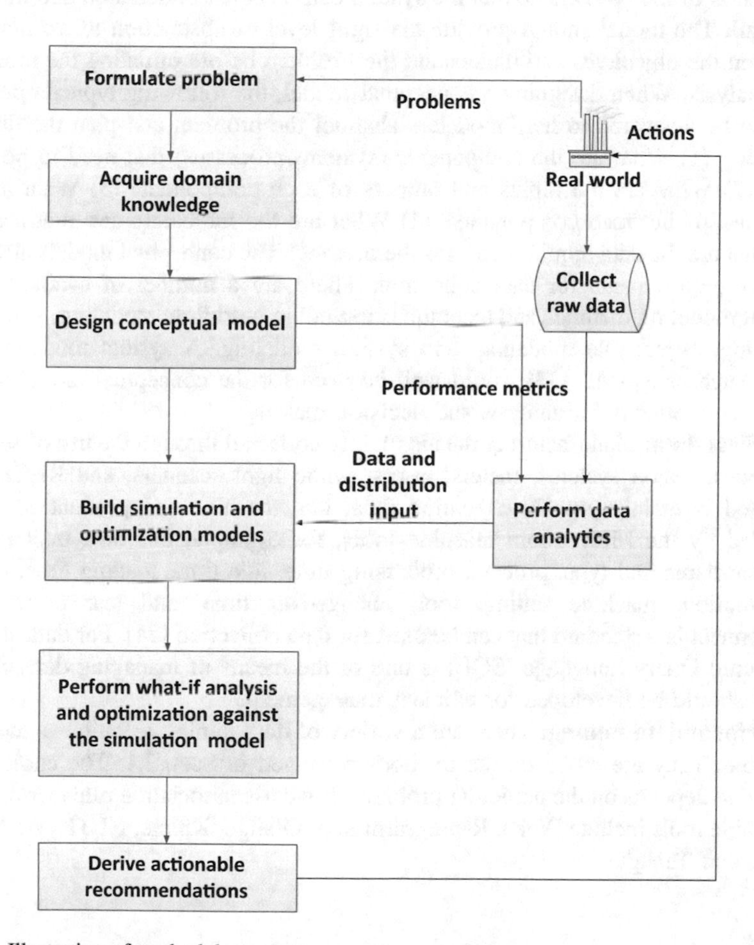

Fig. 5 Illustration of methodology steps

and optimizations. Detailed description of these steps follows in the next paragraphs.

Formulate the problem: This is the definition of the goal and scope of the project. The target plant, work cell, machine, manufacturing operations, or processes are specified at this step. The goal might be, for example, to minimize energy consumption for a foundry shop or to maximize throughput of a machine shop. Relevant resources, operational details, constraints, products, activities, and data collection points are also identified.

Acquire domain knowledge: This is the step to acquire domain knowledge for executing the project. Domain knowledge includes a thorough understanding of the manufacturing processes and system, indicators, metrics, and performance objectives and goals. In addition, knowledge about software (data mining, simulation, and optimization), data collection, communication, and storage are also required.

Design a conceptual model: This is the step to construct a high-level conceptualization of the problem so that the system can be better understood and modeled in detail. The model should provide the right level of abstraction to maintain the focus on the objectives and understand the problem before initiating the modeling and analysis. When designing a conceptual model, the following typical questions need to be answered to help modelers abstract the problem and plan the detailed modeling (1) What are the components (systems/processes) that need to be modeled? (2) What are the inputs and outputs of each component? (3) What are the relationships between components? (4) What are the indicators and metrics? and (5) What are the data requirements for the metrics? The conceptual models also help identify requirements for data collection. There are a number of available conceptual modeling methods and techniques including workflow modeling, workforce modeling, object role modeling, and system modeling. A system modeling language such as SysML [33] would well be used for the conceptual model to represent requirements for analysis and decision making.

Collect data: Manufacturing data is mainly collected through the use of sensors, bar codes, vision systems, meters, lasers, white light scanners, and RFID. Data collected is mainly process execution data, i.e., machine and production events recorded by the MES. From machine tools, for example, this data may include machine name and type, process, processing time, idle time, loading time, energy consumption, machine setting, tool, changeover time, and tear down time. MTConnect is a standard that can be used for data collection [34]. For data storage, Structured Query Language (SQL) is one of the means of managing data. A data model should be developed for efficient management.

Perform data mining: There are a variety of data-mining techniques and tools available. They are based on the methods reviewed in Sect. 2.1. The choice of a technique depends on the particular problem. If we use association rule learning, the applicable tools include Weka, R-programming, Orange, Knime, NLTK, ARMiner, arules, and Tanagra.

Mathematically, the performance indicator, y, e.g., energy, can be represented as a function:

$$y = (x, w),$$

where $x = (x_1, x_2, x_3, \ldots, x_d)^T$ denotes the set of system parameters that are associated with the amount of energy used and w denotes the weight of the parameters. In the work presented in this chapter, y is known and the task of data mining is to determine the system parameters x.

Perform simulation modeling and optimization: The system is represented by a simulation model. Many simulation tools are supplied with optimization modules (as shown in Table 2). Typically, these tools automatically execute multiple runs and systematically compare the results of a current run with past runs to decide on a new set of input values until the optimum is gradually approached. Core manufacturing simulation data (CMSD) standard can be used for representing the input data for the simulation modeling [35].

Derive actionable recommendations: The final step is to derive actionable recommendations by interpreting and translating the output from the optimization process. The users also need to check if the recommended actions conflict with existing knowledge about the system and resolve this conflict if necessary. As Fig. 5 shows, the system performance can be monitored while data is continuously collected so that a new set of decisions can be made when needed.

7 Case Study: Minimizing Energy Consumption and Production Time in Machining Operations

Machining is one of the major manufacturing processes in the metal industry. The process inputs, removal processes, and waste byproducts have a large potential environmental impact. Currently, the relationships among them and their impacts on the environmental have not been fully investigated. As a result, methods for determining control inputs that optimize production objectives have not been fully developed [36]. This section describes how the proposed methodology was applied to a case study that uses data from a machining process for decision making. This case is a first step for understanding and implementing the proposed methodology.

Many machined parts are produced in job shops. The case under study is based on a machining job shop that was used in the research work reported in Kibira et al. [36], and Hatim et al. [37] for simultaneously optimizing process plans and production plans. In this investigation, we use a different part design. The shop consists of the following machine tools: a turning lathe, a milling machine, a drill press, and a boring machine. When orders are received and batched, it can be decided to focus on any or all of these performance objectives (1) minimize costs (e.g., labor, cutting tool, and energy), (2) minimize resource usage (e.g., material, energy, and water),

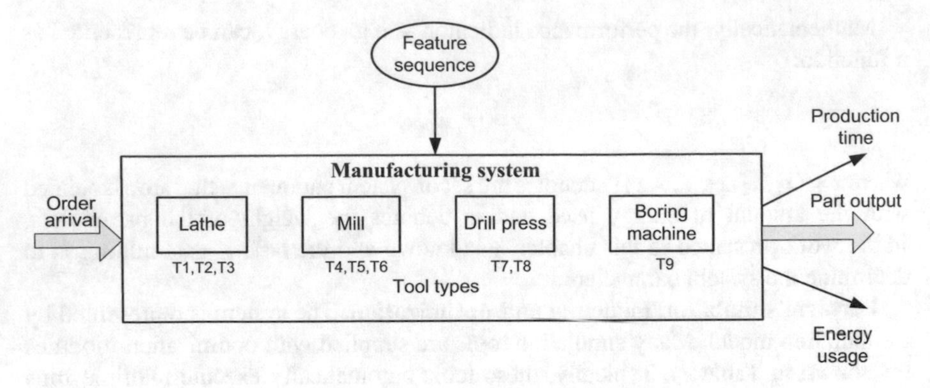

Fig. 6 Conceptual view of inputs and impacts of the machining shop

and (3) maximize production. Figure 6 is a conceptual view of work flow through the machine shop.

Each production batch, or each part in the batch, can potentially have its own process plan because the user can choose different sets of machine and tools to produce a given design feature on a part. We propose three approaches for the sequencing of part-feature production: a predefined, a partially defined, and an unspecified process plan. In the predefined case, each process has a pre-determined machine and cutting tool, determined to optimize a given performance objective such as minimum energy use. In an unspecified case, a machine is selected for processing by a part according to a priority rule such as always choose the machine with minimum number of parts waiting. The partially defined case is a combination of these two. Either of these choices results in a different process plan and hence different energy consumption, production time, and cost. Process and performance data is collected for each batch as it passes through the machine shop. Both types of data depend on resources used for each process within the process plan.

Formulate the problem: The scope and focus is on a machine shop and target product is a grinding head shell, shown in Fig. 7. The manufacturing processes for this part are facing, grooving, threading, spot drilling, and drilling. The objective is to select a sequencing plan, a machine tool, and cutting tools for each process so as to minimize energy consumption and production time.

Acquire domain knowledge: The following expert knowledge was acquired before beginning to model the machine shop operations including production resources, machining processes, energy consumption in machining, machining time, production planning and sequencing in job shop environment, costs of manufacturing processes, performance indicators and metrics, and performance data. Take production resources as an example. Table 3 shows the manufacturing processes to produce a grinding head shell and the machine tools available in the machine shop. These are Computer Numerical Control (CNC) lathe, three-axis vertical milling, press-upright drill, and mills-horizontal boring machine. For each machining operation, one or more cutting tools can be chosen to meet the required specification. Table 4 shows tool types available for each machine. Cutting tools

Fig. 7 Grinding head shell

Table 3 Resource information for manufacturing the grinding head shell

Process	Machine
Facing	Three—axis CNC Lathe
	Three—axis vertical milling machine
Grooving	Three—axis CNC Lathe
	Three—axis vertical milling machine
Threading	Three—axis CNC Lathe
Spot drill	Three—axis CNC Lathe
	Drill press
Drill	Three—axis CNC Lathe
	Drill press
	Boring mills-horizontal boring

Table 4 Tool types for use by each machine tool

Machine	Tool type	Tool description
Three—axis CNC Lathe	T1	Single-point tipped tool
	T2	Form turning
	T3	Drill
Three—axis vertical milling machine	T4	Slot milling
	T5	Mill cutter
	T6	Form milling
Drill press	T7	Center drill
	T8	Reamer
Boring mills-horizontal boring	T9	Boring tool

are as follows: turning (single-point tipped tool, form turning, drill), milling (slot milling, mill cutter, form milling), drilling (center drill, reamer), and boring (boring tool). When a machine may perform a particular operation, each type of tool would perform it differently, which potentially results in different production time and energy use.

Design conceptual model: The conceptual model shown in Fig. 8 is a schematic representation of the problem, activity sequences, and information flow. It includes product design, feature sequence, process selection, machines and tools requirements, and performance indicators that drive the above selections. The part design describes design information, including the features' forms, shapes complexities, dimensions, tolerances, and surface conditions. Alternative networks that describe features' processing precedence during fabrication are described. Next, a set of processes to manufacture a part is determined according to the part's functionalities and design requirements. The combinations of machines and tools that satisfy the design and process requirements are designated. Performance indicators determine the actions that give the machine shop the best chance to meet those objectives.

Model data collection: Based on domain knowledge acquired and the conceptual model developed, mathematical expressions from published literature are used to calculate energy consumption and processing time of the processes [38–44]. The processes in question are turning, milling, and drilling. A matrix of process and prospective machine, and cutting tool to carry out the processing is used to determine the production time and the energy consumed. Three examples are provided to show the expressions employed.

The time to perform a turning operation is given by $T_m = \frac{\pi DL}{vf}$, the time for a drilling process is given by

$$T_m = \frac{\pi D_c(d + 0.5D_c \tan(90 - \frac{\theta}{2}))}{vf},$$

and, the energy consumed by a plain milling operation is given by

$$E = C_z a z D_c^b f^u d^e v T_m.$$

for one specific case used in the model

$$E = \frac{68.2azD_c^{-0.86} f^{0.72} d^{0.86} v}{6120} T_m$$

where

D, L	workpiece diameter and length,
v, f, d	cutting speed, feed rate, and depth of cut,
T_m	machining time,
D_c	diameter of the milling cutter or the drill diameter,

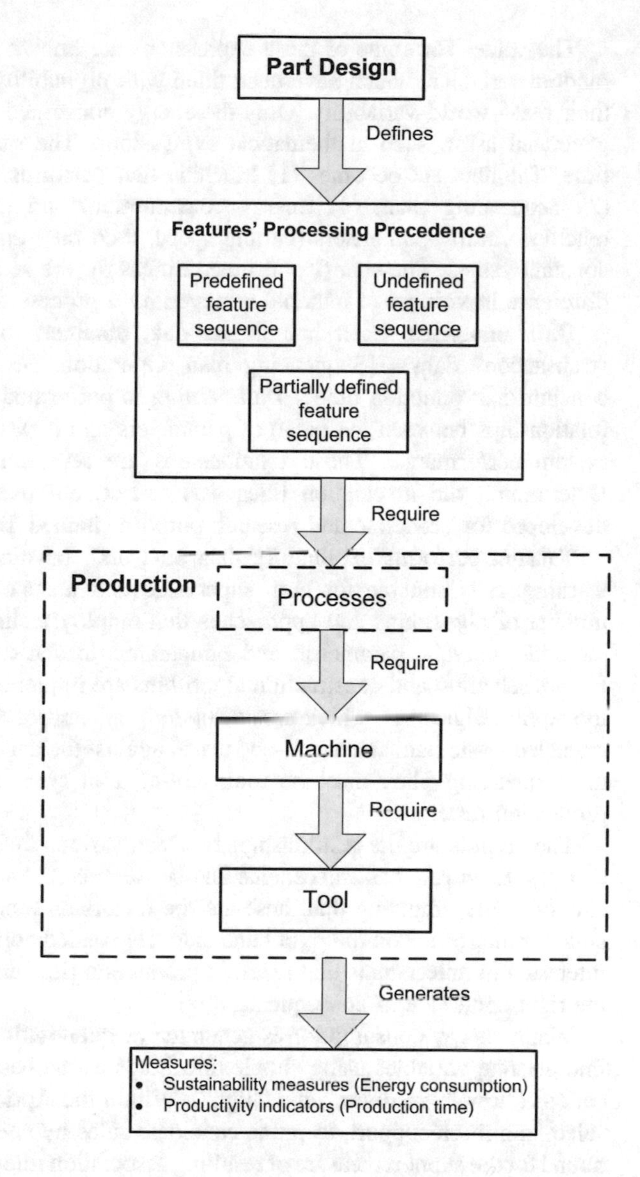

Fig. 8 A conceptual model representing the production process and information flow to a part design

z	No. of teeth in milling cutter or no. of flutes in a tap,
θ	drill point angle,
C_z	constant of the milling operation,
a	milling width,
b, e, u	constants that are determined empirically. These are tabulated for different types of machines and tools.

The values for some of these expressions are known constants. The others are random variables, which have been fitted with probability distributions to simulate their real—world variability. Only the energy consumed during processing can be generated using such mathematical expressions. The same applies to production time. The data set becomes (1) Machine that performs a process, (2) Tool type, (3) Sequencing plan, (4) Energy consumption, and (5) Production time. The machine cutting parameters (cutting speed, feed rate, and depth of cut) are set at constant values. The energy consumed differs for the sequencing plans because of difference in volume of material removed by a process for each plan.

Data analytics: Each line in the data obtained forms a transaction where "transaction" data = {Sequencing plan, Operation, Machine tool, Tool, Energy consumed, Production time}. Data mining is performed to determine the various relationships between the range of parameters and how those relationships impact system performance. Those relationships are represented as association rules. Determining the association rules was carried out using open source software developed for academic and research purposes named Tanagra [45].

Tanagra performs exploratory data analysis, statistical learning, and machine learning. It is suitable for both supervised and unsupervised learning. It uses a number of algorithms and approaches that employ techniques such as clustering, factorial analysis, parametric and nonparametric statistics, association rule, and feature selection and construction algorithms are implemented by Tanagra. We use the Apriori algorithm, which uses a "bottom up" approach and frequent subsets are extended—one item at a time—and tested against the data. The inputs to Apriori are the sequencing plan, machine tool, cutting tool type, energy consumption, and production time.

The outputs are the relationships between various factors expressed in the form of rules. Each rule has antecedents and consequents. Antecedents are the left hand side of a rule, implying that these are the factors and their values that are responsible for the results on the right hand side, also called consequents. As such, we are interested in antecedents that result in production time and energy consumption on the right hand side as consequents.

While energy consumption is generated as quantitative data, it was transformed into discrete variables using simple thresholds as the basis for the discrete classifications ("low," "medium," and "high"). Within the Apriori algorithm, the user can select minimum support, to prune candidate rules by specifying a minimum lower bound for the support measure of resulting association rules. Likewise, the confidence (described in introduction section), is also set. The cardinal is the number of concurring items (itemsets) used in computation. The values chosen are: minimum support is set at 0.16, confidence at 0.6, and minimum cardinal of itemsets is set at 4.

A sample of the derived rules for the demonstration is shown below:

Feature sequence = undefined = > Energy = High
Feature sequence = predefined = > Energy = High
Operation = Spot Drill = > Energy = Low
Operation = Facing = > Energy = High
Machine = M2 = > Energy = High
Machine = M1 = > Energy = Medium
Operation = Drill = > Energy = High
Operation = Threading = > Energy = medium
Operation = Grinding&&Tool = T2 = > Energy = High
Machine = M3&&Tool = T7 = > Energy = Low

The rules show that feature sequencing, operation, machine, and tool are relevant to energy consumption. These factors are included in the simulation model, which generates the performance data. The association rules show relationships between input factors and performance data and they are incorporated into a DES model described.

Simulation and simulation-based optimization: The layout of machines and other details of the system operation were used to develop the DES model using Arena simulation software [17]. The model incorporates intermediate products, work-in-progress, raw materials, lubrication, energy, and operational disturbances. Main Arena modules in the simulation model include part arrival, data requirements for the process, part routing to various machines, part exit, and statistics generation. The manufacturing processes are represented as events, parts as entities, buffers as queues, parts and processes specification data as attributes, and collected data as variables.

The first section of the simulation model deals with part arrival and process data assignment. A part is assigned with information such as design features' dimensions, operation list, and operation orders. The process sequence for the parts can be either of the three options described above. Based on this, the part is then sent to the second section of the model where its operations are decided from the operation matrix developed according to Table 3. A number of combinations of feature-process-machine-tool assignments are implemented in the model. Once an operation is completed, the routing of the part will be decided according to the assigned feature sequencing plan.

Optimization is performed using the OptQuest optimization package supplied by OpTek [46]. It is provided as an option extra with the Arena simulation tool. In OptQuest, resources, such as machine, material, control variables, attributes, constraints, and objective are specified. The user also controls the possible ranges of input variables and set-up inputs for OptQuest. OptQuest uses heuristics known as Tabu search, integer programming, neural networks, and scatter search for seeking within the control (input) space to converge towards the optimal solution.

Table 5 Resulting shop performance due to selected resource combinations

Feature sequence plan	Operation	Resource, R_i	Machining energy (kWh)	Production time (h)
Predefined feature sequence plan	Facing	$R_1 = M1 - T1$	9.676	0.28681
		$R_2 = M2 - T5$	16.961	0.02414
	Grooving	$R_3 = M2 - T4$	16.961	0.02414
	Threading	$R_4 = M1 - T2$	2.902	0.08604
	Spot drill	$R_6 = M1 - T3$	2.580	0.07648
		$R_7 = M3 - T7$	6.484	0.47006
	Drill	$R_6 = M1 - T3$	6.451	0.19120
		$R_9 = M3 - T8$	16.562	1.20068
Partially defined feature sequence plan	Facing	$R_2 = M1 - T1$	8.790	0.22207
	Grooving	$R_3 = M1 - T1$	1.758	0.04441
	Threading	$R_4 = M1 - T2$	2.637	0.06662
	Spot drill	$R_6 = M1 - T3$	2.344	0.05921
	Drill	$R_6 = M1 - T3$	5.860	0.14804
Undefined feature sequence plan	Facing	$R_2 = M1 - T1$	8.790	0.22207
	Grooving	$R_3 = M1 - T1$	1.758	0.04441
	Threading	$R_4 = M1 - T2$	2.637	0.06662
	Spot drill	$R_6 = M1 - T3$	2.344	0.05921
	Drill	$R_6 = M1 - T3$	5.860	0.148047

The results from different scenarios are shown in Table 5. The table also displays the resulting impacts from various system inputs.

Note that the energy consumption as well as production times differ for the same resource set in each plan because of the different sequences in which the design features of the part are produced. Inter-arrival time between successive batch arrivals is set at a constant 120 min. Each batch consists of 15 parts. Data is collected and stored in database.

Determine actionable recommendations: This section discusses the results of various simulation runs from which actionable recommendations are made. Table 5 shows the resources available for each operation. The users can recognize the best process plan, or plans, that minimizes energy consumption and production time (see Table 6. The resource column shows available machine tools for a process; while the indicator columns show the resulting impacts. The table shows the tool-tip energy while the production time displays only the processing time on the machines. The minimum energy consumption is obtained by selecting resources $R_1R_3R_4R_6R_6$.

System users will probably select the partially defined or undefined feature sequencing plans since they have lower energy consumption than the fully predefined sequencing case. At the same time, this sequencing plan would also result in minimum production time for the minimization of energy objective. We note, however, that if the minimum time objective is the one that had originally been set before the table was derived, the production sequence and resource set would have probably been different.

Table 6 Summary of process plans for different feature sequence when minimizing energy consumption

Feature sequence plan	Process plan PP_j	Facing	Grooving	Threading	Spot drill	Drill
Predefined feature sequence plan	PP_1	R_1	R_3	R_4	R_6	R_6
	PP_2	R_1	R_3	R_4	R_7	R_6
	PP_3	R_1	R_3	R_4	R_6	R_8
	PP_4	R_1	R_3	R_4	R_7	R_8
	PP_5	R_2	R_3	R_4	R_6	R_6
	PP_6	R_2	R_3	R_4	R_7	R_6
	PP_7	R_2	R_3	R_4	R_6	R_8
	PP_8	R_2	R_3	R_4	R_7	R_8
Partially-defined feature sequence plan	PP_1	R_1	R_1	R_4	R_6	R_6
Undefined feature sequence plan	PP_1	R_1	R_1	R_4	R_6	R_6

8 Summary, Discussion, and Future Work

Manufacturing industries today collect large volumes of data. Conventional data analysis methods cannot effectively transform this data into knowledge for decision support. Neither can simulation models be applied directly using this data. New approaches are, therefore, needed. This chapter presents a methodology that integrates different methods: data mining, simulation, and optimization for decision-making. This new idea provides the analyst and decision makers with the ability to pinpoint crucial data and prepare model parameters and input data that more effectively help improve performance analysis through simulation optimization. Data mining is first applied to the system data, simulation performs "what-if" analysis for the candidate scenarios, and optimization determines the resource sets, the production plans, and the process plans to optimize a given performance objective. The principal advantage of this methodology over existing approaches is to enable identifying and focusing only on relevant or crucial parameters within collected data. It also helps to reduce the search space for simulation model inputs and optimization by identifying the range of data that significantly affect user-defined system performance.

A case study of a machine shop has been used to demonstrate the methodology. In the case study, we showed how to determine a set of resources and feature sequencing plan that results in minimum tooltip energy during processing. The required prior knowledge can be made available to guide a product specification at the design stage. Similar approaches can be followed for a different objective such as minimum processing time or cost. Data mining to optimize system performance as demonstrated is the first step in developing models for eventually predicting system performance for any part design, machine shop resources, and desired production time.

The methodology involves data collection, model composition, model execution, and result analysis. In practice these activities would be carried out using different tools and models that need to be integrated using standardized interfaces. Therefore, a set of standards are required for the following purposes (1) data collection, (2) data representation, (3) model composition, and (4) system integration. Candidate standards include MTConnect [47] (data collection), CMSD [48] (data representation), Unified Modeling Language (UML) (model composition), and ISA-95 [49] or Open Application Group's Integration Specification (OAGIS) (system integration) [50]. These are briefed next.

MTConnect standard facilitates the organized retrieval of process information from numerically controlled machine tools through continuous data logging. It provides a mechanism for system monitoring, process, and optimization with respect to energy and resources. This standard needs to be extended to collect other data besides CNC machine tools. The CMSD is a standard for integrating simulation applications with other manufacturing applications. CMSD uses a neutral data format to facilitate exchanging both simulation input and output data across supply chain partners. Among CMSD goals are supporting the construction of manufacturing simulators and the testing and evaluation of manufacturing software. More standardization efforts are needed especially for data collection. Currently, data collected is still limited to machine tool data.

For model conceptual design and composition, UML is a standard language for specifying, visualizing, constructing, and documenting the artifacts of software systems. An example of a diagramming method based on the UML is SysML, which supports management of system requirements along with the system development and operation.

The ISA-95 standard defines interfaces between enterprise and shop floor activities while OAGIS establishes integration scenarios for a set of applications including ERP, production scheduling, MES, and capacity analysis. However, OAGIS and ISA-95 were not intended to provide interfaces with simulation systems nor with each other.

Future work includes the definition and description of a framework for data collection and interface for input to data mining and simulation tools; investigation of data mining standards for the methodology; the requirements analysis for extension of existing standards for interfacing between tools for data mining, simulation, optimization, and manufacturing system monitoring; and conducting industrial size case studies.

Acknowledgements This effort has been sponsored in part under the cooperative agreement No. 70NANB13H153 between NIST and Morgan State University. The work described was funded by the United States Government and is not subject to copyright. Qais Y. Hatim, formerly a Guest Researcher at the National Institute of Standards and Technology, contributed to the research of this chapter.

References

1. AEI (American Enterprise Institute). http://www.aei.org.
2. Kibira, D., Jain, S., & McLean, C. (2009). A system dynamics modeling framework for sustainable manufacturing. In *Proceedings of the 27th Annual System Dynamics Society Conference*, July 26–30, Albuquerque, NM. http://www.systemdynamics.org/conferences/2009/proceed/papers/P1285.pdf.
3. ISO/IEC 9075-1. *Information Technology—Database Languages—SQL—Part 1: Framework* (SQL/Framework). http://www.iso.org/iso/catalogue_detail.htm?csnumber=45498.
4. UNECE (United Nations Economic Commission for Europe). http://www.unece.org.
5. Gröger, C., Niedermann, F., Schwarz, H., & Mitschang, D. (2012). Supporting manufacturing design by analytics: Continuous collaborative process improvement enabled by the advanced manufacturing analytics platform. In *Proceedings of the 2012 IEEE 16th International Conference on Computer Supported Cooperative Work.*
6. Ding, S. X., Yin, S., Peng, K., Hao, H., & Shen, B. (2013). A novel scheme for key performance indicator prediction and diagnosis with application to an industrial strip mill. *IEEE Transactions on Industrial Informatics, 9*(4), 2239–2247.
7. Better, M., Glover, F., & Laguna, M. (2007). Advances in analytics: Integrating dynamic data mining with simulation optimization. *IBM Journal of Research and Development, 51*(3/4), 477–488.
8. Brady, T., & Yellig, E. (2005). Simulation data mining: A new form of computer simulation output. In M. E. Kuhl, N. M. Steiger, F. B. Armstrong, & J. A. Joines (Eds.), *Proceedings of the 2005 Winter Simulation Conference* (pp. 285–289). New Jersey: Institute of Electrical and Electronics Engineers Inc.
9. Fayyad, U., Piatetsky-Shapiro, G., & Smyth, P. (1996). From data mining to knowledge discovery in databases. *American Association for Artificial Intelligence, AI Magazine, 17*(3), 17–54.
10. Agard, B., & Kusiak, A. (2005). Data mining in selection of manufacturing processes. In O. Maimon & L. Rokach (Eds.), *The data mining and knowledge discovery handbook* (pp. 1159–1166). Springer.
11. Choudhary, A. K., Harding, J. A., & Tiwari, M. K. (2009). Data mining in manufacturing: A review based on the kind of knowledge. *Journal of Intelligent Manufacturing, 20,* 501–521.
12. Agrawal. R., & Srikant, R. (1994). Fast algorithms for mining association rules in large databases. In J. B. Bocca, M. Jarke, & C. Zaniolo (Eds.), *Proceedings of the 20th International Conference on Very Large Data Bases (VLDB), Conference*, Santiago, Chile (pp. 487–99).
13. Chen, W.-C., Tseng, S.-S., & Wang, C. Y. (2005). A novel manufacturing defect detection method using association rule mining techniques. *Expert Systems with Applications, 29,* 807–815.
14. Agrawal, R., Mannila, H., Srikant, R., Toivonen, H., & Verkamo, I. (1996). Fast discovery of association rules. In U. Fayyad, G. Piatetsky-Shapiro, P. Smyth, & R. Uthurusamy (Eds.), *Advances in knowledge discovery and data mining*. Cambridge, Mass: AAAI/MIT Press.
15. Ng, M. A. (2003). Parallel tabu search heuristic for clustering data sets. In *Presented at the International Conference on Parallel Processing Workshops (ICPPW''03)*, Kaohsiung, Taiwan.
16. Kerdprasop, K., & Kerdprasop, N. (2013). Cluster-based sequence analysis of complex manufacturing process. In *Proceedings of the International Multi-Conference of Engineers and Computer Scientists*, IMECS 2013, Hong Kong (Vol 1).
17. Kelton, D. W., R. P. Sadowski, & Swets, N. B. (2010). *Simulation with Arena* (5th ed.). McGraw-Hill Book Company, International Edition.
18. Law, A., & Kelton, D. (2007). *Simulation modeling and analysis* (2nd ed.). McGraw-Hill, International Editions.

19. Negahban, A., & Smith, J. S. (2014). Simulation for manufacturing system design and operation: Literature review and analysis. *Journal of Manufacturing Systems, 22,* 241–261.
20. Rogers, P., & Gordon, R. J. (1993). Simulation for real-time decision making in manufacturing systems. In G. W. Evans, M. Mollaghaseni, E. C. Russell, & W. E. Biles (Eds.), *Proceedings of the 1993 Winter Simulation Conference* (pp. 866–874).
21. Zhang, H., Jiang, Z., & Cuo, C. (2009). Simulation based optimization of dispatching rules for semi-conductor wafer fabrication system scheduling by the response surface methodology. *International Journal of Advanced Manufacturing Technology, 41,* 110–121.
22. Skoogh, A., Michaloski, J., & Bengtsson, N. (2010). Towards continuously updated simulation models: Combining automated raw data collection and automated data processing. In B. Johansson, S. Jain, J. Montoya-Torres, & E. Yucesan (Eds.), *Proceedings of the 2010 Winter Simulation Conference* (pp. 1678–1689).
23. Shao, G., Shin, S. -J., & Jain, S. (2014). Data analytics using simulation for smart manufacturing, In A. Tolk, S. Y. Diallo, I. O. Ryzhov, L. Yilmaz, S. Buckley, & J. A. Miller (Eds.), *Proceedings of the 2014 Winter Simulation Conference* (pp. 2192–2203).
24. Azadivar. F. (1999) Simulation optimization methodologies. In P. A. Farrington, H. B. Nembhard, D. T. Sturrock, & G. W. Evans (Eds.), *Proceedings of the 1999 Winter Simulation Conference* (pp. 93–100).
25. Brady, T., & Bowden, R. (2001). The effectiveness of generic optimization routines in computer simulation languages, In *Proceedings of the Industrial Engineering Research Conference*, Dallas, Texas.
26. Amaran, S., Sahinidis, N. V., Sharda, B., & Bury, S. J. (2014). Simulation optimization: A review of algorithms and applications. *Journal of Operational Research Society, 12,* 301–333.
27. Carson, Y., & Maria, A. (1997). Simulation optimization: methods and applications. In S. Andradottir, K. J. Healy, D. H. Withers, & B. L. Nelson (Eds.), *Proceedings of the 1997 Winter Simulation Conference* (pp. 118–126). Piscataway, New Jersey: Institute of Electrical and Electronics Engineers Inc.
28. Fu, M., Glover, F. W., & April, J. (2005). Simulation optimization: A review, new developments, and applications. In M. E. Kuhl, N. M. Steiger, F. B. Armstrong, & J. A. Joines (Eds.), *Proceedings of the 2005 Winter Simulation Conference.*
29. Phatak, S. J. Venkateswaran, G. Pandey, S. Sabnis, & Pingle, A. (2014). Simulation based optimization using PSO in manufacturing flow problems: A case study. In A. Tolk, S. D. Diallo, I. O. Ryzhov, L. Yilmaz, S. Buckley, & J. A. Miller (Eds.), *Proceedings of the 2014 Winter Simulation Conference* (pp. 2136–2146).
30. Remondino, M., & Correndo, G. (2005). Data mining applied to agent based simulation. In *Proceedings of the 19th European Conference on Modeling and Simulation*, Riga, Latvia.
31. Bogon, T., Timm, I. J., Lattner, A. D., Paraskevopoulos, D., Jessen, U., Schmitz, M., et al. (2012). Towards assisted input and output data analysis in manufacturing simulation: The EDASim approach. In C. Laroque, J. Himmelspach, R. Pasupathy, O. Rose, & A. M. Uhrmacher (Eds.), *Proceedings of the 2012 Winter Simulation Conference* (pp. 257–269). Piscataway, New Jersey: Institute of Electrical and Electronics Engineers Inc.
32. Dudas, C. (2014). Learning from multi-objective optimization of production systems: A method for analyzing solution sets from multi-objective optimization. *PhD Thesis,* Stockholm University, Sweden.
33. Friedenthal, S., Moore, A., & Steiner, R. (2015). *A practical guide to SysML: The systems modeling language* (3rd ed.). Elsevier Inc: Morgan Kaufmann.
34. MTConnect Part 1. (2011). The Association for Manufacturing Technology, "Getting Started with MTConnect: Connectivity Guide", *White Paper*, MTConnect.
35. SISO (Simulation Interoperability Standards Organization). *Core Manufacturing Simulation Data (CMSD) Standard.* https://www.sisostds.org/DesktopModules/Bring2mind/DMX/Download.aspx?Command=Core_Download&EntryId=36239&PortalId=0&TabId=105.
36. Kibira, D., Hatim, Q., Kumara, S., & Shao, G. (2015). Integrating data analytics and simulation methods to support manufacturing decision making. In L. Yilmaz, W. K. V. Chan,

I. Moon, T. M. K. Roeder, C. Macal, & M. D. Rossetti (Eds.), *Proceedings of the 2015 Winter Simulation Conference* (pp. 2100–2111).

37. Hatim, Q., Shao, G., Rachuri, S., Kibira, D., & Kumara, S. (2015). A simulation based methodology of assessing environmental sustainability and productivity for integrated process and production plans. *Procedia Manufacturing, 1,* 193–204.
38. Zhao, M., & Zhu, S. (1995). *Mechanical engineering handbook*. Beijing, China: China Machine Press.
39. Iwata, K. (1972). A probabilistic approach to the determination of the optimum cutting conditions. *Journal of Manufacturing Science and Engineering, 94*(4), 1099–1107.
40. Sonmez, A., Baykasoglu, A., Dereli, T., & Filiz, I. (1999). Dynamic optimization of multipass milling operations via geometric programming. *International Journal of Machine Tools and Manufacture, 39*(2), 297–332.
41. Sardinas, R. Q., Reis, P., & Davim, P. (2006). Multi-objective optimization of cutting parameters for drilling laminate composite materials by using genetic algorithms. *Composites Science and Technology, 66*(15), 3083–3088.
42. Congbo, L., Ying, T., Longguo, C., & Pengyu, L. (2013). A quantitative approach to analyze carbon emissions of CNC-based machining systems. *Journal of Intelligent Manufacturing, 4,* 34–46.
43. Jeswiet, J., & Kara, J. (2008). Carbon emission and CES in manufacturing. *CIRP Annals-Manufacturing Technology, 57,* 17–20.
44. Malakooti, B., Wang, J., & Tandler, E. (1990). A sensor-based accelerated approach for multi-attribute machinability and tool life evaluation. *Internatial Journal of Production Research, 28*(12), 2373–2392.
45. Rakotomalala, R. (2005). TANAGRA: A free software for research and academic purposes. In *Proceedings of European Grid Conference 2005*, RNTI-E-3 (Vol. 2, pp. 697–702). Amsterdam.
46. OpTek, http://www.opttek.com/products/optquest/.
47. AMT. (2013). Getting Started with MTConnect: Monitoring Your Shop Floor—What's In It For You?, *AMT—The Association for Manufacturing Technology.* Retrieved February 2, 2015, from http://www.mtconnect.org/media/39437/gettingstartedwithmtconnectshopfloormonitoringwhatsinitforyourevapril4th-2013.pdf.
48. SISO. (2012). *SISO-STD-008-01-2012: Standard for Core Manufacturing Simulation Data—XML Representation*. Orlando, F L: Simulation Interoperability Standards Organization.
49. ANSI/ISA 95. *ANSI/ISA–95.00.03–2005—Enterprise-control system integration: Part 3: Activity models of manufacturing operations management.*
50. OAGi. Open Application Group's Integration Specification (OAGIS). (2014). http://www.oagi.org/dnn2/DownloadsandResources/OAGIS100PublicDownload.aspx, Inc.

Part II
Simulation Optimization Study Cases

Improving Airport Performance Through a Model-Based Analysis and Optimization Approach

M. Mujica Mota, P. Scala and D. Delahaye

Abstract Traditionally airport systems have been studied using an approach in which the different elements of the system are studied independently. Until recently scientific community has put attention in developing models and techniques that study the system using holistic approaches for understanding cause and effect relationships of the integral system. This chapter presents a case of an airport in which the authors have implemented an approach for improving the turnaround time of the operation. The novelty of the approach is that it uses a combination of simulation, parameter analysis and optimization for getting to the best amount of vehicles that minimize the turnaround time of the airport under study. In addition, the simulation model is such that it includes the most important elements within the aviation system, such as terminal manoeuvring area, runway, taxi networks, and ground handling operation. The results show clearly that the approach is suitable for a complex system in which the amount of variables makes it intractable for getting good solutions in reasonable time.

1 Introduction

Air global transportation is in continuous growth, looking at the most recent statistics European flights have increased by 0.7% in May 2015 compared with the same month of the last year and it was above the forecast, furthermore preliminary data for June 2015 say that there will be a 1.2% of flights increase compare to June 2014 [10]. The majority of nations in Europe have seen a growth in their local flight, there are reports that mention the levels of congestion the airports in Europe are facing [10, 12]. The direct effect of congestion in the airports is delays that

M. Mujica Mota (✉) · P. Scala
Aviation Academy, Amsterdam University of Applied Sciences,
Amsterdam, The Netherlands
e-mail: m.mujica.mota@hva.nl

D. Delahaye
Ecole Nationale de L'Aviation Civile, Toulouse, France

M. Mujica Mota and I. Flores De La Mota (eds.), *Applied Simulation and Optimization 2*, DOI 10.1007/978-3-319-55810-3_4

correlate with the increasing traffic. The numbers of EUROCONTROL (the European organization for the safety or air navigation) [11] illustrate how the percentage of delayed flight in December 2016 increased by approximately 7% when compared to the same month in the previous year. These situations make evident that capacity in airports is being chocked with the increase on traffic, and this situation might become dramatic if the forecasts of Boeing and EUROCONTROL are correct [10]. For this reason scientific community has paid a lot of efforts for developing tools, new paradigms and novel infrastructure that alleviates the different congestion problems that arise when the traffic increases. These solutions range from optimization tools, re-allocation paradigms or the design of novel infrastructures that have flexibility among their characteristics [8].

1.1 Case Study: Lelystad Airport

Amsterdam Schiphol (AMS) is the main airport in the Netherlands and it was the fifth busiest airport in Europe in 2014 in terms of passenger traffic [1]. Furthermore AMS is also the main hub for KLM, which provided 54% of the seats available at the airport in 2013, and a major airport for the SkyTeam alliance, whose members—including KLM—are responsible for 66.3% of the airport traffic in terms of ATM [27]. Its role as a hub, by airport management and government, is central to the airport strategy, especially considering the small size of the domestic market in the Netherlands and the airport's role as economic engine for the region. However due to environmental reasons, the capacity is limited to 510,000 air traffic movements per year (landings and departures). In 2015 there were 450,679 movements at the airport, 91% of the imposed cap [29]. Since the operation is approaching to the limits, Schiphol Group would like to support the airport strategy by redistributing traffic non-related to the hub development to other airports in the Netherlands. The objective of this action is to relieve capacity and at the same time continuing providing support for the development of the region. The preferred alternative is to upgrade Lelystad Airport (LEY) to attract commercial flights of European cities and regions [28], putting focus on tourist destinations. In that way LEY will take an important role in the multi airport system of the Netherlands composed currently by Schiphol, Rotterdam and Eindhoven.

In recent years Low Cost Carriers (LCCs) in Europe have put focus on short-haul point-to-point leisure traffic, in addition they have been targeting business travellers more actively, and some of them even offer interline connectivity using simple hub structures. This means that the development process at Lelystad should consider not only the type of passengers and airlines that are desired but also the performance parameters the airport should have in order to become attractive for these types of carriers since the airports cannot force the activity in it, instead they make the airport attractive through the offering of incentives economical and operative ones.

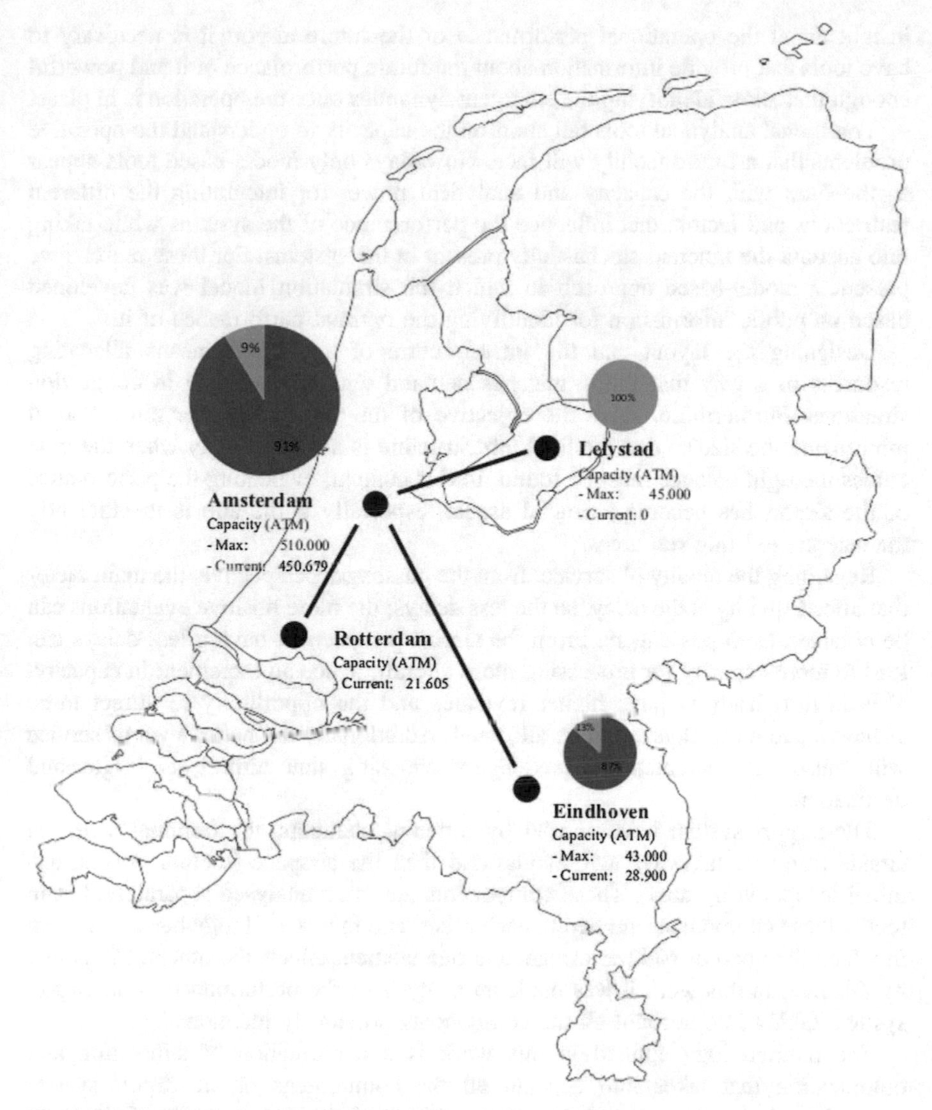

Fig. 1 Lelystad and the multi airport system of the Netherlands

Lelystad is the largest airport for general aviation traffic in the Netherlands. It is located 56 km from central Amsterdam, about 45 min by car to the east. The airport is fully owned by the Schiphol Group, which also owns Rotterdam airport (RTM) and a 51% stake in the Eindhoven airport (EIN), both in the Amsterdam Multi-Airport System (see Fig. 1).

In order to attract airlines, especially LCCs, Lelystad would need to provide differentiation factors: availability of slots; low aeronautical charges; incentive programs and quick aircraft turnaround [15]. Therefore, in order to have better insight about the operational performance of the future airport it is necessary to have tools that provide information about the future performance of it and powerful enough that allow identifying the emergent dynamics once the operation is in place.

Traditional analytical tools fall short on the capacity to understand the operative problems that a future facility will face. Nowadays only model-based tools appear as the ones with the capacity and analytical power for integrating the different restrictions and factors that influence the performance of the systems while taking into account the inherent stochasticity present in the systems. For these reasons we present a model-based approach in which the simulation model was developed based on public information for identifying the optimal performance of it.

Designing the layout and the infrastructure of an airport means allocating resources in a way that traffic matches demand without incurring in congestion situations. Furthermore, since the objective of the airport operator aims also at minimizing the size so that the final infrastructure is not half empty when the time comes the right balance must be found. In this context, evaluating the performance of the airport has become a crucial aspect, especially if the aim is to efficiently manage the existing resources.

Regarding the quality of service, from the passenger perspective, the main factor that affects quality is the delay, so the less delays; the more positive evaluations can be obtained from passengers. From the airport perspective, having less delays can lead to more capacity for processing more aircraft, hence an increment in capacity. This in turn leads to have higher revenues and the opportunity to attract more airlines since more slots could be allocated. Additionally the better level of service will cause the increment in passengers choosing that airport as origin and destination.

The airport system is composed by different elements, the terminal area, the airside (runway, taxiways and stands) and then the airspace (sectors, routes, terminal manoeuvring area). These components are often analysed separately, but in reality these components are tied to each other, and they act all together as a system in which the good or bad performance in one element affects the others. Motivated by this fact, in this work it was made an analysis of the performance of an airport system, taking into account all the components previously mentioned.

The methodology applied in this work is a combination of simulation and optimization that takes into account all the components of an airport system (ground + airspace), and evaluates the airport performance in terms of the turnaround time (TAT). The simulation paradigm used in this work is a Discrete-event Simulation (DES) in a program called SIMIO [31]. The optimization approach is a simulation-based optimization in which the search space is the domain of the Cartesian product of the values of the main factors that affect the objective function; the search is performed by an embedded tool called OptQuest [23] that has different heuristics for optimizing the search.

With the use of the simulation model, different configurations of resources were evaluated paying attention to the TAT. The use of design of experiments (DOE) was carried out employing a multi-level factorial design with the purpose of evaluating the effect of the factors and their interactions for the system response.

Moreover, with the study of the Analysis of Variance (ANOVA), the main factors that affect the objective function were determined. Finally, for optimizing the TAT, we used the information of the identified factors for making the optimization search more efficient than the one that could be done without the analysis.

1.2 Previous Work

Optimization of airport resources is a subject that was faced by researchers in many studies; most of them treated the airport as a two separate entities, from one side airspace and from the other ground side. In this context, many techniques that aimed at improving airport performance were employed, taking into account different variables. Concerning the airspace, specifically for the Terminal Manoeuvring area (TMA) many studies focused on the sequencing and merging problem and scheduling problem. The former is concerned in finding the best sequence for aircraft flow in order to determine conflict-free situations [18, 33, 34], the latter is about scheduling of aircraft flow in order to minimize the deviation between the scheduled landing time and the actual landing time [3–6, 21].

The techniques most utilized were from the operations research arena in which some of the solutions used stochastic optimization models [2], however, due to the complexity of the problems, for many of them heuristics were implemented in order to find sub optimal solutions. Just to mention some, the aircraft scheduling problem was studied extensively by Beasley et al. [4–6] this work focused on developing a mixed-integer one-zero problem and then the authors employed two heuristics respectively for the static and dynamic case. Other relevant work is the one from Balakrishnan et al. [3], which uses constrained position shifted (CPS) for improving the sequence of aircraft by changing the position of the aircraft in order to minimize the makespan. Hu and Chen [14] proposed a receding horizon control (RHC) technique where the scheduling and sequencing problem were treated in a dynamic way; they introduced a genetic algorithm for solving it.

Regarding the ground side, most of the studies are related to the optimization of gate assignment, the scheduling of departing aircraft and taxiing operations, with the objective of avoiding congestion situations and favouring a smooth flow of aircraft in the taxiways. For instance, in the work of Dorndorf [9] the authors present a survey about the techniques used to cope with the gate assignment problem, among others we can find the work of Bolat [7] in which a branch and bound algorithm was combined with two heuristics. A Coloured petri net (CPN) technique was proposed by Narciso and Piera [22] in order to calculate the number of stands needed to absorb the traffic. In other studies pushback control strategies were proposed in order to determine the best sequence of departures without incurring in congestion situations [16, 24, 30].

As it can be seen for the previous review, the most implemented techniques refer to analytic and heuristic models, and there is a clear distinction between airspace and ground side. In this work the problem is treated from a holistic view in which both airspace and ground side are analysed together for making a more complete study. Additionally, a methodology has been followed that permits optimizing airport performance following a structured way. The approach focuses in performance measured as turnaround time which is the key for determining the amount of resources an airport needs in order to improve throughput and reduce delays due to congestion.

The chapter continues in the following way, in Sect. 2 the proposed methodology is presented. Sect. 3 presents how the methodology is applied in a particular case, and finally in Sect. 4 the correspondent conclusions are presented.

2 Methodology

The approach uses first Discrete Event Simulation (DES) together with statistical techniques for identifying the most influencing factors in the performance of the airport under study. After performing an analysis of the different factors that influence the performance, they are disaggregated for making a more refined selection of those. The identification of the ultimate ones allows the reduction of the search space of the optimization tool embedded in the simulation program used.

DES is an approach that is used in many applications like logistic and manufacturing [17]. Recently DES was also applied to the aviation field with the scope of modelling the airport operation for both airspace and ground, even inside the terminal [19]. Using this approach, it has been possible to make an initial analysis and evaluation of airports performance [20, 25, 26]. The methodology uses statistical tools like Design of experiments and the ANOVA for identifying the factors that impact the system the most and a selection of parameters is done which at the final stage will be used to optimize the values of the most influential elements of the system.

The methodology applied works in phases, in the first phase it performs the identification of the factors that affect the performance of the airport using an objective function of the turnaround time. During this phase the significance of the different factors that affect such performance are identified and then a combination of DOE with ANOVA is performed for making a more refined selection of the elements that affect the indicator.

In a second phase the model is combined with an optimization algorithm for performing the improvement of the system under study in which the decision variables are the ones that affect the objective function.

Figure 2 illustrates the different phases of the methodology.

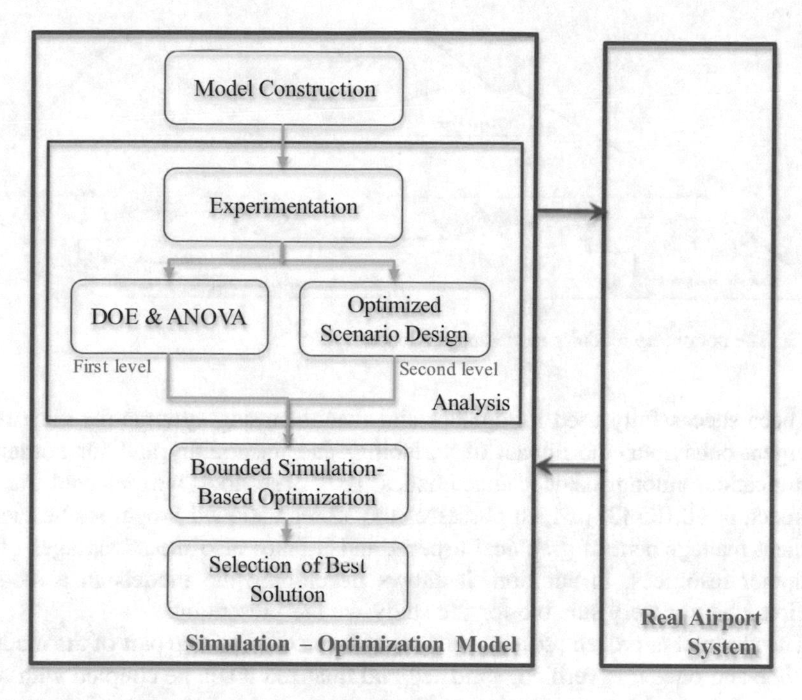

Fig. 2 Methodological approach for airport optimization

2.1 Development of the Airport Model and Identification of Variables

Until recently, scientific community has been taken simulation as a key tool for evaluating systems performance during the planning phase of facility development. In the aviation field the studies concerning systems performance and capacity evaluation are quite recent but its potential has been recognized by international institutions and also as consultants which are becoming keen for the use of simulation for performing studies [19, 25, 26].

The first phase focuses on the development of a simulation model of the system under study and the identification of the main variables. In the case of the airport of this work, we used the DES approach. This is an approach that is used for modelling systems of dynamic nature in which there is strong interaction between the different processes of the system and stochasticity is one of the characteristics that define them. In comparison with other approaches, the time advances as events are happening in the model, so the number of calculations is much less than the ones required for agent-based technology for instance. As with most of the simulation approaches, it allows the identification of emergent dynamics within the system and it has the full potential for integrating the inherent stochasticity which in some situations hinders the smooth behaviour of the system under study. This approach

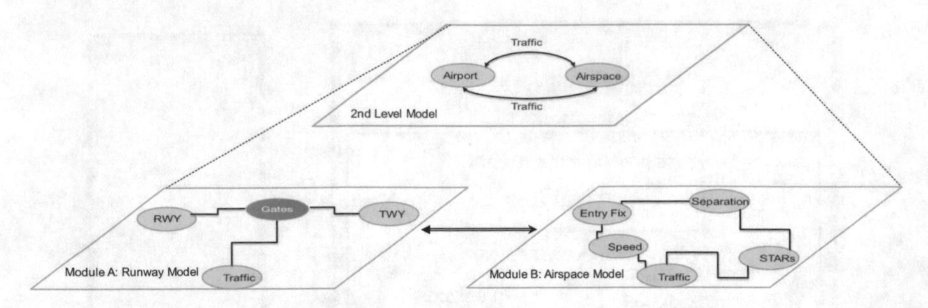

Fig. 3 The bottom-up modular modelling approach

has been successfully used in logistics and manufacturing systems for identifying emergent behaviour, the impact of variability and uncertainty and for bottleneck identification among other characteristics [20]. The tool we selected for the approach is SIMIO [31] which possesses the aforementioned properties besides an efficient management of graphical aspects and it takes also the advantages of the computer resources. In addition, it allows developing the models in a modular fashion which is very suitable for the study we are presenting.

The modular approach [26] allows putting focus only on one part of the model at the time and once it is verified, validated and finalized it can be coupled with other modules for obtaining the final one. The authors suggest strongly this approach since the reliability of the final model is higher if we perform a bottom-up modelling approach. Figure 3 illustrates the proposed approach for this phase.

For the developed model, the main components of airside and airspace were included like:

- Runway system
- Taxiway system and stands
- Approaching and departing routes
- Airspace

The architecture of the different modules' models is illustrated by Fig. 4.

The airside is made by a coupling of two modules: the runway model and the turnaround model. The runway model integrates the main characteristics and restrictions of the utilization of any runway in an airport system such as wake vortex separations, speed limitations, taxi speed and limitation of the runway.

The turnaround model is made by a model in which all the services required by an aircraft at the gate are implemented. These services are performed by a number of vehicles dedicated to providing them. Table 1 illustrates the different implementations and restrictions of the runway and turnaround model.

The last module that composes the integral model is the one of the airspace close to the airport, in particular the area known as terminal manoeuvring area, which is composed by a radius of approximately 40 nautical miles. This area is important since it is within its limits that the sequencing of arrivals is performed.

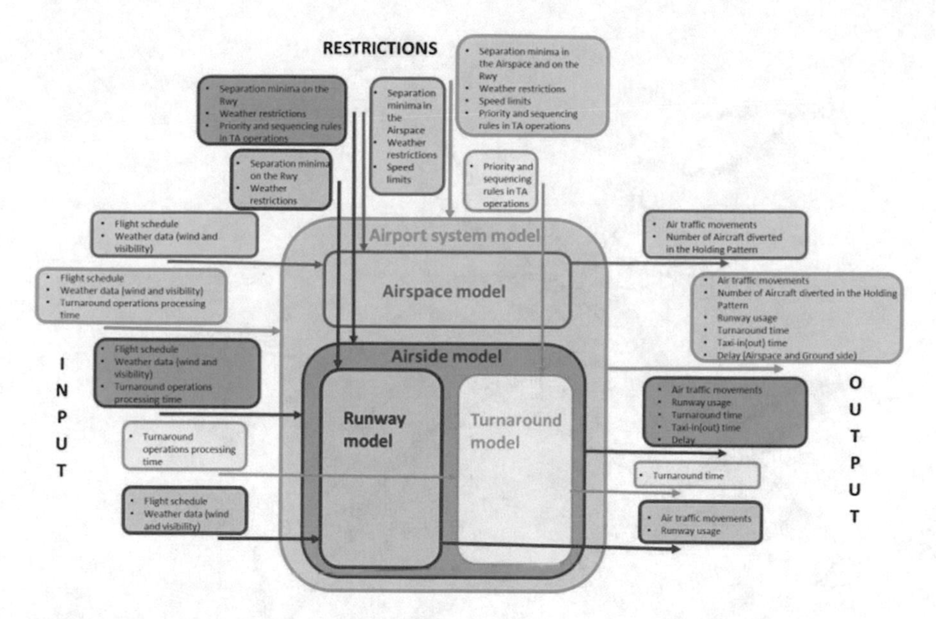

Fig. 4 Modular architecture of the airport model

Table 1 Characteristics of the runway and turnaround model

Parameter	Value
Number of runways	1
Number of exit ways	1
Taxiway type	Parallel
Number of stands	16
Aircraft speed Taxi in	[45 Knot…24 Knot]
Aircraft speed Taxi out	[19 Knot]

The modelling approach uses a network of nodes located at different altitudes and positions in the space in which the connections between them represent the airways followed by the aircraft in their routes to the airport. Due to the differences of scale in the airport and airspace model, the airport itself just represents one segment of the network. Figure 5 illustrates the network created for modelling the airspace.

The acronyms present in the figure refer to *Initial Approach Fix (IAF)* and *Final Approach Fix (FAF)* which are the segments in which the controllers guide the aircraft for their final route to landing [32]. The basic restrictions that must be taken care of for the development of the simulation model are the separations that need to be respected by the aircraft to land. These separations are for safety reasons which ensure the minimization of the risk of collision or interactions between the aircraft in the area surrounding the airport.

Tables 2 and 3 present the description of the different restrictions and parameters that compose the model.

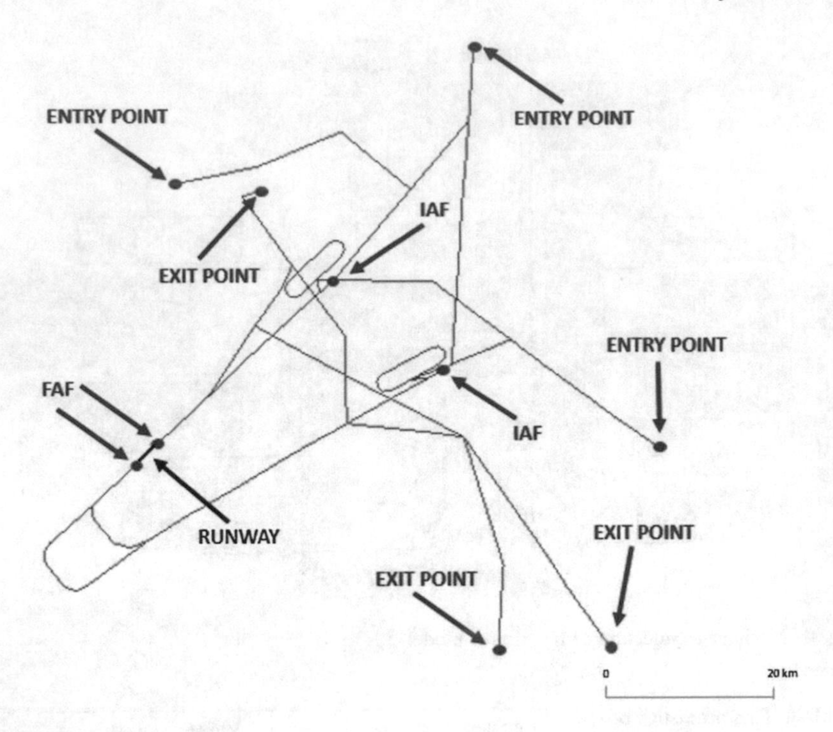

Fig. 5 TMA airspace network

Table 2 Characteristics of the TMA model

Parameter	Value	Value
Entry point (Speed)	250 Knot	160 Knot
Initial approach fix (Speed)	160 Knot	130 Knot
Final approach fix (Speed)	–	130 Knot
Holding pattern (number and speed limit)	1 for each route, 200 Knot	
Aircraft mix	Code C (B737–A320)	

Table 3 Separation minima in nautical miles (ICAO)

		Leading aircraft		
		Heavy	Medium	Light
Trailing aircraft	Heavy	4	3	3
	Medium	5	3	3
	Light	6	4	3

Once the three models have been developed and validated against the expected speed and relevant variables, they were merged model that represents the airport system (airside and airspace). The different modules interact with each other in such a way that it is possible to evaluate the behaviour of different performance indicators (PI) and the emergent dynamics which would not be possible to perceive if

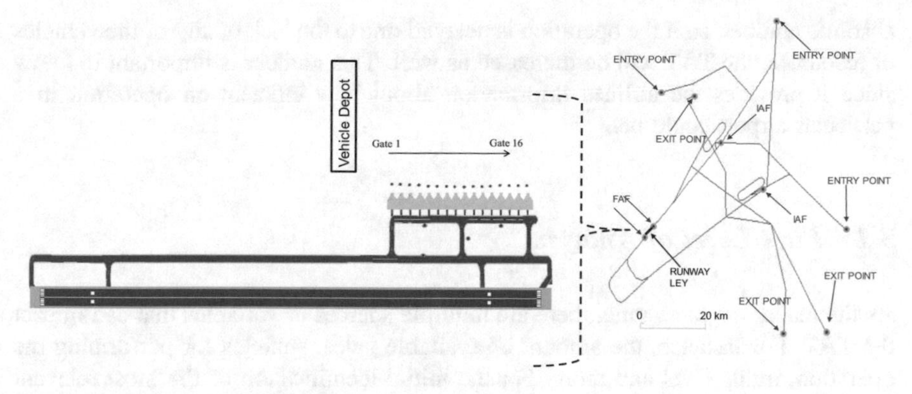

Fig. 6 Complete model of the airport of Lelystad

the models were analysed independently. Figure 6 depicts the complete model, in which the entities first are generated in the airspace model, then they are sequenced for landing and the landing is performed. For the landing process they get out of the airspace model and enter the airside model. In the airside all the landing and taxiing is performed until the aircraft gets to the gate in which the turnaround operation is performed by the ground handling vehicles.

With the complete model, it was possible to evaluate performance indicators (PIs) like number of movements, data about turnaround time and about delay under different scenarios. As an initial approach focus was put on the analysis of the TAT which is very important for understanding the potential of the airport under study.

3 Getting More Insight: Design of Experiments

Design of experiments is a technique that permits to identify the main parameters, or factors that affect the performance of a system. With this technique, it is possible to evaluate what the main effects of the factors involved are, and also the effect of their interactions. This technique allows identifying the main effects for each factor. To that end, for each factor different values were assigned, called levels. For each combination of factor level a response is evaluated and an analysis is performed in order to identify if the factor is statistically significant for the studied variable.

This phase focuses on developing structured experiments with the model for identifying the most relevant factors.

For the example we present, we applied recurrently the technique in the simulation model to make an identification of the variables that affect the outcome of the PIs. For the first and second level analysis we put the focus on the objective under study: Turnaround Time.

Turnaround time (TAT): This parameter is the time measured from the moment the aircraft parks in the stand until it is ready for taxing out to the runway. This is an absolute number, so if the operation is delayed due to the lack of any of the vehicles or resources, the TAT will be increased as well. This number is important to know since it provides the airlines information about how efficient an operation in a particular airport could be.

3.1 First Level of Analysis

As the reader might assume, there are multiple sources or variables that can impact the TAT. For instance, the amount of available gates, vehicles for performing the operation, traffic level and more. For the initial identification of the most relevant variables, we applied DOE in categories that group some factors. This selection was based on expert opinion and the selected ones were: air traffic, available vehicles for the turnaround, and stand allocation. Using these factors we performed a *multi-level full factorial design*. Table 4 illustrates the different categories of factors we evaluated for the design.

For the first and the second factors we set three levels and two levels for the last one. In addition, 50 replications were made for each level.

The evaluated levels for the three factors followed the following logic:

- Incoming Flow of aircraft. As it has been mentioned, this study deals with the evaluation of a future airport in the Netherlands. The public information states that the amount of expected traffic is approximately 50,000 ATMs per year. Thus the Level 2 is approximately this value so this traffic is considered the one expected by the airport. The other two levels explored the situation in which 30% more and 30% less traffic than expected is received in the airport.
- Number of Vehicles. The number of vehicles refers to the *sets* of vehicles that can be used for the operation. Without economical limitations we can estimate that we might use one complete set per aircraft, thus the initial set is of 9 vehicles. One set itself is composed by 1 fuel, 2 passenger bus, 1 water, 2 bulk trucks, 2 stair trucks and 1 loader. The other two levels are used for evaluating the reduction in vehicles so that it is possible to perceive when the turning point is (if there is) of performance due to the lack of vehicles.
- Apron's entering mode. For this factor, only two levels were evaluated, they concerned with how the aircraft were allocated in the available stands. The two levels are, from left to right, assuming a first-in first-served allocation putting

Table 4 Evaluated category factors

Factors	Level 1	Level 2	Level 3
A—Incoming flow of aircraft (flights/day)	92	132	190
B—Number of vehicles	2	5	8
C—Apron's entering mode	Left–right	Center-out	–

Analysis of Variance

Source	DF	Adj SS	Adj MS	F-Value	P-Value
Model	7	790.29	112.899	3.82	0.028
Linear	3	578.34	192.779	6.53	0.010
Traffic	1	120.25	120.251	4.07	0.071
Vehicles	1	432.73	432.726	14.66	0.003
Stand Allocation	1	25.36	25.361	0.86	0.376
2-Way Interactions	3	234.45	78.149	2.65	0.106
Traffic*Vehicles	1	221.89	221.891	7.52	0.021
Traffic*Stand Allocation	1	12.31	12.315	0.42	0.533
Vehicles*Stand Allocation	1	0.24	0.242	0.01	0.930
3-Way Interactions	1	22.87	22.866	0.77	0.399
Traffic*Vehicles*Stand Allocation	1	22.87	22.866	0.77	0.399
Error	10	295.24	29.524		
Total	17	1085.54			

Fig. 7 Analysis of variance for the factors of TAT

priority in the stands closest to the left part of the apron while center-right assumes hat the priority is put in the central stands.

3.1.1 Analysis of Variance (ANOVA)

Once we have run the full factorial analysis, we performed the one-way ANOVA for identifying the impact of the different categories evaluated. Figure 7 presents the results obtained with the ANOVA test.

From the ANOVA we could identify that the most significant category is the vehicles set assuming the standard *p* value of 0.05. This category is followed by traffic; however statistics cannot support conclusively this assumption. In addition we can also perceive that the interaction of traffic with vehicles is significant. So, as expected, traffic itself affects, however that is a variable in which we cannot manipulate to get a better or worse performance. For this reason we paid attention to the amount of vehicles in order to going further in the analysis.

Figure 8 depicts the Pareto chart of standardized effects to graphically illustrate the diverse effects of the different categories evaluated.

Once we selected as vehicles as the most influential and controllable factor, the next question that arise is what the right mix of vehicles would be for a smooth and efficient operation.

3.2 *Second Level of Analysis*

We run the *second level* DOE in which the factors were the different categories of vehicles and their levels were the number of them. As the reader might note, the combinatorial challenge make it impossible to run a full factorial design, which in

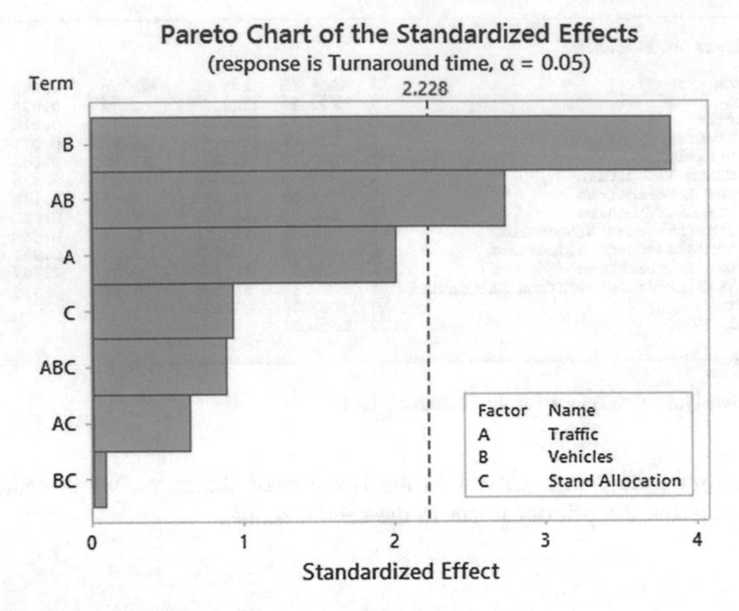

Fig. 8 Plot of the standardized effects for turnaround time

this case it might imply to run at least 19 683 different scenarios. For that reason, we implemented the Federov algorithm [13] which allowed the reduction of the amount of configurations to evaluate by doing and intelligent selection of them. After implementing this algorithm, the number of scenarios to evaluate was reduced to only to 27 as Table 5 presents.

Table 5 has been encoded for the different number of vehicles, it corresponds to −1 as 2 vehicles, 0 corresponds to 5 vehicles and +1 corresponds to 8 vehicles.

After running the 27 scenarios, we performed again the ANOVA for identifying which vehicles were the most influential for the objective pursued. In this case and due to the few amounts of points for the analysis it was not possible to consider the 2nd order interactions. Therefore we could only make an analysis of the first order interactions or the direct effect of the use of vehicles.

Figure 9 presents a scatter plot that together with ANOVA helps identifying the influence in the TAT of some parameters which later would be used for improving the optimization search.

From the scatter plot we could identify that some values of vehicles minimize the turnaround time, namely Stairs1, Stairs2, Bulk2, fuel truck, Bus2. This result was also used for the last phase of the optimization.

In addition to this analysis, we performed the ANOVA for identifying which vehicles were the most influential for the TAT. We identified that the main factors

Table 5 Design of experiments based on Federov's algorithm

Scenario number	Loader	Bulk1	Bus1	Stairs1	Stairs2	Water service	Bulk2	Fuel truck	Bus2
219	1	−1	−1	1	1	−1	−1	−1	−1
723	1	−1	1	1	1	1	−1	−1	−1
4609	−1	−1	1	1	1	−1	−1	1	−1
4867	−1	1	−1	−1	−1	1	−1	1	−1
4941	1	1	1	1	−1	1	−1	1	−1
5049	1	1	1	−1	1	1	−1	1	−1
5077	−1	−1	−1	1	1	1	−1	1	−1
5232	1	−1	1	0	0	−1	0	1	−1
5851	−1	−1	1	−1	−1	−1	1	1	−1
5894	0	1	−1	1	−1	−1	1	1	−1
5968	−1	−1	−1	1	0	−1	1	1	−1
6019	−1	1	1	−1	1	−1	1	1	−1
8202	1	−1	1	−1	1	−1	1	−1	0
12555	1	1	1	1	0	−1	1	1	0
13123	−1	−1	−1	−1	−1	−1	−1	−1	1
13131	1	1	−1	−1	−1	−1	−1	−1	1
13443	1	0	1	1	−1	0	−1	−1	1
13687	−1	1	1	1	−1	1	−1	−1	1
14312	0	−1	−1	1	1	0	0	−1	1
14367	1	−1	−1	0	−1	1	0	−1	1
15087	1	−1	1	−1	−1	1	1	−1	1
15136	−1	1	0	1	−1	1	1	−1	1
15255	1	1	1	−1	1	1	1	−1	1
17760	1	−1	1	−1	−1	0	−1	1	1
18007	−1	1	1	−1	−1	1	−1	1	1
19012	−1	0	−1	1	−1	−1	1	1	1
19029	1	−1	1	1	−1	−1	1	1	1

that affect the turnaround time was firstly the fuel truck, and then the use of the stairs. Figure 10 illustrates the outcome of the ANOVA analysis.

This result is very important, since it suggests that when someone is interested in improving the TAT of this particular airport, he should ensure that there are enough amounts of fuel trucks and stairs. The Pareto chart of standardized effects in turn can also illustrate the impact of the fuel truck and the stairs as the reader can see in Fig. 11.

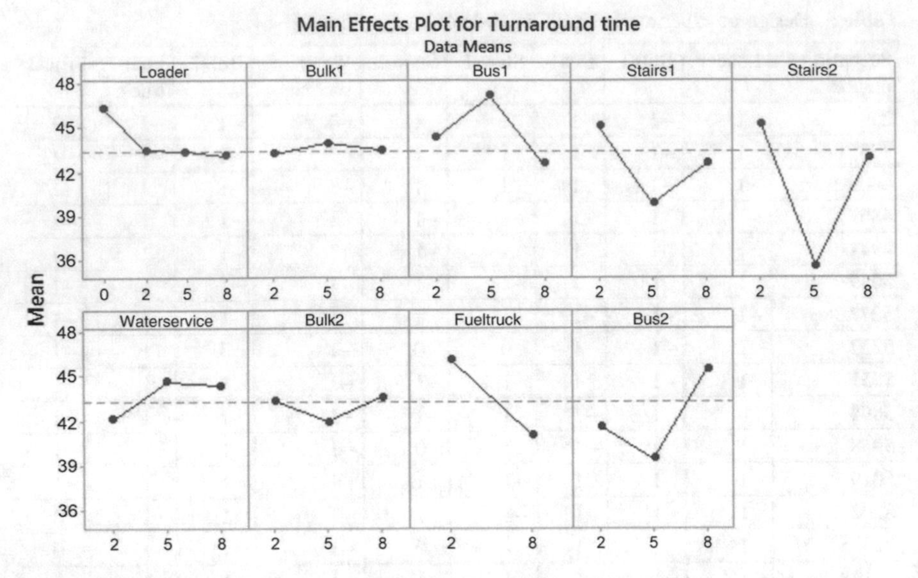

Fig. 9 Dependency of TAT on the modification of vehicle numbers

Analysis of Variance

Source	DF	Adj SS	Adj MS	F-Value	P-Value
Model	9	295.964	32.885	2.99	0.025
Linear	9	295.964	32.885	2.99	0.025
Loader	1	16.066	16.066	1.46	0.243
Bulk1	1	0.078	0.078	0.01	0.934
Bus1	1	2.848	2.848	0.26	0.617
Stairs1	1	36.351	36.351	3.30	0.087
Stairs2	1	36.745	36.745	3.34	0.085
Waterservice	1	7.392	7.392	0.67	0.424
Bulk2	1	7.287	7.287	0.66	0.427
Fueltruck	1	115.960	115.960	10.54	0.005
Bus2	1	3.275	3.275	0.30	0.592
Error	17	187.046	11.003		
Total	26	483.010			

Fig. 10 ANOVA for the 1st order interaction in the TAT analysis

3.3 *Optimizing the Model Response*

In the next phase of the analysis, we used the previous results for making a more informed search over the solution space of the simulation model.

In most of the commercial simulation tools there are programs embedded that perform a simulation-based optimization. This optimization is performed by

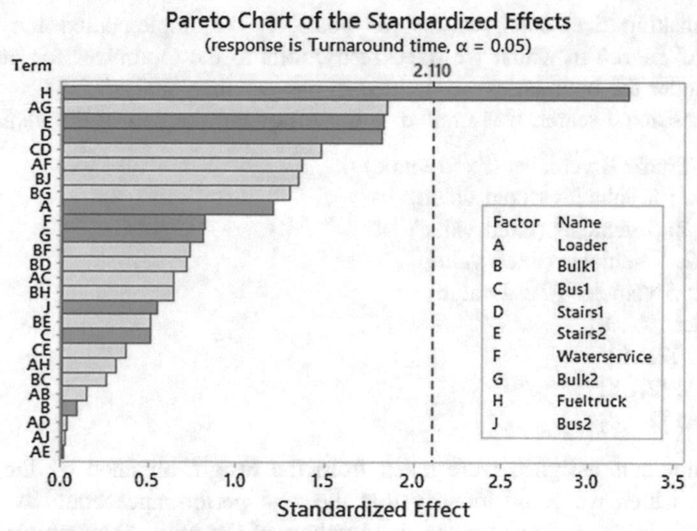

Fig. 11 Standardized effects for the TAT evaluation

parameterizing the simulation model and then undergoing a search in the domain of the parameters' values. The search is done in most of the cases using a brute-force approach in which the program just tests different values and make several replications of the model in order to find the best values for the objective function. As the reader might infer, the more parameters and the higher the range of the domain the more time consuming the search becomes. For this reason, it is necessary to give support to the search, otherwise the required time to get to a good solution could take a lot of time, and sometimes it would become unfeasible to wait for a solution.

For the previous reason, in the next stage, we used the information obtained from the previous analysis for restricting the domain of the search in the algorithm of the optimization program embedded in SIMIO.

3.3.1 Optimization Phase

The final phase of the methodology focuses on getting the optimal values for the Turnaround Time which is the factor analysed in this study.

OptQuest is an optimization tool present in SIMIO, and it allows the user to specify the objective function(s), domains, independent variables which will define the search space, and it will use the simulation model for performing the evaluation of the objective function. As the reader might know, the search over a high dimensional space takes from some minutes to even days, for that reason it is important to define wisely the boundaries and objectives of the optimizer.

For making the search as fast as possible, we implemented the so-called *Restricted Search* in which we fixed restrictions to the Optimizer for making the search under the boundaries we defined in the previous analysis.

The restricted search was limited to the following domain of the vehicles:

- Fuel Truck: 8 vehicles (fixed value)
- Stairs1: 5 vehicles (fixed value)
- Stairs2: 5 vehicles (fixed value)
- Bulk2: 5 vehicles (fixed value)
- Bus2: 5 vehicles (fixed value)
- Loader: [2…8]
- Bus1: [2…8]
- Bulk1: [2…8]
- Water: [2…8]

The numbers assigned were taken from the insight obtained by the previous phase in which we could identify that the best performance could be achieved somewhere in the region near the fixed values of the initial five vehicles. For the remaining vehicles we relaxed the search so that the algorithm of the optimizer can search freely on the complete domain.

For the sake of comparison we also performed the same optimization but letting OptQuest make the search on the domain of the relaxed variables, for this reason we called it as *Free Search*.

The obtained results for both searches are presented in the following table.

For limiting the speed of calculation and time to get the results it is also necessary to establish some limits for the allowed number of combinations for providing the solution. In our example, for making a comparison between the free and the restricted search, we set the limits to 50 and 100. In addition we also set another limit for the free search just for having an idea of the improvement that can be achieved if the analyst had enough time to let the model run.

3.3.2 Maximum Combinations 50

Regarding the performance of the approach, when we pay attention to the scenarios, the first one is the limited by 50 permutations.

Table 6 illustrates that in terms of Turnaround time, the free search provides a slightly better solution than the restricted one, however the restricted search finds a similar solution with only 5 combinations and a smaller number of vehicles than the free search (56 compared to 65).

When we check the solution with the minimum number of vehicles, we identified that after two permutations the free search provides a solution with 18 vehicles and a turnaround time of 42.46 min, while the restricted one provides a solution of 36 vehicles but with a shorter turnaround time of 37.01 min. These results supports the premise that limiting the search space based on the results of the DOE & ANOVA it will provide a better starting point for the search.

Table 6 Analysis of the optimized search

Type of search		Free search			Restricted search	
Maximum number of combinations		50	100	300	50	100
Solution with minimum TAT	Turnaround time (min)	29.41	29.23	29.40	29.56	29.56
	Number of vehicles	65	67	58	56	56
	Number of combinations	17	37	204	5	5
Solution with minimum number of vehicles	Turnaround time (min)	42.46	42.46	42.46	37.01	29.56
	Number of vehicles	18	18	18	36	56
	Number of combinations	2	2	2	2	5

3.3.3 Maximum Combinations 100

Regarding the turnaround time, for this amount of maximum number of permutations we can appreciate that the achieved Turnaround times are very similar, however the restricted search finds a solution which is less costly since it uses only 56 vehicles while the free search 67. In addition, the restricted search finds it with a minimum amount of permutations.

If we wanted to pay attention to a solution of minimum vehicles, the free search finds a suitable solution of 18 vehicles while the restricted one finds one of 36 vehicles but with a better turnaround time in the same amount of permutations which is in line with the previous example.

Regarding the free search with a limit of 300 permutations, we can appreciate that the results are not necessarily better, they can be even worse than a more restricted search. This can be noted in the turnaround time when we let it make a free search on a more relaxed fashion. This result also indicates the complexity of the solution space of this system.

4 Conclusions

Managing an airport system is a complex task in which the decision involves many variables, thus the decision makers require decision-support tools that provide them insight of the consequences of taking particular decisions.

In this work we presented a case of the analysis of an airport in the Netherlands which is currently under construction. For the decision makers it is important to identify what the most influential variables are in order to improve the performance.

This is key for them since the more efficient the airport, the most attractive for airlines to move there.

In this work we illustrated how a structured methodology can help identifying the most influential decision variables for the system in place. With the identification of them, it is possible to use simulation together with optimization for finding the values of the decision variables that improve the performance of the airport under study; in this case we put focus on the turnaround time. The results illustrate that certainly the methodology successfully drives the search space into a region of good solutions so we could obtain very good values without performing a time-consuming search.

The methodology has been implemented in the case of an integral airport model developed in SIMIO using OptQuest as the optimization tool. However this methodology can be easily implemented in a different area using a different simulation tool and a different optimizer.

References

1. ACI. (2014). https://www.acieurope.org/component/content/article/29-article/103-world-airport-traffic-data.html.
2. Arias, P., Guimarans, D., Boosten, G., & Mujica, M. (2013). A methodology combining optimization and simulation for real applications of the stochastic aircraft recovery problem'. In *Proceedings of the EUROSIM'13*, Cardiff, U.K.
3. Balakrishnan, H., & Chandran, B. (2010). Algorithms for scheduling runway operations under constrained position shifting. *Operations Research, 58*(6), 1650–1665.
4. Beasley, J. E., Krishnamoorthy, M., Sharaiha, Y. M., & Abramson, D. (2000). Scheduling aircraft landings—the static case. *Transportation Science*, 180–197.
5. Beasley, J. E., Krishnamoorthy, M., SharaihaY, M., & Abramson, D. (2004). Displacement problem and dynamically scheduling aircraft landings'. *Journal of the Operational Research Society, 55,* 54–64.
6. Beasley, J. E., Sonander, J., & Havelock, P. (2001). Scheduling aircraft landings at London Heathrow using a population heuristic. *Journal of the Operational Research Society, 52,* 483–493.
7. Bolat, A. (2000). Procedures for providing robust gate assignments for arriving aircrafts. *European Journal of Operational Research, 120*(1), 63–80.
8. DeNeufville, R., & Scholtes, S. (2011). *Flexibility in engineering design*. MIT Press.
9. Dorndorf, U., Drexlb, A., Nikulin, Y., & Peschc, E. (2007). Flight gate scheduling: State-of-the-art and recent developments. *Omega, 35*(3), 326–334.
10. Eurocontrol. (2013). Challenges of growth 2013. Technical report.
11. Eurocontrol. (2016, December). Coda Report: All-causes Delay to Air Transport in Europe. Technical report.
12. Eurocontrol. (2015). Industry Monitor, Issue N 174. 01/07/2015.
13. Federov, V. V. (1972). *Theory of optimal experiments*. In W. J. Studden & E. M. Klimko (Eds). New York: Academic Press.
14. Hu, X., & Chen, W. (2005). Receding horizon control for aircraft arrival sequencing and scheduling. *IEEE Transactions on Intelligent Transportation Systems*, *6*(2).
15. Jimenez, E. (2014). Airport strategic planning in the context of low-cost carriers ascendency: insights from the European experience. Ph.D. thesis, Universidad do Porto.

16. Khadilkar, H., & Balakrishnan, H. (2014). Network congestion control of airport surface operations. *AIAA Journal of Guidance, Control and Dynamics, 7*(3), 933–940.
17. Longo, F. (2013). On the short period production planning in industrial plants: A real case study. *International Journal of Simulation and Process Modelling, 8,* 17–28.
18. Michelin, A., Idan, M., & Speyer, J. L. (2009). Merging of air traffic flows. In *AIAA Guidance, Navigation, and Control Conference*, 10–13, Chicago, Illinois.
19. Mujica, M. (2015). Check-in allocation improvements through the use of a simulation-optimization approach. *Transportation Research Part A*, 320–335.
20. Mujica, M., de Bock, N., Boosten, G., Jimenez, E., & Pinho, J. (2015). Simulation-based turnaround evaluation for Lelystad airport. In *Air Transport Research Society World Conference*, 2–5 July, Singapore.
21. Murca, M. C. R., & Müller, C. (2015). Control-based optimization approach for aircraft scheduling in a terminal area with alternative arrival routes. *Transportation Research Part E, 73,* 96–113.
22. Narciso, M. E., & Piera, M. A. (2015). Robust gate assignment procedures from an airport management perspective. *Omega, 50,* 82–95.
23. OptQuest Webpage http://www.opttek.com/products/optquest/.
24. Pujet, N., Delcaire, B., & Feron, E. (1999). Input-output modeling and control of the departure process of congested airports. In *A collection of technical papers: AIAA Guidance, Navigation and Control Conference and Exhibit*, Part 3 (pp. 1835–1852), 9–11 August, Portland, Oregon, USA.
25. Scala, P., Mujica, M., & Zuniga, C. A. (2015). Assessing the future TMA capacity of Lelystad airport using simulation. In *Air Transport and Operations Symposium*, 20–22 July, Delft, The Netherlands.
26. Scala, P., Mujica, M., & De Bock, N. (2015). Modular approach for modelling an airport system. In *EMSS European Modeling & Simulation Symposium*, 21–23 September, Bergeggi, Italy.
27. Schiphol Magazine. (2014). http://trafficreview2014.schipholmagazines.nl/air-transportmovements.html#atmmainairlinesa.
28. Schiphol Group. (2014). Ondernemingsplan Lelystad Airport, March 2014.
29. Schiphol Group. (2016). http://www.schiphol.nl/SchipholGroup1/Onderneming/Statistieken/VerkeerVervoerCijfers1.html.
30. Simaiakis, I., & Balakrishnan, H. (2014). A queuing model of the airport departure process. *Transportation Science* (accepted, July 2014).
31. SIMIO Web Page http://www.simio.com.
32. van Baren, G., & Treve, V. (2015). The current practice of separation delivery at major European airports. http://www.atmseminar.org/seminarContent/seminar11/papers/466-Van%20Baren_0126150311-Final-Paper-5-7-15.pdf.
33. Zuniga, C. A., Delahaye, D., & Piera, M. A. (2011). Integrating and sequencing flows in terminal maneuvering area by evolutionary algorithms. In *DASC 2011, 30th IEEE/AIAA Digital Avionics Systems Conference*, Seattle, United States.
34. Zuniga, C. A., Piera, M. A., Ruiz, S., & Del Pozo, I. (2013). A CD & CR causal model based on path shortening/path stretching techniques. *Transportation Research Part C, 33,* 238–256.

Airport Ground Crew Scheduling Using Heuristics and Simulation

Blaž Rodič and Alenka Baggia

Abstract International airports are complex systems that require efficient operation and coordination of all their departments. Therefore, suitable personnel and equipment scheduling solutions are vital for efficient operation of an airport as a system. Many general solutions for fleet scheduling are available; however, there is a lack of scheduling solutions for airport ground crews, especially for work groups with overlapping skills. In the presented case, a scheduling solution for airport ground crew and equipment in a small international airport is described. As analytical methods are unsuitable for the system in question, the proposed scheduling solution is based on heuristics. A combined agent based and discrete event simulation model was developed to validate and improve the heuristic algorithms until they produced acceptable schedules and shifts. The algorithms first compute the requirements for workforce and equipment based on flight schedules and stored heuristic criteria. Workforce requirements are then optimized using time shifting of tasks and task reassignments, which smooth the peaks in workforce requirements, and finally the simulation model is used to verify the generated schedule. The scheduling procedure is considerably faster than manual scheduling and allows dynamic rescheduling in case of disruptions. The presented schedule generation and optimization solution is flexible and adaptable to other similar sized airports.

1 Introduction

In this chapter, we describe the development of a scheduling solution for airport ground crew and equipment in a small international commercial airport. Similar to other service providers, airports are facing constant competition. To attract airlines

B. Rodič (✉)
Faculty of Information Studies, Novo mesto, Slovenia
e-mail: blaz.rodic@fis.unm.si

A. Baggia
Faculty of Organizational Sciences, University of Maribor, Kranj, Slovenia
e-mail: alenka.baggia@fov.uni-mb.si

M. Mujica Mota and I. Flores De La Mota (eds.), *Applied Simulation and Optimization 2*, DOI 10.1007/978-3-319-55810-3_5

and passengers, they must offer efficient and high quality services for airlines and passengers. At the same time, labour, equipment and other costs need to be kept low enough to generate profits. This presents airports with a difficult optimization problem.

In order to provide efficient and high quality services, airport's resources used for passenger and airplane services need to be scheduled. An airport is a complex logistics system. Analysis and optimization of processes at an airport can be a tedious and time-consuming work since the processes are interleaved, cannot be analysed separately and are usually too complex to be modelled with an exact mathematical approach.

The entire airline industry faces a range of different, yet typically complex scheduling problems, from aircraft or fleet scheduling [1–4], ground crew scheduling [5], disruption management [6], aircraft landing sequence scheduling [7, 8] to personnel training scheduling [9].

An important factor in the optimization of air traffic logistics are delays and delay costs. While 50% of flight delays are caused by the carriers, 19% of delays are caused by airport operations [10]. Depending on contracts between airlines and airports, the cost of these delays can be transferred to the airport. The analysis of tactical delay costs with network effect [11] shows that delays cost airlines from €90.80 to €110.50 per minute, depending on the plane status and other factors. The steep costs of delays highlight the importance of optimization in airport operations.

However, most of the related research in recent years is focused on optimization of airport surface operations, from ground movement, runway scheduling and gate assignment [12] and aims to efficiently utilize the resources and lower the impact on the environment. Further, most of the research on personnel scheduling problems in the airline industry focuses on cabin crew scheduling, whereas airport ground crew scheduling has only gained the attention of researchers in recent years. Ground crew scheduling is as important to the airports as cabin crew scheduling is for the airlines, and is vital to ensure security, safety and quality of airport service.

Airport ground crew operations and tasks can be divided into passenger-related tasks and aircraft-related tasks, where the latter include maintenance, cargo, baggage, loading, cleaning, catering, towing and operations [5].

Ground crew scheduling is a complex problem since in addition to common constraints of personnel scheduling, the required equipment and skills of the crewmembers have to be considered. Interconnections between work groups and overlapping of ground crew skills increase the number of constraints and possibilities, which have to be considered in developing scheduling algorithms. Since professional ground crew scheduling solutions may be prohibitively expensive for smaller airports, solutions that require a lot of manual work are still in place, presenting great possibilities for improvement.

Most of the research on ground crew scheduling offers partial solutions for individual work groups e.g. check-in [13–15], baggage handling [14, 16], security [17], or runway [12, 18, 19]. Although mathematical (linear programming) models can be used to resolve rostering problems of a specific work group type, it cannot be applied to a complex system, therefore other techniques need to be employed [20].

Staff scheduling is usually carried out in several stages, where the demands are calculated first, followed by the generation of work shift plans [21].

The airport in question is an international airport located in the southeast Europe with over 30.000 flights and over 1.400.000 passengers per year. From the year 2004 the airport grew substantially with the introduction of low-cost carriers, followed by the construction of a new passenger terminal and renovation of runway. The growth of passenger and freight traffic still continues, and better information technology (IT) support of the internal processes will be required. The airport has a single 3300 m long runway equipped with CAT III/B Instrument Landing System, a 23 m wide taxiway, and 25 independent parking positions. Airport's Aerodrome Reference Code (International Civil Aviation Organization) is 4E. The terminal capacity is 500 passengers per hour, with 13 check-in counters and 2 baggage claim conveyors. The total area of the airport is 320 hectares.

The arrival or a departure of an aircraft requires the execution of a series of ground crew tasks. The scheduling of these tasks and the workforce and equipment requirements were performed manually using spreadsheets; however, the procedure was too lengthy to allow dynamic rescheduling in case of flight schedule changes, and did not adequately address the variation of workforce requirements during peak and off-peak times.

While research such as deals with fixed shifts with repetitive peak time and static demands (e.g. [13]), our project's end goal is to develop an automated work-force scheduling and shift generation system, that would produce floating shifts adjusted to variation of workforce requirements throughout the day in a fraction of the time needed for manual schedule preparation, and would allow dynamic rescheduling in case of unforeseen events or disruptions. In order to achieve optimal workforce deployment, we needed to minimize the criteria of personnel costs and aircraft delay costs. In this chapter, we describe the development of a solution for the optimization of number of workers present in work groups covering individual types of tasks throughout the working day.

1.1 Ground Crew Scheduling Problem Description

Ground crew scheduling problem at the considered airport is confined with the arrival and departure of the aircraft i.e. the presence of the aircraft at the airport. Depending on the type of aircraft, airline (carrier) and other attributes, a set of tasks has to be performed in a predefined time sequence. The general scheduling rules at this airport are described in the following paragraphs.

The workforce and equipment requirements for each task and their scheduling depend on the flight schedules and parameters of each flight, e.g. destination, aircraft type, and carrier. Tasks can be performed by work groups that have appropriate skills. To simplify scheduling, skills are arranged into skills groups, and every employee is a member of one or more skill groups. Employees that belong to a skill group can perform one or more types of tasks. While employees from almost any skill group can perform simple tasks such as luggage handling, specialized tasks such as supply control can be performed only by employees from a single skill group. After the workforce requirements during a day are defined, shifts are constructed according to business rules (e.g. minimum duration of shifts, allowed shift start times, maximum number of shifts per day per employee), legal limitations (e.g. maximum duration of shifts) and other limitations (e.g. available workers in a skill group, available equipment). Each work group performs only one type of task per shift, and therefore all workers in this work group are selected from the same skill group.

The tasks are performed by the three airport service departments:

- Aircraft supply service,
- Passenger service and
- Technical service (including the Fire department).

Each service department consists of personnel with different skills, matched to specific tasks. There are two types of tasks: "fixed" tasks, which are performed by a constant number of workers that work in fixed shifts, and are thus not a part of the optimization problem, and the "operational" tasks. Workforce requirements for the operational tasks and thus shifts vary according to the number and type of events (flight arrivals and departures) at the airport. Aircraft supply service has three fixed tasks (Shift manager, Crew bus driver, Trolley collector) and nine operational tasks (Load balancer, Supply controller, Group leader, Sorter, Baggage handling worker A and B, Tractor driver, Cleaner/Driver and Cleaner). Passenger service has four fixed tasks (Ground attendant type 1, Call centre, Information desk, Business lounge counter) and five operational tasks (Check-in, Gate, Transfer, Guidance, Lost and found desk). The technical service has two fixed tasks (Fireman, Shift manager), and ten operational tasks (Follow me driver, Bus driver, Power unit operator, Water tank driver, Aircraft towing, Flatbed operator, Deicing, Disabled people van driver, Air-start system operator and Aircraft cabin and engine blades heating operator).

In the past, planners (usually heads of service departments), human resource (HR) department and IT department were involved in the process of ground crew scheduling. HR department provided up to date information on personnel skills and availability, while the IT department provided latest flight schedules. Based on expert knowledge of planers, schedules were generated using a spreadsheet application 14 days in advance, with the final confirmation of the schedule at least 24 h before the execution of the tasks.

In order to prepare schedules, planners needed to know what are the workforce requirements for every type of task throughout the day. Their decision criteria included the type of aircraft, carrier, length of stopover and type of flight. Based on flight data and their expert knowledge, shifts of workers were generated and gathered in a schedule.

Based on the decision criteria and heuristic rules identified from manual scheduling procedures we have documented the following attributes of flights as the scheduling criteria:

- Type of stopover (arrival or departure),
- Flight type (charter, scheduled or transfer),
- Aircraft type (320, CRJ, SH3 etc.),
- Carrier (9 carriers are currently using the airport),
- Destination.

For each criteria type, e.g. "Aircraft type", multiple criteria can be defined, i.e. the aircraft type of a particular flight can influence the requirement and parameters of several different tasks, e.g. number of cleaners, requirement of an auxiliary power unit to start the engines etc.

The main scheduling issue in the presented case arose from the big difference in workforce needs in peak time and outside of peak-time for some of the skill groups, which lead to difficult and inefficient rostering of employees. While according to heuristic rules additional workers were needed during peak times, manually prepared schedules did not schedule additional workforce, as the peak times are much shorter than a minimum shift duration. In the past, the airport instead employed students working part time during peaks, but this solution was not sustainable. The discrepancy between the number of available manually scheduled workers and the number of workers required according to heuristic rules is shown in Fig. 1. Here we can see that outside of peak hours, there are more workers available than required, but at peak time, the task requirements exceed the number of available workers, leading to overload and potential errors and delays.

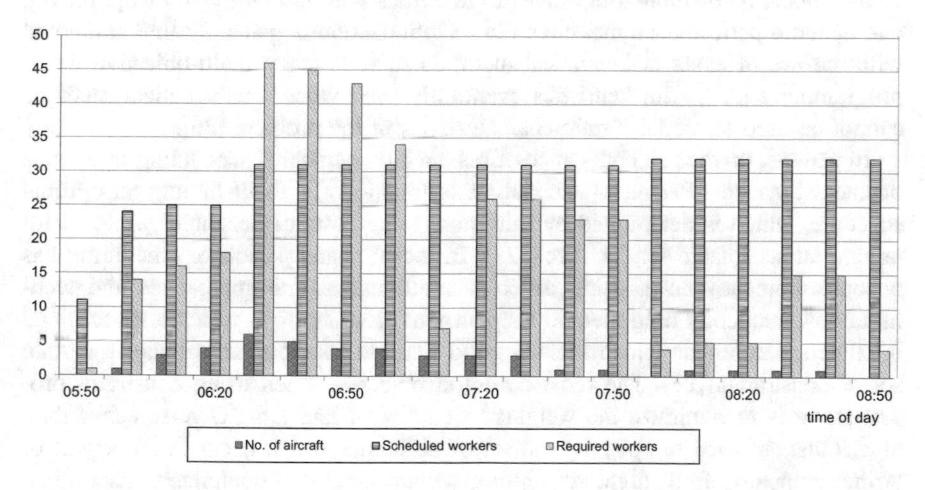

Fig. 1 The discrepancy between the number of available manually scheduled workers and the number of workers required according to heuristic rules

Additionally, ad hoc work schedule changes are a constant, especially due to the changes in flight schedules and other interruptions caused by different events (e.g. flight delays due to weather on route and changes at other airports, flight cancelation due to adverse weather or malfunction, aircraft change due to malfunction or airline decision). These changes can require different airport services with additional personnel and equipment for different aircraft, or rescheduling of a task to peak time when personnel and equipment are not available. The response to the ad hoc flight changes needs to be accurate and efficient, and should not disrupt the flight schedule. In the past, planners had varying success adapting the schedules to unexpected changes.

Although the process of manual schedule generation was not seen as a major issue at the airport, it was quite time consuming and stressful. The scheduling problem increased in complexity after an additional passenger terminal was built and the runway has been upgraded, resulting in an increase of airport traffic. Peak-time ground crew scheduling was partially alleviated using part time (student) workforce for a while, however, a new work legislation has reduced the availability of part time workers, and situations with redundant workforce during low traffic and a lack of workforce during peak time have become more common.

1.2 Literature Review

Scheduling is described as the allocation of activities or actions on a timeline to resources, according to specific performance criteria [22]. Moreover, scheduling is a decision-making process with the goal of optimizing one or more parameters [23] or the allocation of limited resources to activities with the objective of optimizing one or more performance measures [24]. Optimization is generally first attempted with the use of exact mathematical methods such as fuzzy multi-objective linear programming [25], with heuristics eventually used where mathematical methods cannot be used to model certain characteristics of the problem [26].

In general, there is an abundance of research on personnel scheduling in various business branches. Personnel scheduling is traditionally about finding an optimal schedule, which is determined by minimizing the costs of personnel while maintaining an acceptable service level [27]. In recent years, personnel scheduling has become more advanced, using different multidimensional approaches and techniques. A frequently used method is constraint programming, which is an artificial intelligence technique which seeks a good feasible solution that satisfies a certain set of constraints [23]. The most common objective when using constraint programming is to minimize the weighted quantity of late jobs. Diverse scheduling algorithms are used to optimize the set of schedules which occur in an airport or airline company, from flight scheduling to personnel and equipment scheduling, and researchers often tend to use heuristics instead of exact solution techniques [9].

Heuristics are often used because in advanced personnel scheduling solutions, skill requirements and shift definition significantly contribute to the complexity of the scheduling problem.

In most airlines, several departments are involved in the scheduling process [10]. On the other hand, ground crew schedules are usually handled by a division of the airport management [5]. For a single work group of employees with different skills, workforce demands can be calculated and a memetic algorithm can be used to evaluate the schedule [17]. The study of aircraft maintenance staff with the time constraint and different skill requirements is presented in [9], where the first step of schedule generation is the definition of optimal skill mix, and the second step is the optimization of training costs. In the case of check-in and baggage handlers scheduling [14], each day is divided into time blocks with different constraints defined and a required number of employees given. At the end, employees are scheduled for work in three different shifts, 8 h each. Goal programming also proved to be efficient for generating shift duties for baggage services section staff [16].

A simulation study focused on aircraft maintenance, uses a classification of aircrafts according to the time of stay at the airport [14]. Depending on the length of stay, maintenance programs are scheduled for technicians and total technician requirements are calculated for each sub-shift of the day. Using the stochastic methods, delay costs in air traffic can be calculated [28]. Attempts were even made to influence the schedule of aircraft landings in order to balance the workload of ground staff [29].

In addition to optimization methods and heuristics, discrete event simulation (DES) can be used to improve aircraft ground handling performance [30]. Agent based modelling (ABM) was used to simulate and optimize the complex socio-technical air transportation system [31], while [32] used ABM simulation to predict the airport capacity. Based on the research of simulation methods used in personnel scheduling problems [30–33], we can conclude that the combination of DES and ABM methods proved to be more flexible and accurate than using DES alone. While airport operations can certainly be modelled with DES alone, the addition of ABM components to a DES model of airport traffic allows us to model the activities of ground crew and the communication between ground crew groups, their supervisors and aircraft with less abstraction and more detail, thus improving simulation accuracy and transparency and comprehension by the client. Modelling of decision and work processes at the level of individual agents (groups, individuals and their equipment) will also allow us to introduce elements such as the effect of personal work efficiency, fatigue and equipment malfunction with less abstraction, while the spatial aspect of agents allows us to model the movements of aircraft and ground crew on the tarmac and measure the time a group of workers spends travelling between aircraft.

2 Methodology

2.1 Scheduling Problem Definition

The process of scheduling does not depend on the type of scheduling problem. In case of the production scheduling, the scheduling algorithm searches for the most appropriate machine(s) to process an order, while in case of personnel scheduling, the algorithm searches for the most appropriate person(s) to perform a task. The scheduling algorithm could therefore be defined as a search procedure to find an optimal solution among all possible solutions aligned with criteria function.

According to the previous research on scheduling (see e.g. [34]), the scheduling problem can be defined as a general search procedure, where only one variable's value is changed when different types of scheduling problem are addressed. A general example of such a procedure is shown in Algorithm 1, where the variable can be a machine, person, equipment, etc., depending on the type of scheduling problem. Each solution is evaluated based on the defined criteria function to find an optimal solution.

```
FOR each position in the time frame
    FOR each set of requirements
          FOR each variable
                EVALUATE possible solution
          END FOR
          SELECT optimal solution
    END FOR
    INSERT optimal solution in the time frame
END FOR
```

Algorithm 1: General search procedure

Since the equivalence between different scheduling problems can be observed, every scheduling domain (SD) can be described with four basic elements as presented in Eq. 1: object type (O), parameter (P), syntax (S) and algorithm (A) [35], where different types of relations between object types describe the behaviour of the scheduling problem. Object types, parameters and the syntax represent the input for the scheduling algorithm, which defines the logic to generate a schedule.

$$SD \subseteq O \times P \times S \times A \tag{1}$$

In line with the diversity of the scheduling problems addressed in the air transport industry, diverse scheduling approaches were used, from combinatorial optimization problems [20], linear programming [13], multi objective genetic algorithms [36], etc. The ground crew scheduling problem addressed in this research can be aligned with the multiple machine scheduling problems [37], although in case of ground crew scheduling, the sequence of operations or tasks is not fixed and the sequence of orders is known in advance. Ground crew scheduling operations can be scheduled independently to some extent (e.g., there is no dependence between aircraft cleaning and passenger guidance). Although the sequence of operations is well known in advance (flight schedule), it is often subject to various disruptions and rendering of the schedule. Another specific of the ground crew scheduling is the limitation of the time window for the execution of a task, which is limited by the presence of the aircraft at the airport.

Approaches to shift planning and crew assembly often use limited validity assumptions [21] or deal with simplified problems such as partial solutions for individual work groups [13]. Therefore, in the presented case of a small international airport, a heuristic approach was used to generate work schedule and shift generation.

2.2 *Simulation and Modelling*

In simulation and modelling of logistics systems, three different simulation methods are generally used, and are selected depending on the complexity and abstraction level of the discussed system:

- System dynamics (SD) is a form of continuous simulation of complex systems with a high level of abstraction, using stocks, flows, feedback loops and time delays to model flows and levels of materials, people, funds etc. [38, 39]. It does not allow the modelling or tracking of individual entities.
- Discrete event simulation (DES), uses a low level of abstraction to model systems as a series of events or instants in time when a stage-change occurs [40]. In DES, the system is modelled as a process, with a sequence of operations that are performed on entities or transactions. A DES model requires that the data which describe the processes are obtained, analysed, extracted and prepared in a suitable format for the model. Integration of simulation software and operational databases can preserve the model accuracy even after minor changes in processes.
- Agent based modelling (ABM) allows experimentation with models, composed of agents that interact within an environment [41]. ABM allows different levels of abstraction, making it more flexible than SD or DES. The main attribute in an ABM simulation model is an object (agent) and its individual behaviour [42]. An agent can be an individual person or object (e.g. pedestrian, worker, aircraft, and truck) or a group of persons or objects with a common decision mechanism.

A conventional approach to modelling the ground crew processes at an airport would involve DES methodology to model the set of tasks as separate processes, with workers and equipment as resources and the steps in a process as delay elements. While formally correct, this approach can be too rigid to model the dynamics of an airport. The addition of ABM adds more flexibility, as workers and work groups can be modelled as agents that move from plane to plane performing tasks, and make autonomous decisions according to a set of rules and assignment of tasks. Work group movements, communication between entities, pre-emption, change of tasks, travel delays and other real events and conditions are easier to model using agents. Such a model can be more realistic and more flexible than a conventional DES model, while still allowing the monitoring of resource utilization and other statistics. ABM is a better choice than DES to model processes that are dynamic and must quickly adapt to changing requirements and events on a real-time basis [43]. ABM allows the inclusion of descriptive models of how people actually make decisions within the modelled system, and modelling of the effects of all decision makers within the system. In contrast, DES models take a normative approach, i.e., indicating what should be done rather than how the system really works. In addition, ABM is better than DES when it is important that individual agents have spatial or geo-spatial aspects to their behaviours (e.g., agents move over a landscape or between parked aircrafts). However, the development and validation of ABM can be considerably more difficult than SD or DES models.

3 Heuristic Algorithm Development

Planners in service departments have previously depended on their expert knowledge (i.e. heuristics) to generate manual schedules. We have collected the data on their heuristics to define the constraints and scheduling requirements used in the heuristic approach. Similar to [44], we have used a two-step approach to generate a feasible scheduling solution for ground crew scheduling, with the first step defining the work force requirements and the second step constructing shifts based on the work force requirements, business rules and legal limitations.

In the first part of the solution, the requirements and constraints for all tasks on flights within a selected time frame are identified to generate a feasible skill group and equipment schedule. This schedule defines the number of personnel for each skill group and the required equipment for every minute within the time frame, and does not include shifts or employee names.

A flowchart diagram representation of the skill group scheduling algorithm is presented in Fig. 2. First, the planner defines the time frame for the schedule (usually 14 days, starting after the end of current schedule), and thus defines the range of flight data to be transferred from the Flight Information System (FIS). The algorithm then analyses the flight data, and determines the time window, within which all flight related tasks need to be completed. This time window limits the optimization of workforce requirements via the time shifting of tasks. According to flight characteristics and stored heuristic rules, the criteria and requirements for this flight are determined and the number of personnel, their required skill groups and required equipment are recorded. Information about the skill profiles are stored in the Human Resources System (HRS). Finally, the personnel/workforce

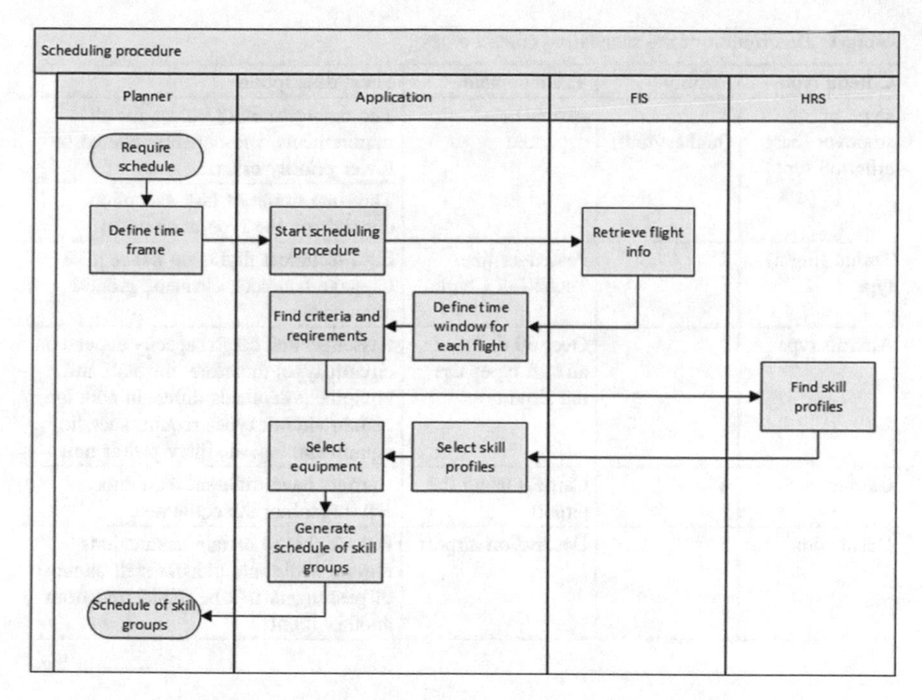

Fig. 2 The flow of the airport ground crew skill profile scheduling procedure

requirements per skill group and equipment requirements from all flights are summed for every minute of time within the selected time frame.

As the availability of the equipment was determined not to be a constraint of the scheduling problem, the equipment requirements are modelled, but not subject to optimization in the presented solution. The focus is therefore on personnel scheduling optimization.

The scheduling criteria types, presented in Table 1, stated by the airports experts were included in the heuristics algorithm: type of stopover, traffic (flight) type, aircraft type, carrier and destination.

These criteria are used to determine the tasks that need to be performed per flight and their parameters. There are four basic scheduling parameters per each task:

- Skill required,
- Start of task,
- Duration of task, and
- Number of workers per task.

Advanced task parameters, tied to skill groups, include the possibility of time shifting and skill groups allowed to perform the task.

The criteria have different priorities, i.e. they must be used in a prescribed sequence to arrive at the solution. The first criterion to be used is *type of stopover* with *priority* value 1. This is the base criterion, which sets the default requirement

Table 1 Description of the scheduling criteria types

Criteria type	Priority	Value domain	Short description
Type of stopover (base criterion type)	1 (highest/first)	Arrival or departure	The default/starting values for all requirements, subsequently altered by lower priority criteria.
			The time frame of task execution depends on the type of stopover
Traffic (flight) type	2	Passenger line, Technical, Charter	E.g., technical flights do not require baggage handlers, cleaners, ground attendants
Aircraft type	3	Over 30 different aircraft types use the airport	Passenger and cargo capacity depend on aircraft type; therefore, the staff and equipment demands differ. In addition, certain aircraft types require specific equipment, e.g. auxiliary power unit
Carrier	4	Carriers using the airport	Carriers have different requirements, esp. regarding the equipment
Destination	5	Destination airport	E.g., flights to certain destinations require additional transfer staff as most of passengers will be transferred from another flight

values, which can later be altered by the subsequent criteria (with *priority* value 2 or more). In addition, two types of criteria exist: relative and absolute. Relative criteria will reduce or increase a scheduling parameter (e.g. required number of workers or task duration) while absolute criteria, if defined for the given flight parameters, will set the scheduling parameter to a predefined value. For example, certain carriers have a fixed demand for the number of ground attendants at check-in. Therefore, all previously calculated workforce demand at check-in are overridden with a fixed number.

Figures 3 and 4 describe the process of aircraft supply service (fixed tasks are not included) for the arrival and departure of an aircraft of type C (e.g. Airbus 321). The aircraft supply service department included nine different operational tasks which were mapped to skill groups with the same name, listed in Fig. 3. Most of the skill groups required one person to be assigned to the skill group. The two exceptions are the skill groups Cleaner, where according to the requirements, two workers should be assigned to the task, and Baggage sorter type A, where only half of person (i.e. a half of worker's full time utilization) is assigned to the task. The assignment of "half persons" per task in combination with time shifting of tasks allows a degree of workforce requirements optimization of certain types of tasks. The required number of individual workers in specific task group is not shown in the graphical presentation. As it can be seen from Figs. 3 and 4, most of the tasks overlap and can be performed simultaneously. The only task which requires a strict sequence is the task of baggage sorting (A and B) in Fig. 4, which has to be finished before the tractor driver drives the baggage to the aircraft.

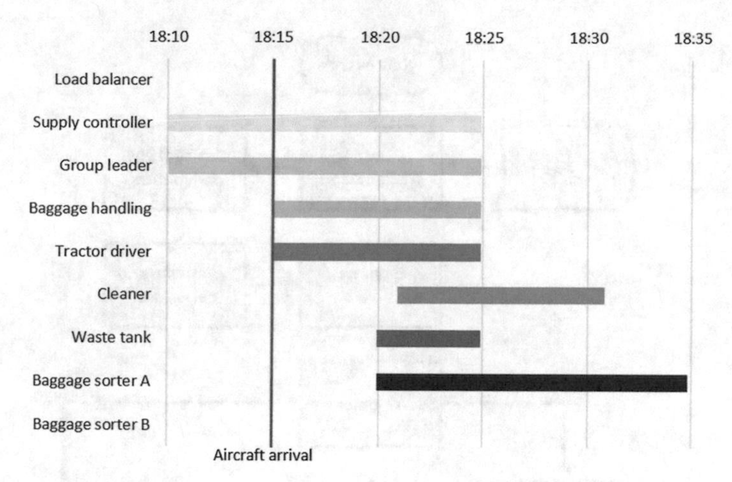

Fig. 3 The process of aircraft supply service for the arrival of type C aircraft

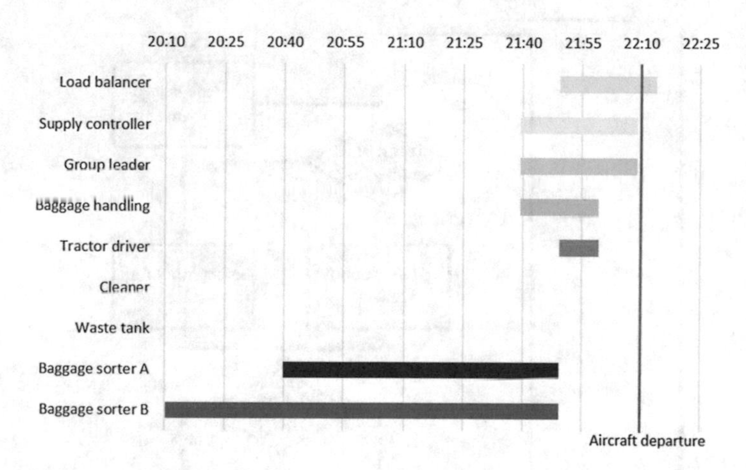

Fig. 4 The process of aircraft supply service for the departure of type C aircraft

There are two types of tasks: fixed and operational. Fixed tasks do not depend on the airport traffic and are always performed in the same manner. These tasks (e.g. 1 manager per shift, 1 person in call centre) cannot be optimized, and were therefore not modelled in our solution. For operational tasks, the required number of staff varies with the airport traffic.

Based on the recorded heuristics, the algorithm was coded in a software program to generate a timeline of workforce requirements for all tasks. The algorithm uses flight information and documented criteria to calculate workforce requirements for every flight, per task. The required number of workers for every type of task is then calculated for each minute within a given time frame (the end and start date of the schedule), producing a timeline of heuristic (ideal) workforce requirements.

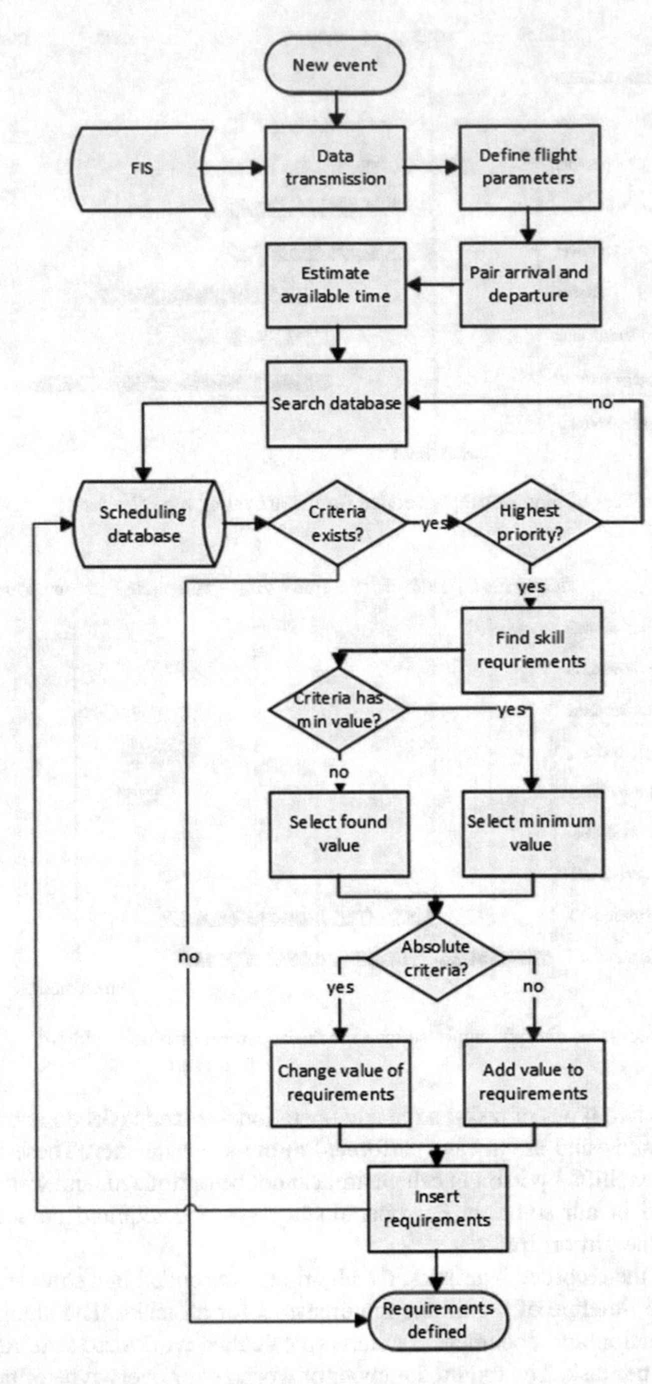

Fig. 5 The first version of the workforce requirements scheduling algorithm

Figure 5 shows the first version of the workforce requirements scheduling algorithm, which computes the ideal workforce requirements.

3.1 Optimized Heuristic Algorithm

Upon further examination, workers were found to temporarily change task assignments to help overloaded colleagues, and several types of tasks were performed either sooner or later than defined in the heuristics in order to avoid peak worker overload. Furthermore, the most overloaded skill groups were found to be temporary overloaded during peak times, and performing two tasks in the same time period.

The main issue was therefore to smooth out workforce requirements for daily peak times without causing flight delays due to the workforce overload. Seasonal peak times are not considered problematic since they can be planned in advance and additional workforce can be employed during the season period. In contrast, daily peak times are problematic as they are mostly shorter than minimum shift length (2 h).

Therefore, additional heuristic rules were implemented to reproduce the in-field optimization behaviour of the examined system. Main improvement of the algorithm, described in the following section, was the "smoothing" of requirement peaks, implemented by shifting the execution of a task to a time, where more workers are available. A task can be shifted to an earlier or a later time, according to limitations defined by airport planners and implemented in the algorithm. Further improvement is the temporary reassignment of workers, which is implemented by the option of a worker being assigned to several tasks simultaneously inside a short time frame (e.g. a tractor driver usually helps as a baggage sorter when he stops the tractor, although he is formally still busy waiting to drive the tractor back). The maximum duration of an overlapping activity is limited by the duration of peak time requirements (typically less than 30 min) and the end of a shift. This version of algorithm improves on the manual schedules by reducing workforce requirements in the off-peak times, however the peak time optimization is still not satisfactory, as the algorithm replicated the overloading of personnel during peak times.

Diagram of the optimized scheduling algorithm is presented in Fig. 6.

The optimized workforce requirements scheduling algorithm was implemented as a standalone application in Java, using an Oracle database to store the data on flights and heuristic rules. 24 relational tables were used to describe the criteria and the demands of the airport ground crew scheduling problem. Before the start of scheduling procedure, flight and personnel data is transferred from the FIS and HRS.

Table 2 shows an example of parameters stored in the database, assembled from the FIS into a single table using an SQL query. DD1 defines the date of the flight, *FLTNO_A* and *FLTNO_D* describe the aircraft's arrival and departure code. The type of traffic (e.g. C—charter passenger only, F—scheduled cargo/mail, S—

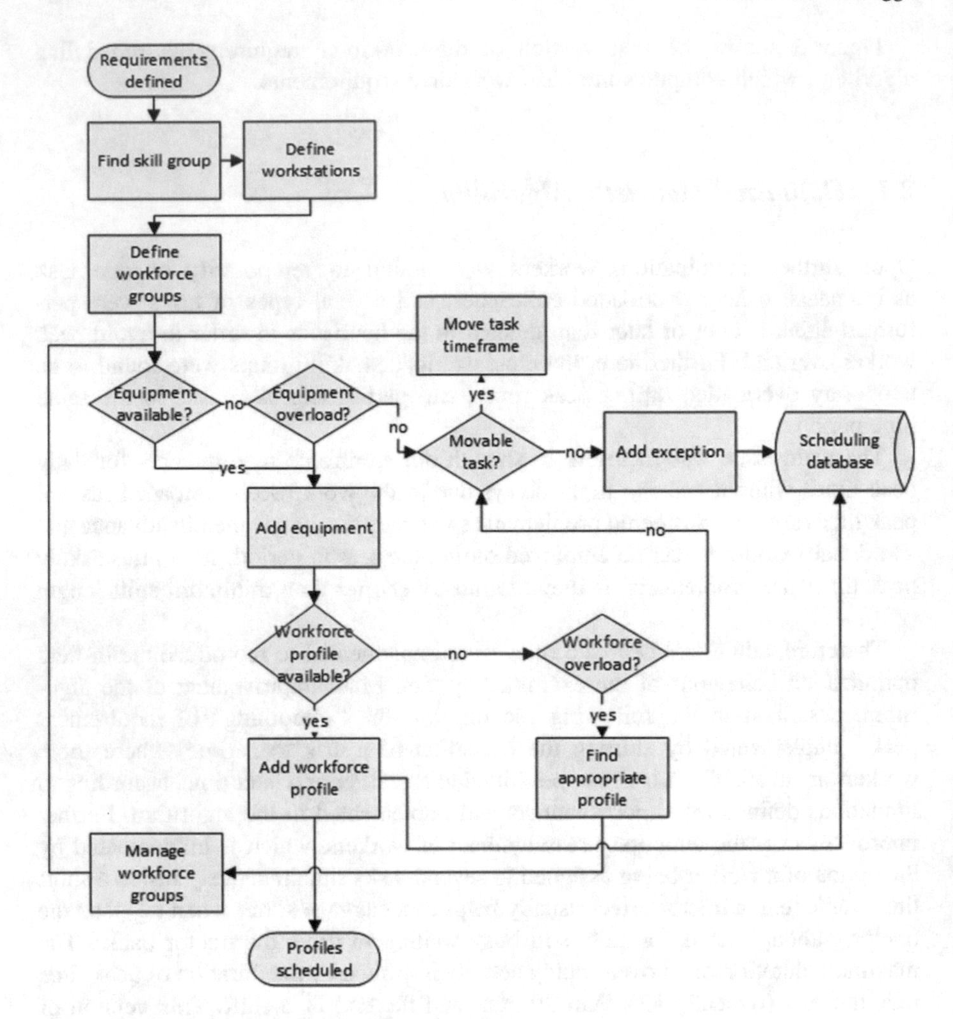

Fig. 6 Version 2 (optimized) scheduling algorithm

scheduled passenger) is defined in column *TRFTYP*, *ST_A* and *ST_D* show the time of arrival or departure, with *ROUTE_A* and *ROUTE_D* as arrival or departure airport. *ACTYP* defines the type of aircraft.

According to the criteria types and detail information on considered flight schedule, heuristic rules stored in the database can be used to generate a schedule. Table 3 shows an example of heuristic rules stored in the database. The column *ID_CT* stands for identification number of criterion type, *DM* for demand (number of workers to be set, added or subtracted according to criterion value), *ST* for task start time, *DR* for duration of task, *ABS* for the distinction between absolute and relative criteria, and *CV* for criteria value, i.e. the condition where this particular rule/line is used, *ID_S* for identification number of skill, i.e. which skill group this rule applies to, *MVB* for indication of movable tasks, i.e. tasks with flexible start or

Table 2 Flight schedule stored in database

DD1	FLTNO_A	FLTNO_D	TRFTYP	ST_A	ST_D	ROUTE_A	ROUTE_D	ACTYP
01.05.2016		JP648	S		00:20:00		IST	735
01.05.2016	JP395		S	01:30:00		BRU		CRJ
01.05.2016	JP299		S	02:40:00		CPH-BCN		320
01.05.2016	FAH6972	6972	F	06:25:00		VIE		F27
01.05.2016		OK827	S		06:40:00		PRG	AT4
01.05.2016		JP376	S		06:45:00		BRU	CRJ
01.05.2016	JP649		S	06:50:00		IST		735
01.05.2016		JP102	S		06:50:00		MUC	CRJ
01.05.2016		JP938	S		07:00:00		WAW	CRJ
01.05.2016	JP687		S	07:05:00		IST		735

Table 3 Heuristic rules stored in database

ID	ID_CT	PARAMETER	SKILL	DM	ST	DR	ABS	CV	ID_S	MVB	LEBD	A_D
401	24	EVENT TIME	CHECKIN	1	30	−30	R	360	1			D
1	7	TRFTYP	CHECKIN	2	120	100	A	S	1			D
253	10	CARRIER	CLEANER ORD	−2			R	EZY	19		20	A
250	6	TRFTYP	CLEANER ORD	2	6	10	A	C	19		20	A
278	10	CARRIER	CLEANER DRV	−1			R	EZY	20		20	A
279	10	CARRIER	CLEANER DRV	−1			R	W	20		20	A

end time, *LEBD* for latest allowed time for task completion before departure and *A_D* for the arrival/departure label. Absolute criteria require setting the work demand/start/duration to their value, thus replacing the current value, while the relative criteria require adding their value to the current values for work demand/start/duration.

The heuristic rules stored in database define the flow of the scheduling procedure in detail. The base criteria for scheduling the tasks is the stopover type: arrivals and departures require different skills and require a different time frame calculation method, i.e. in case of a departure, the task start times are to be subtracted from the departure time, while for arrivals the task start times are added to the arrival time.

For example, the criterion in Table 3, row 2 applies if the flight type (*A_D*) is *D* (Departure), *TRFTYPE* value is *S* (scheduled passenger flight), and *CV* (carrier) is *S*. In that case, the *CHECKIN* skill demand (*DM*) is 2 (two workers). Therefore, two ground attendants with skill ID number 1 (*ID_S*) should start their work 120 min (*ST*) prior to the flight departure (*A_D*), and should perform their task for 100 min (*DR*). Since the value for *ABS* is *A*, the *DM* (demand/number of workers) value of 2 is absolute, thus it overrides any previously set value for *DM* for this skill type. Since the task is not movable (*MVB* is empty), we cannot change the start time of the task and therefore no value for the latest start before departure is given. *CV* criterion defines specific rules for certain carriers, which may for example require a check-into complete earlier.

For example, if the flight carrier has the *CV* value "360" an additional criterion with a lower priority also applies (Table 3, row 1). This criterion type is R (relative), which means that the start time is 30 min (prior to the departure of the aircraft), 1 ground attendant is added to the check-in counter, and the relative event duration is -30 min, which means the duration of this task is reduced by 30 min (i.e. check-in ends 30 min sooner, but there is an additional ground attendant present).

Figure 7 shows the ideal workforce requirements for load balancers estimated by the first heuristics algorithm (solid line), and the requirements "smoothed" by the optimized heuristic algorithm (dashed line). Short periods of peak times are clearly visible.

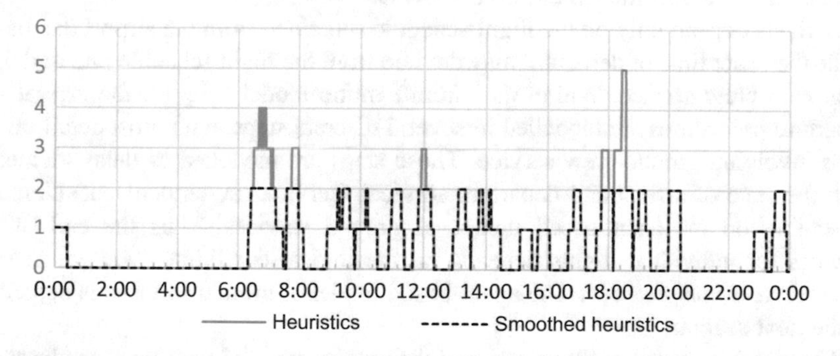

Fig. 7 Workforce requirements of first and second versions of the algorithm for the load balancer skill group

4 Simulation Model

We have decided to validate the generated workforce schedules using a model of the airport operations. We have used AnyLogic as the main a general-purpose simulation and modelling tool. While AnyLogic is built in Java, and the models are translated into Java code, most of the modelling can be performed using the Visual Interactive Modelling (VIM) approach, which allows fast development and intuitive comprehension of model operation for the stakeholders without previous knowledge of simulation methods. Java code of the model is accessible within AnyLogic and can be modified during model development. A unique characteristic of AnyLogic from version 7 on is that DES entities, resources, and agents use the same object type, allowing easier integration of DES and ABM models. A combination of DES and ABM methods can be used in several different situations, from implementation of a DES server as an agent, an entity as an agent or different combinations of agent usage to introduce messages into the model [42].

Our model is divided into two parts: the DES based aircraft traffic model, which models the arrivals and departures of aircraft, and the ABM based ground crew work group model. The models are linked via passing of messages between aircraft and work group agents. The DES based aircraft traffic also implements several ABM-specific features in order to allow simulation of movement on the airport and the passing of messages.

4.1 Aircraft Traffic Model

In the first step of the model development we have identified the entities present in the system and selected the modelling method appropriate for the required level of autonomy and abstraction. We have determined that a classic DES model will be sufficient to model the aircraft traffic, i.e. the arrivals and departures of aircraft, however we have supplemented it with ABM elements to allow the modelling and animation of aircraft movement on the tarmac.

Arrivals depend only on the flight schedule (obtained from the airport database), while the exact time of departure may deviate from the flight schedule due to delays in ground crew service. Within the aircraft traffic model (Fig. 8), the arrival and departure procedures are modelled in several discrete steps, with most detail on the steps involving ground crew service. These steps are modelled as delay elements, with the state of arrival and departure services (serviceArr, serviceDept) elements depending on the agent-based model of ground services. Thus the end of the services for individual flights depends on the logical condition: *"Are all ground crew services completed?"*. The agent-based model of ground services is described in the next section.

The physical layout of the airport and the aircraft taxi and parking procedures are also modelled, with the main purpose of improving model comprehension and

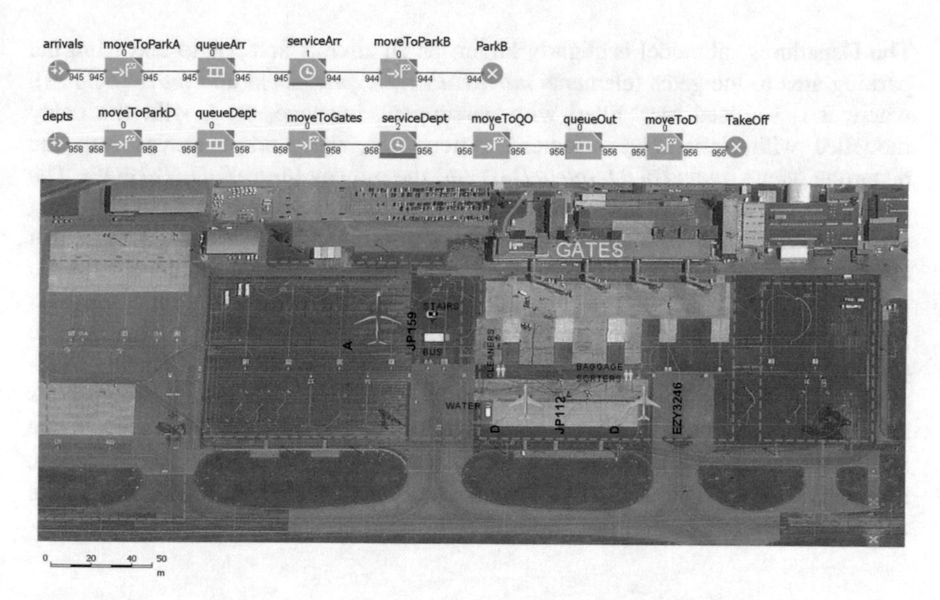

Fig. 8 Main view of the simulation model with DES model of aircraft traffic and airport layout

acceptance by the end users. According to our client, taxi and parking logistics do not influence the ground crew service performance or delays in departures, and are therefore not included in model statistics. While the arrival services are not problematic from the aspect of flight delays, as they are completed long before the departure of the aircraft, they affect the availability of ground crew groups and equipment.

The Aircraft traffic model is implemented with two DES submodels: the Arrivals submodel, and the Departures submodel. The separation into submodels reflects the business rules of the airport and the limitations of the FIS database: individual aircraft are not tracked in the FIS after the arrival procedure is complete, thus the Turnaround process cannot be modelled. After the arrival tasks are completed, an aircraft is removed from the model by parking them at "Parking B" (exit point in the model is *ParkB*). The Departures submodel in turn assumes that the aircraft is present (parked) at the airport.

The starting point in the Arrivals submodel is the *arrivals* element, which generates aircraft in the submodel according to the arrivals schedule in the FIS. At this point, the aircraft also appears in the animation seen below the DES submodels. Subsequently, the aircraft taxies to the gates (DES elements *moveToParkA* and *queueArr* (wait on the apron for gate assignment)), waits to be serviced in the *ServiceArr* element, and is removed from the model via the *moveToParkB* and *ParkB* elements.

The departure traffic is also based on the FIS database, with the task start times based on the planned (in FIS) departure times. The actual (modelled) departure times depend on the execution of tasks, which allows us to calculate flight delays.

The Departures submodel is slightly larger, as an aircraft first has to taxi from the parking area to the gates (elements *moveToParkD*, *queueDept* and *moveToGates*), where it is serviced (e.g. filled with passengers, baggage, etc.), with the delay modelled with *ServiceDept* element. Afterwards, the aircraft moves into the departure queue (*moveToQO*, *queueOut*) and the runway (*moveToD*, *TakeOff*). The modelled queues serve to keep aircraft waiting until a taxiway or a gate or a parking area is available. The elements *arrivals* and *depts* are linked to a local database that contains data on every aircraft from the FIS and the service requirements for each flight. The service requirements are assigned to every aircraft at the moment of its generation/entry in the model according to the ideal heuristic requirements (i.e. the requirements are based only on flight data and do not take into account the availability of workers). The service requirements are subsequently passed as messages from an aircraft to the relevant ground crew work groups, which then add the service request to their internal queue. Message passing is an ABM specific feature, however as AnyLogic models all DES entities as agents, the addition of this feature to the DES aircraft traffic model was straightforward.

4.2 Ground Crew Work Group Model

In DES models, services/stations are often modelled as static resources that entities/transactions (e.g. products, patients) travel through on a fixed path, in a predefined sequence, and an entity cannot be serviced by multiple stations. While it is possible to model the ground services processes using a classical DES model, the required abstraction would in our opinion make the model less comprehensible and rigid in comparison to the actual processes. ABM however allows us to model the entities and processes in a way that is closer to reality, i.e. the ground service work groups have the role of service stations, however they travel to the aircraft and not vice versa; the sequence of services depends on the availability of service work groups, and the place of an aircraft within a service work groups' internal queue; and perhaps most important, an aircraft can be serviced by several work groups simultaneously.

We have modelled the work groups performing tasks as agents. All work group agents have an internal state chart model of their task process, shown in Fig. 9. The starting point is the state *Waiting*, where the agent waits for a message from an arriving or departing aircraft requesting services from this work group and specifying service requirements (number of workers, start time, end time). These requests are added to the internal queue (an array data structure). Currently, the requests are processed according to the FIFO rule, but the implementation of priority based service, e.g. according to available number of workers or available equipment could be easily implemented. If there is at least one request in the queue, the work group agent proceeds to the relevant aircraft at the specified service start time, and performs the service (modelled as a delay) for the required length of time.

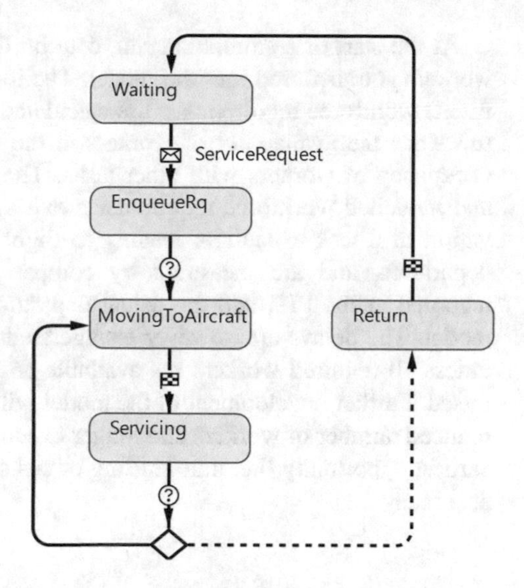

Fig. 9 State chart of the agent based model of a ground crew group

After the servicing is complete, the work group agent passes a *ServiceComplete* message to the aircraft and either proceeds to the next aircraft in its internal queue or returns to the waiting area.

4.3 Model Operation

The start, end and work requirements of tasks are determined by the aircraft to be serviced by the ground crew (i.e. ideal heuristic requirements), while the availability of workers is determined by the workforce requirements timeline generated and optimized by the heuristic algorithm (version 2 of the algorithm). By combining the timeline of optimized workforce numbers and ideal requirements of a flight, we can verify the effects of a generated workforce requirements timeline in practice and foresee the potential flight delay costs.

Flight schedule from the FIS is used to generate arrivals and departures at the airport. Using the first version (ideal requirements) of heuristics, each aircraft is assigned the values of service parameters on entry in the model.

Workforce requirements generated by the heuristic method described in previous section are used to vary the availability of workers during the simulation run. Workers are modelled as resources and divided into work groups. Each work group performs only one type of tasks. A work group is then modelled as an agent. The duration of a simulation run for a month of simulated time is approximately 20 s on a Windows 10 computer with an Intel i5-4200 M CPU and 8 GB of RAM.

At the start of a simulation run, data on the flight schedules and availability of workers is transferred into the model. The ideal (i.e. calculated separately for each flight) workforce requirements are calculated. Flight delays appear in a simulation run since the availability of workers in the model is subject to the schedule and occupancy of workers with other tasks. Therefore, the discrepancy between ideal and modelled workforce requirements exists. The discrepancy results in a prolongation of a task execution, leading to flight delays. Delays are only possible for departures, and are measured by comparing the scheduled departure time as recorded in the FIS, and the actual departure time as recorded by the simulation model. The delays are however exaggerated because the start of a task is delayed unless all required workers are available and the start times of tasks are not optimised. Further development of the model will include the execution of tasks with a reduced number of workers and longer execution time and the execution of tasks at earliest opportunity (i.e. time shifting of tasks) and should model flight delays more accurately.

5 Results and Discussion

The first version of the algorithm produced a workforce requirements timeline with pronounced peaks, but with ideal numbers of workers available at every minute of the day to perform the flight dependent tasks. On the other hand, the second version of the algorithm produced a workforce requirements timeline with less peaks, but with a higher chance of flight delays and human errors during peak time.

While the end user is satisfied with the current solution, as it produces schedules in a fraction of the time required for manual schedule development (minutes vs. hours), and the schedules are better than manually produced schedules at least during off-peak times, there is still potential for optimization during peak-time.

In order to achieve optimal workforce deployment, we would need to minimize the criteria of personnel costs and aircraft delay costs. Next step is the development of a simulation model based optimization solution, which would find the optimum between the ideal numbers of workers and the smoothed workforce requirements.

5.1 Further Optimization Possibilities

Personnel costs grow linearly with the number of workers present in work groups and can be estimated as the number of work hours' multiplied by average hourly costs. Delay costs however are not linear: here we experience diminishing returns with the increase of number of workers. The behaviour and the nature of the variables here closely resembles the U-curve optimisation problem, which is common in product development and inventory management [45]. According to [45] the linear component represents the rising cost of services (e.g. number of

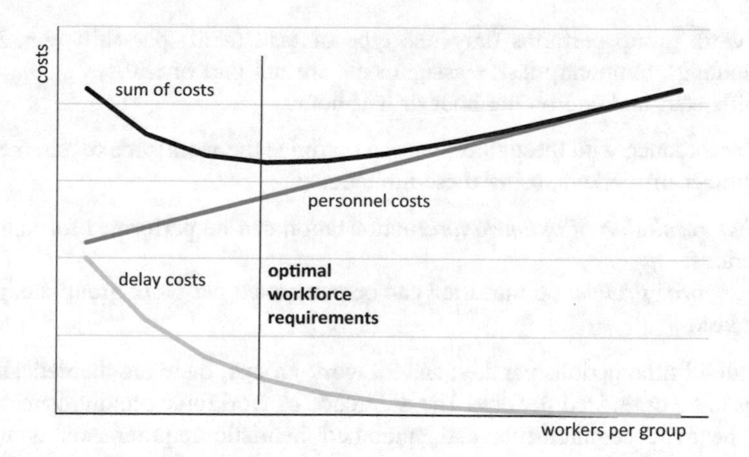

Fig. 10 U-curve optimisation problem of personnel and flight delay costs (*source* own)

workers) or carrying/holding costs, and the non-linear component, which can be approximated as a negative exponential curve, represents delay costs or ordering/release costs. An example of a U-curve optimisation problem in aviation is described in [46].

Figure 10 illustrates the U-curve optimisation problem in the case of personnel and flight delay costs. The goal of optimization is to minimize the sum of both types of costs. The optimization problem in our case is complicated by the fact that there are several work groups to be optimized, and that the workload changes over time.

The criteria function (Eq. 2) in our case has two main variables: the personnel costs (p) for the simulation period (t), and the costs of flight delays (d) per simulation period. The objective is to minimize the criteria function

$$F = \sum_{i=1}^{p} p_i + d_i \quad (2)$$

The costs of personnel (p) depend on the number of workers (w) and costs (c) of worker per shift period in skill groups (g) as presented in Eq. 3.

$$p = \sum_{i=1}^{t} (w_i * c_i) * g_i \quad (3)$$

The costs of flight delays depend on flight schedules and available number of workers in various skill groups.

Workforce scheduling and shift rules at the airport prescribe that:

1. workers are assigned to work groups,
2. a work group is created at the start of a shift, and disbanded at the end of a shift,
3. all workers in a work group start and end their shifts at the same time,

4. a work group performs only one type of task (skill) per shift (e.g. luggage handling), temporary task reassignments are not part of shifts,
5. shifts start and end on the hour or half-hour.

In accordance with these rules, we can optimize the workforce requirements, i.e. the number of workers, using these limitations:

- *Time resolution of optimization:* optimization can be performed for half-hourly periods;
- *Workforce groups:* optimization can be performed per work group, i.e. per type of task.

With 48 time periods per day, and 29 work groups, there are theoretically 1392 values to be optimized per day. The difference of workforce requirements between ideal heuristic requirements and smoothed heuristic requirements is up to 4 workers, therefore there are 5 different possible optimal values. Theoretically, there are 29×5 scenarios to be tested per each hour, or $29 \times 5 \times 48 = 6960$ scenarios per day to be tested in order to find the optimal workforce numbers.

However, there are several factors that reduce our optimization problem:

- Out of 29 skill groups, only 7 skill groups from all three service departments can require more than 1 person per shift: Load balancer, Baggage handler, Cleaner, Baggage sorter type A, Baggage sorter type B, Gate stewardess, and the "Follow me" car driver,
- There is a limit on the number of workers available per work group,
- The difference of workforce requirements between heuristic and smoothed versions is usually only 1 or 2 workers (heuristic results will be used as the upper bound of the optimization),
- Differences in workforce requirements between heuristic and smoothed versions appear only in peak hours, i.e. in up to approximately 10 h per day for all work groups together,
- There is sufficient time from aircraft arrival to departure to perform all arrival related tasks, therefore additional workforce requirements optimization (apart from smoothing) is not necessary. Optimization can be done for departures tasks only.

Due to these alleviating factors, we only have to generate up to approximately $7 \times 3 \times 20 = 420$ scenarios, i.e. half hour long simulation runs per day. With a half hour simulation run duration of 1 s, a day's workforce requirements could be optimized in less than 7 min. AnyLogic can perform optimization of a set of parameters using a user defined criteria function. Part of the optimization procedure could therefore be automated using the built-in AnyLogic optimizer, removing the need to manually prepare and execute the simulation runs.

6 Conclusions

International airports are complex systems that require efficient operation and coordination of all their departments. Therefore, suitable personnel and equipment scheduling solutions are vital for efficient operation of an airport as a system.

Based on the experiences from the presented ground crew optimization project, we can conclude that the problems of airport ground crew scheduling are more demanding than general machine or order scheduling problems found in literature and encountered in our previous projects, even at smaller international airports. Mathematical scheduling models were not applicable in described projects, therefore customized heuristic algorithms were to be developed.

Our work in this project has so far resulted in two versions of heuristic work-force requirements scheduling algorithms, a shift construction algorithm, and a simulation model of airport operations used for verification and future optimization of workforce requirements, which combines DES and ABM. The algorithm for generation of floating shifts and assignment of individuals to shifts is described in [34]. The shifts are generated according to the generated workforce requirements and demands about shift length.

The heuristic work-force requirements scheduling algorithms and the shift construction algorithm are currently implemented in a software package that is undergoing testing at the client. While the manual scheduling takes several hours, the automated scheduling can be completed within minutes, allowing dynamic rescheduling in case of changes in flight schedules. The presented schedule generation and optimization solution is flexible and adaptable to other similar sized airports.

Our future work on the project will involve model-based optimization of workforce requirements as outlined in the previous section and the adaptation of the entire scheduling solution to the airport's development of infrastructure. Whereas competitiveness is definitely the main reason for the optimization of airport operations, sustainability issues also need to be considered. Efficient airport ground operations are one of the key aspects towards sustainable air transportation [36].

References

1. Bian, F., Burke, E., Jain, S., & Kendall, G. (2005). Measuring the robustness of airline fleet schedules. In G. Kendal, E. Burke, S. Petrovic, & M. Gendreau (Eds.), *Multidisciplinary scheduling theory and applications* (pp. 381–392). New York: Springer Science + Business Media.
2. El Moudani, W., & Mora-Camino, F. (2000). A dynamic approach for aircraft assignment and maintenance scheduling by airlines. *The Journal of Air Transport Management, 6,* 233–237. doi:10.1016/S0969-6997(00)00011-9.

3. Yan, S., Tang, C. H., & Fu, T. C. (2008). An airline scheduling model and solution algorithms under stochastic demands. *European Journal of Operational Research, 190,* 22–39. doi:10.1016/j.ejor.2007.05.053.
4. Gurtner, G., Bongiorno, C., Ducci, M., & Miccichè, S. (2016). An empirically grounded agent based simulator for the air traffic management in the SESAR scenario.
5. Clausen, T. (2011). Airport ground staff scheduling.
6. Løve, M., Sørensen, K. R., Larsen, J., & Clausen, J. (2002). Applications of evolutionary computing. In S. Cagnoni, J. Gottlieb, E. Hart, et al. (Eds.), *EvoWorkshops 2002: EvoCOP, EvoIASP, EvoSTIM/EvoPLAN Kinsale*, Ireland, April 3–4, 2002 Proceedings (pp. 315–324). Berlin: Springer.
7. Tavakkoli-Moghaddam, R., Yaghoubi-Panah, M., & Radmehr, F. (2012). Scheduling the sequence of aircraft landings for a single runway using a fuzzy programming approach. *The Journal of Air Transport Management, 25,* 15–18. doi:10.1016/j.jairtraman.2012.03.004.
8. García Ansola, P., García Higuera, A., Pastor, J. M., & Otamendi, F. J. (2011). Agent-based decision-making process in airport ground handling management. *Logistics Research, 3,* 133–143. doi:10.1007/s12159-011-0052-y.
9. De Bruecker, P., den Bergh, J., Belien, J., & Demeulemeester, E. (2014). A two-stage mixed integer programming approach for optimizing the skill mix and training schedules for aircraft maintenance. In *International Conference on Operations Research.*
10. Burke, E. K., De Causmaecker, P., De Maere, G., et al. (2010). A multi-objective approach for robust airline scheduling. *Computers & Operations Research, 37,* 822–832. doi:10.1016/j.cor.2009.03.026.
11. European Organisation for the Safety of Air Navigation. (2015). Standard Inputs for EUROCONTROL Cost-Benefit Analyses. Brussels.
12. Weiszer, M., Chen, J., & Stewart, P. (2015). Preference-based evolutionary algorithm for airport runway scheduling and ground movement optimisation. In *IEEE Conference on Intelligent Transportation Systems Proceedings,* ITSC 2015–Octob:2078–2083. doi:10.1109/ITSC.2015.336.
13. Lin, D., Xin, Z., Huang, Y. (2015). Ground crew rostering for the airport check-in counter. In *2015 IEEE International Conference on Industrial Engineering and Engineering Management* (pp. 2–6). Singapore.
14. Bazargan, M. (2004). *Airline operations and scheduling*. Aldershot: Ashgate Publishing Limited.
15. Stolletz, R. (2010). Operational workforce planning for check-in counters at airports. *Transportation Research Part E Logistics and Transportation Review, 46,* 414–425. doi:10.1016/j.tre.2009.11.008.
16. Chu, S. C. K. (2007). Generating, scheduling and rostering of shift crew-duties: Applications at the Hong Kong International Airport. *European Journal of Operational Research, 177,* 1764–1778. doi:10.1016/j.ejor.2005.10.008.
17. Abdoul Soukour, A., Devendeville, L., Lucet, C., & Moukrim, A. (2013). A memetic algorithm for staff scheduling problem in airport security service. *Expert Systems with Applications, 40,* 7504–7512. doi:10.1016/j.eswa.2013.06.073.
18. Bennell, J. A., Mesgarpour, M., & Potts, C. N. (2011). Airport runway scheduling. *4OR, 9,* 115–138. doi:10.1007/s10288-011-0172-x.
19. Soomer, M. J., & Franx, G. J. (2008). Scheduling aircraft landings using airlines' preferences. *European Journal of Operational Research, 190,* 277–291. doi:10.1016/j.ejor.2007.06.017.
20. Qi, X., Yang, J., & Yu, G. (2004) Scheduling problems in the airline industry. In *Handbook of scheduling algorithms, models and performance analysis.*
21. Herbers, J. (2005). Models and algorithms for ground staff scheduling on airports.
22. Spyropoulos, C. D. (2000). AI planning and scheduling in the medical hospital environment. *Artificial Intelligence in Medicine, 20,* 101–111. doi:10.1016/S0933-3657(00)00059-2.

23. Pinedo, M. L. (2012). *Scheduling: Theory, algorithms, and systems*. New York: Springer.
24. Leung, J. Y. T. (2004). Introduction and notation. In *Handbook of scheduling algorithms, models and performance analysis.*
25. Peidro, D., Díaz-Madroñero, M., & Mula, J. (2010). An interactive fuzzy multi-objective approach for operational transport planning in an automobile supply chain. *WSEAS Transactions on Information Science and Applications, 7,* 283–294.
26. Cafuta, K., Klep, I., & Povh, J. (2012). Constrained polynomial optimization problems with noncommuting variables. *SIAM Journal on Optimization, 22,* 363–383. doi:10.1137/110830733.
27. Castillo, I., Joro, T., & Li, Y. Y. (2009). Workforce scheduling with multiple objectives. *European Journal of Operational Research, 196,* 162–170. doi:10.1016/j.ejor.2008.02.038.
28. Kleinman, N. L., Hill, S. D., & Ilenda, V. A. (1998). Simulation optimization of air traffic delay cost. In *Proceedings of the 1998 Winter Simulation Conference* (pp. 1177–1181). Washington DC.
29. Boysen, N., & Fliedner, M. (2011). Scheduling aircraft landings to balance workload of ground staff. *Computer and Industrial Engineering, 60,* 206–217. doi:10.1016/j.cie.2010.11.002.
30. Piera Eroles, M. À., Ramos, J. J., & Fernandez Robayna, E. (2009). Airport logistics operations. In Y. Merkuryev, G. Merkuryeva, À. M. Piera, & A. Guasch (Eds.), *Simulation-based case studies in logistics education and applied research* (pp. 209–228). London: Springer. doi:10.1007/978-1-84882-187-3_12.
31. Bouarfa, S., Blom, H., & Curran, R. (2016). Agent-based modeling and simulation of coordination by airline operations control. *IEEE Transactions on Emerging Topics in Computing, 4,* 1–1. doi:10.1109/TETC.2015.2439633.
32. Peng, Y., Wei, G., Junqing, S., & Bin, S. (2014). Evaluation of airport capacity through agent based simulation. *International Journal of Grid and Distributed Computing, 7,* 165–174.
33. Sabar, M., Montreuil, B., & Frayret, J. M. (2012). An agent based algorithm for personnel shift-scheduling and rescheduling in flexible assembly lines. *Journal of Intelligent Manufacturing, 23,* 2623–2634. doi:10.1007/s10845-011-0582-9.
34. Rodič, B., & Baggia, A. (2013). Dynamic airport ground crew scheduling using a heuristic scheduling algorithm. *International Journal of Applied Mathematics and Informatics, 7,* 153–163.
35. Baggia, A., Leskovar, R., & Kljajić, M. (2008). Implementation of the scheduling domain description model. *Organizacija, 41,* 226–232. doi:10.2478/v10051-008-0024-4.
36. Weiszer, M., Chen, J., & Locatelli, G. (2014). An integrated optimisation approach to airport ground operations to foster sustainability in the aviation sector. *Applied Energy, 157,* 567–582. doi:10.1016/j.apenergy.2015.04.039.
37. Kofjač, D., & Kljajic, M. (2008). Application of genetic algorithms and visual simulation in a real-case production optimization. *WSEAS Transactions on Systems and Control, 3,* 992–1001.
38. Forrester, J. W. (1961). *Industrial dynamics*. Cambridge, MA: MIT Press.
39. Sterman, J. D. (2000). Business dynamics: Systems thinking and modeling for a complex world. *Management*. doi:10.1108/13673270210417646.
40. Stewart, R. (2004). Simulation: the practice of model development and use, 2nd ed. *Journal of Simulation*. doi:10.1057/palgrave.jos.4250031.
41. Gilbert, N. (2007). Agent-based models. In *Agent-based model (Quantitative applications in the social sciences)* (pp. 1–20). doi:10.1146/annurev-polisci-080812-191558.
42. Borshchev, A. (2013). *The big book of simulation modeling*. AnyLogic North America.
43. Siebers, P. O., Macal, C. M., Garnett, J., et al. (2010). Discrete-event simulation is dead, long live agent-based simulation! *Journal of Simulation, 4,* 204–210. doi:10.1057/jos.2010.14.

44. Brucker, P., Qu, R., & Burke, E. (2011). Personnel scheduling: Models and complexity. *European Journal of Operational Research, 210,* 467–473. doi:10.1016/j.ejor.2010.11.017.
45. Reinertsen, D. G. (2012). *The principles of product development flow: Second generation lean product development, kindle*. Redondo Beach: Celeritas Publishing.
46. Awad, M. S. (2009). Allocate the right aircraft capacity. In *13th Air Transport Research Society World Conference*.

Optimization of Take-Off Runway Sequences for Airports Under a CDM Framework

Roland Deroo and Alexandre Gama

Abstract With the regular growth of air traffic, airports are becoming the most critical part of the aircraft path. Improving ground operations to absorb the delays generated is becoming a necessity. This chapter presents a new departure sequencing algorithm based on operation research methods in the context of the CDM implementation over the European airports. This algorithm is described and results and benefits are demonstrated using data from Paris Charles de Gaulle airport. The performance of the algorithm is also investigated using a fast-time simulation tool.

1 Introduction

As nodes of the air transportation network, airports are (and will always be) crucially impacted by the growth of air traffic [1]. Over the last few years, important efforts have been made in order to improve operations and increase the overall capacity of the system. One of these major projects is the Collaborative Decision Making program (CDM) initiated about ten years ago by EUROCONTROL [2], the European network manager. The main idea of this program is to get airport stakeholders to work together in an efficient and transparent way and in particular to share data in order to provide the network manager with the most reliable information concerning the incoming traffic as early as possible.

The technical core of the CDM implementation on an airport is the conception of a Pre-Departure Sequencer (PDS) whose role is to compile all the information coming from the different stakeholders in order to calculate for each departing flight a Target Start-up Approval Time (TSAT) which is supposedly its optimum off-block time (being the time when an aircraft leaves it stand to reach the runway).

R. Deroo (✉) · A. Gama
Service Technique de l'Aviation Civile, Bonneuil-sur-Marne, France
e-mail: roland.deroo@aviation-civile.gouv.fr

A. Gama
e-mail: alexandre.gama@aviation-civile.gouv.fr

M. Mujica Mota and I. Flores De La Mota (eds.), *Applied Simulation and Optimization 2*, DOI 10.1007/978-3-319-55810-3_6

The CDM implementation at Paris Charles de Gaulle airport (CDG) is used as a case study in this paper to compare potential benefits of various PDS algorithms.

The goal of this chapter is to present a new method for the sequence calculation based on research operation methods. By using more input parameters and more powerful algorithms to deal with the increased complexity of the calculations, it is possible to compute a more accurate departure sequence. The quality of the information that is sent to the network manager is therefore improved and the overall operational capacity can be increased.

This paper will be divided into five main parts. In the first part, the overall context is explained and related papers are discussed. The second part aims at describing the basic inputs of a departure sequencer. The third part presents the core of the proposed new algorithm and additional features are investigated in the fourth part. Finally results are shown in the fifth part.

2 Context and Literature Review

2.1 A-CDM Concept

Initiated by EUROCONTROL in the 2000s, the CDM program implementation process is described in [3]. The main steps can be summarised in five main actions that have to be achieved (cf. Fig. 1).

The expected benefits of the CDM program are various and affect all the airport stakeholders. For instance, airlines can expect a reduction of the fuel consumption of their aircraft due to reduced runway waiting times and this directly translates to decreased operating costs for them. This reduction also leads to environmental benefits for the airport operator thanks to reduced pollutants emissions. Meanwhile, EUROCONTROL can expect a better prevision of the oncoming traffic as take-off times are more accurate. Capacity buffers in the airspace can thus be reduced and the overall operational capacity of the network is improved.

This paper focuses on the step 3 and 4 of the implementation process which concern the technical part of the overall CDM process.

2.2 Operation Research Techniques for Departure Sequencing

The scheduling/sequencing problem is a well-documented topic in the operation research area as can be seen in [4, 5] or [6]. There are various methods from the operation research field that can be used for the specific airport sequencing problem. Some of them have already been described in several papers and a rather exhaustive summary can be found in [7]. In [8–11] the sequencing problem is addressed but a

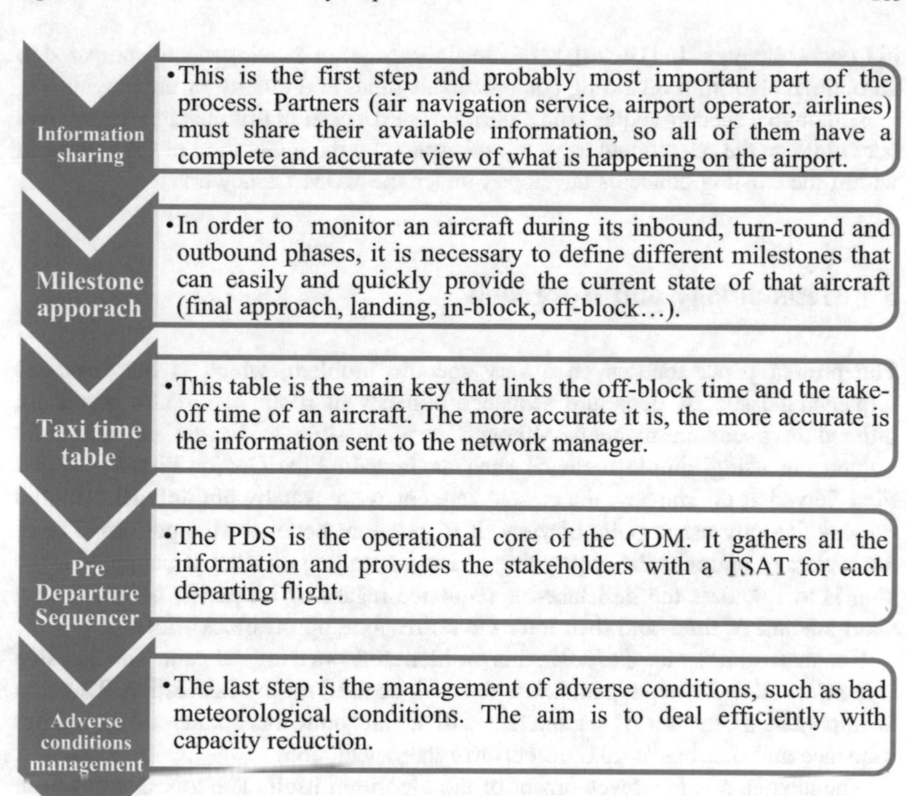

Fig. 1 CDM implementation process

rather different approach is used as the sequencer is designed as a decision aid tool for the air traffic controllers and not as an overall ground operation optimization tool. A different method is used in [12, 13]. The problem is modelled by a constraint satisfaction program, focusing more on the terminal manoeuvring area, as the sequence is calculated based on the available runways on the airport and the standard departure routes used. Results showed capacity improvement but some work remained to be made in order to improve the calculation time for an important airport such as CDG. In [14, 15], operation research methods are used to treat a more global problem that includes departure, arrival and conflicts avoidance during taxiing phases. Another interesting approach is described in [16, 17] where a Constrained Position Shifting (CPS) approach for scheduling is used to indicate how much an aircraft may be moved from an initial position in the sequence.

Also the constraints are not the same as with the departure case, the arrival scheduling problem is also showing some similarities. Operations research techniques and more specifically heuristic algorithms are widely used as well to solve this problem. In [18], a population heuristic algorithm is applied to solve the problem in London airspace. In [19, 20], heuristic algorithms are developed and compared to an optimal algorithm regarding computational times and quality of the outputs.

Although a specific sequencing algorithm is proposed in this chapter, the focus is here more on the operational aspects, and especially the integration of the algorithm within the existing concepts developed under the CDM framework.

3 Methodology and Approach

The present paper focuses on a very specific problem, which is the departure sequencing issue. A departure sequence consists of a set of aircraft which are ordered in a certain manner. Although it is possible to handle the departure sequencing using simple methods such as First-In First-Out or First-Scheduled First-Served at the runway, the created sequences are usually not optimal from the runway capacity usage point of view. It is therefore better for the performance of the system to optimize the sequencing phase: depending on a set of parameters, the goal is to calculate the best take-off sequence regarding a specific objective in a short amount of time, and then infer the corresponding off-block times.

The methodology used to solve this problem follows a logical path. The first step is to select all the relevant parameters that will be used for the calculation. This step is important as the set of parameters will define both the quality of the output sequence and the time it takes to perform the calculation.

The next step is the development of the algorithm itself. The tool that has been developed is a heuristic algorithm, with an initialization phase that calculates a first sequence, and an optimization phase where the initial sequence is modified to reach a better one based on an objective function. This structure meets the two aforementioned requirements. Once the initial sequence is defined, new sequences are created with small modifications (by switching two aircraft in the sequence at a time) of the current reference sequence. When a better sequence is found (based on the objective function), it becomes the new reference, and the process starts over. The tool will keep looking for better sequences until the time limit is reached. Because the initial sequence is the one that is built with the current PDS algorithm, the final sequence is at least as good as the one that is used in the actual operations presently.

To evaluate the performance of the tool, two different approaches are considered. The first one, which is fast and easy to deploy is to perform tests on static data samples. The idea here is to freeze the time at a certain point, and to compare the sequence created by the current PDS algorithm, and the sequence created by the new algorithm at this specific time using the overall delay (which is defined as the sum of all individual delays). Such a comparison is possible, because the data available at any time are the same in both cases. This evaluation approach is implemented in part 7.

The second evaluation approach is to use a simulation tool, to replicate real life, and therefore make a dynamic comparison which is mandatory to truly evaluate the performance of the algorithm as real-life operations on an airport present a lot of contingencies to which adaptation are required. This testing phase using an airside simulation tool is presented in part 8.

3.1 Inputs and Constraints

As core of the system, the PDS gathers all the information from the different stakeholders in order to compile them and provide the desired TSATs. These inputs are described in the following paragraphs, sorted by stakeholder.

3.1.1 Airline Parameters

Airlines and their ground assistants provide the departure time of the aircraft. Two pieces of information are available:

- Scheduled Off-Block Time (SOBT): This is the time written on the passengers tickets. This information is known in advance.
- Target Off-Block Time (TOBT): This is an updated off-block time provided by the airline. If the airline does not modify the departure time, then the TOBT matches the SOBT.

3.1.2 Network Manager Parameters

The Network Manager Operations Center (NMOC) is the network manager's unit. Its main mission regarding our problem is to determine and allocate take-off slots to regulate the traffic in the airspace. As an example, if the expected arrival time in an airspace block (an airspace of defined dimensions in space and time within which air navigation services are provided) of an aircraft leads to a demand greater than the block's capacity, the system will delay the aircraft so it arrives later, when the demand is lower. This traffic management method is implemented through the Calculated Take Off Time (CTOT) allocation. A CTOT is an important time constraint as the effective take-off has to occur between [CTOT – 5 min] and [CTOT + 10 min].

3.1.3 Airport Operator Parameters

The airport operator provides the taxi time table. This table which contains the estimated taxi times links all the airport stands to the departure runways entries, and thus allows the calculation of the TSAT from the take-off time.

3.1.4 ATC Parameters

These are the parameters that are set by the ATC (air traffic control) concerning the usage of the airport. There are three values:

- Runway capacity: the ATC set the maximum number of take-off within one hour. This parameter might be different for each runway of the same airport.
- Runway pressure: this parameter defines the maximum amount of queuing allowed at the runway entry. It can be set as a waiting time, or as a number of waiting aircraft.
- Runway configuration: depending mostly on the wind, the ATC specifies the runways in use. It will affect which taxi time values from the taxi time table are used.

All these inputs and their providers are summarized in Fig. 2 Summary of the PDS parameters and their providers. These parameters are the only ones needed for a simple departure sequencer fulfilling the basic CDM PDS requirements. The following set of constraints describes the job of such a PDS. They must be fulfilled by each flight.

$$TSAT \geq SOBT \tag{1}$$

$$TSAT \geq TOBT \tag{2}$$

$$TTOT \geq CTOT - 5' \tag{3}$$

$$TTOT \leq CTOT + 10' \tag{4}$$

$$TTOT = TSAT + EXOT + ERWT \tag{5}$$

$$ERWT \leq RWY\ PRESSURE \tag{6}$$

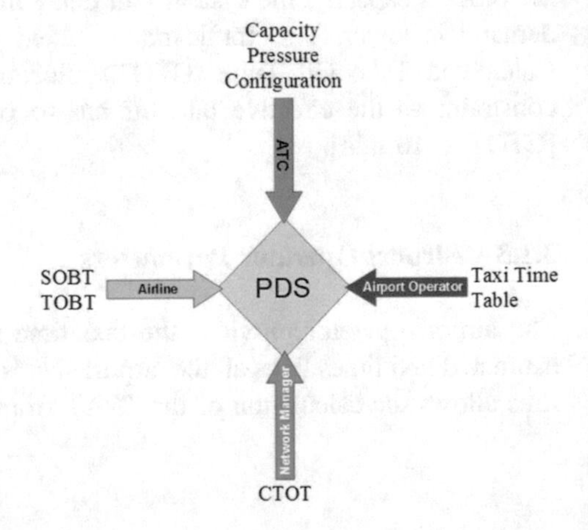

Fig. 2 Summary of the PDS parameters and their providers

where:

- EXOT is the Estimated Taxi Out Time. This is the value taken from the taxi time table between the stand and the runway.
- ERWT is the Estimated Runway Waiting Time. This is the calculated queuing time at the runway.
- TTOT is the Target Take-Off Time.

Equations (1) and (2) ensure that the off-block time is set after the aircraft is ready. Equations (3) and (4) are here to make sure the aircraft takes off within the NMOC slot. Equation (5) links the runway schedule and the off-block schedule. Finally, (6) limits the runway waiting time at its maximum value, the runway pressure. The capacity parameter which does not appear directly in those equations defines the maximum of take-off slots that can be allocated in an hour in respect of the aforementioned equations.

Although these basic inputs and constraints might define a departure sequencer, it is possible to improve and optimize the calculated sequence by using operation research methods, additional parameters and objective functions.

3.2 *Heuristic Algorithm*

This part will present a sequencing algorithm which aims at maximizing the runway capacity by optimizing the overall sequence and more specifically the relative position of the aircraft within the sequence. This sequencer consists of three different parts, each of them having its own function:

- An initialization phase
- A heuristic algorithm that improves the initial sequence
- Several additional modules to add flexibility.

3.2.1 Initialization Phase

In order to initiate the optimization phase, it is important to build a feasible first sequence. Basic requirement of a pre-departure sequencer can be found in [21]. If the initial sequence can already be considered as "good" according to an objective function, then the optimization phase can be run faster and may offer better results. This initial sequence is computed in the following way:

(1) *Take-off slots*: the capacity parameter provided by the ATC is used. An hour is divided in as many take-off slots as the capacity value. As an example, if the capacity is set to 30 departures per hour, there is then 1 slot every 2 min. The departure sequencer's task is to allocate these slots to the departing flights.

(2) *Aircraft sorting*: in order to allocate efficiently each aircraft to one runway slot, the aircraft have to be sorted by order of priority. Thus, if several aircraft are able to use the same departure slot, the latter will be allocated to the highest aircraft in the priority list which is based on the SOBT.
(3) *Available slots computing*: once the slots are defined and the aircraft sorted, the next step is to calculate for each aircraft its earliest take-off time. This determines for each flight the first available runway slot. This value is calculated as the maximum value between the potential CTOT and the sum of the TOBT (or SOBT) and the EXOT.
(4) *Slots allocation*: the next step is to allocate one runway slot to each flight. Based on their first available slot and priority order, the departure sequencer is able to provide a TTOT (Target Take-Off Time) to each aircraft.
(5) *TSAT calculation*: once the TTOTs are defined, the TSATs are calculated based on the taxi time table and the runway pressure value. Providing the TSATs to all the stakeholders is the final step of the departure sequencer.

The different phases of this initialization phase are summarized in the Algorithm 1.

Algorithm 1 PDS Initialization

Input: SOBT, TOBT, CTOT
Capacity, Pressure, Configuration

1: define runway slots

2: sort aircraft by priority

3: **for** each flight, calculate earliest take-off time

4: **for** each flight, calculate TTOT

5: **for** each flight, calculate TSAT

Output: Initial off-block sequence

3.2.2 Optimization Phase

After the computation of the initial sequence, the next step is to optimize it using a dedicated algorithm. Details regarding this optimization phase are presented in this paragraph.

3.2.3 Wake Vortex Separation

Once the initial departure sequence has been computed during the first phase, a new parameter is included in order to have a more accurate sequence. In the present algorithm, this parameter is the true wake vortex separation between two successive take-offs.

The capacity parameter provided by the ATC is an aggregation of the different separation between the departures. It has to be evaluated by the controllers depending on the traffic. If the traffic diversity (in terms of aircraft wake vortex categories) is important, the capacity might be reduced as the average separation increases. However, for a given set of flights, the actual runway throughput may vary significantly depending on the order in which the aircraft take-off. Lowering the capacity settings according to the average separation does therefore not ensure that the runway throughput will always match the demand for departures.

Figure 3 shows different sequences for the same three aircraft (a medium, a light and a heavy). In the depicted scenario, the "First Come First Served" sequence is not the optimal one. This example shows that in some cases, delaying one flight by reorganizing the sequence allows increasing the overall capacity.

The main objective of this optimization phase is to test modifications of the order in which the aircraft are within the departure sequence and look for improvements of the departure time of the last sequenced aircraft.

Using the true separation time between two departures instead of a mean value (the ATC capacity) can be considered as a real time capacity adaptation.

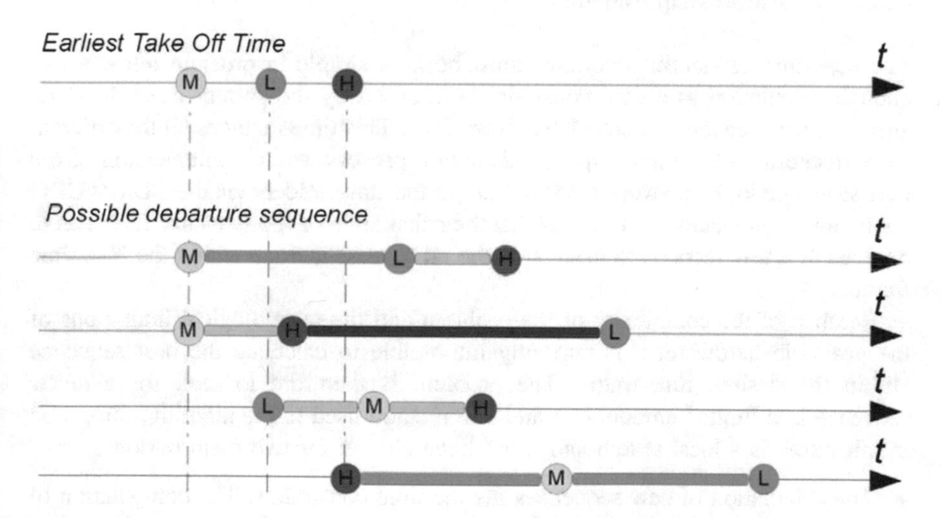

Fig. 3 Possible sequences for three departing aircraft taking into account the wake vortex separation

3.2.4 Objective Function

In order to evaluate the quality of the sequences explored during the optimization phase, an objective function must be defined. In our algorithm, we developed two objectives functions:

- Improving the overall capacity: this objective can be seen as minimizing the departure time of the last aircraft.
- Reducing the overall off-block delay: this objective is to minimize the sum of the difference between the TOBT and the TSAT (see Eq. (7)).

$$min \sum_{f} (TSAT(f) - TOBT(f) \;\forall\, flights\, f\; in\; the\; sequence \tag{7}$$

In order to evaluate the performance of a sequence, these two objectives are used and both aspects are tested as the different stakeholders might consider differently the quality and the performance of a sequence. Therefore, a new sequence is considered as better when either:

- the departure time of the last aircraft is improved and the overall off-block delay is not increased;
- the overall off-block delay is improved and the departure time of the last aircraft is not deteriorated.

3.2.5 Algorithm Requirements

The departure sequencer algorithm must be kept simple in order to allow a fast enough calculation as a time constraint is imposed by the system itself. In CDG airport, a new sequence is calculated every 30 s. This time includes all the different steps (reception of the new inputs, calculation process, and communication of the new sequence to the network manager and to the stakeholders via the CDM@CDG communication channels). It means that the calculation of the sequence itself has to be done in a few seconds to make sure the all process is done within the 30 s time frame.

Because of the complexity of the problem and the technological limitations of the available hardware, it is currently impossible to calculate the best sequence within the desired time-frame. The problem is therefore to seek for a *better* sequence in a limited amount of time. The method used in the algorithm proposed in this paper is a local search and it has been chosen for two main reasons:

- The calculation of new sequences fits the time constraints. The computation of the neighboring sequences and the evaluation of their performance are done in a short amount of time which allows exploring a very large number of sequences within the time-frame.

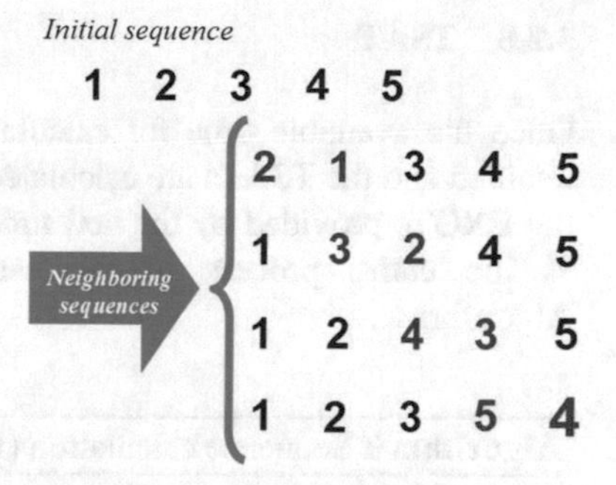

Fig. 4 Neighborhood structure used in the local search

- The objective is not to find the optimal sequence, but to search for a good sequence in a short time interval. Therefore, local optimization is a good match as it has proved quite successful for combinatorial optimization problems [22].

3.2.6 Neighborhood Structure

The local search algorithm needs a neighborhood structure to create the new sequences to explore from the initial one. In order to keep a fast-time calculation, a unique and simple structure is used. This is a switch structure, where new sequences are obtained by switching two successive aircraft in the sequence. A sequence of n aircraft has therefore a neighborhood consisting of n-1 sequences (see Fig. 4).

The method used for the neighborhood exploration is a best improve search. It means that all the sequences are tested, according to the two objective described previously and the best is kept.

3.2.7 Sequence Feasibility

For each sequence its feasibility is also tested, in particular regarding to the CTOT (if applicable). Indeed, a CTOT given by the network manager has a small tolerance interval (−5 min; +10 min) and also has a high delay penalty when it is not complied with. This means that all the aircraft for which a CTOT has been issued have to take-off as much as possible within that time interval. Therefore, when a neighboring sequence is computed, a test is made to ensure that all the aircraft with a CTOT have a runway slot within the interval.

3.2.8 TSAT

Once the available time for calculation is elapsed, the current best sequence is retained and the TSATs are calculated based on the TTOTs from that sequence and the EXOTs provided by the taxi time table.

The entire process of the sequence calculation is summarized on the Algorithm 2.

Algorithm 2 Sequence calculation (for *n* aircraft)

Input: SOBT, TOBT, CTOT
Capacity, Pressure, Configuration
Aircraft Type, Wake vortex separation
Neighbourhood structure, objective function

1: *t* = *current_time*
2: Compute initial sequence
(see Alg. 1)

3: Add wake vortex separation
current_sequence = *initial_sequence*
current_quality = *f(current_sequence)*

4: **while** *current_time* < *t* + *10s*
Create *n-1* neighbouring sequences
$n\_sequence_1$, $n\_sequence_2$, ...
for *i* from *1* to *n-1*
if $f(n\_sequence_i)$ > *current_quality* and *sequence_feasibility* = *true*
then *current_sequence* = $n\_sequence_i$
current_quality = $f(n\_sequence_i)$
i=*i*+*1*
else *i*=*i*+*1*
endif
endfor
endwhile

5: for each flight, calculate TSAT

Output: Off-block sequence

---

4 Additional Features

In order to deal with the other objectives of the CDM implementation on an airport, additional modules have been investigated concerning two main fields:

- Adverse conditions and more specifically deicing operations during winter
- Taxi time and therefore environmental impact reduction

Both items will be discussed in this part, focusing more on the changes and the complexity added to the departure sequencer.

4.1 Deicing Operations

Dealing with adverse conditions and especially deicing operations is one of the five steps to achieve in the CDM program implementation on an airport as said in the first part of this paper. Deicing operations are particularly problematic for the sequencing phase for the two following reasons:

- The taxi time is modified if the aircraft is deiced on a deicing pad. It means that in order to create an accurate runway sequence, taxi times from the stands to the pads and from the pads to the runway must be well estimated as well as the deicing phase in itself, depending on the meteorological conditions, the aircraft type and the pilot needs.
- The TOBT is modified if the aircraft is deiced on its stand. The deicing time also has to be well estimated in order to predict an accurate runway sequence.

Moreover, depending on the weather conditions, all aircraft might not have to deice before taking-off, and may use different types of deicing facilities (deicing on stand, or remote deicing on a specific area of the airport). The authors propose the following way to deal with this problem with two sequencers working together:

- A deicing sequencer: its job is to allocate a deicing slot (starting at the Estimated Commencement of Deicing Time (ECZT) and ending at the Estimated End of Deicing Time (EEZT)) and a deicing resource to the aircraft, and to modify its Earliest Take-Off Time.
- The usual departure sequencer with the modified information from the deicing sequencer that provides the TSAT.

With this method, the departure sequencer can work in the exact same way as in normal conditions. A deicing tool is simply connected to update all the necessary information. The main issues come from the accuracy of the information and its predictability. As shown in [23], the deicing problem is complex, and is part of the *tactical* phase (which means that decision are taken few minutes before take-off). The objective in this paper is not to propose a complete integration of it in the

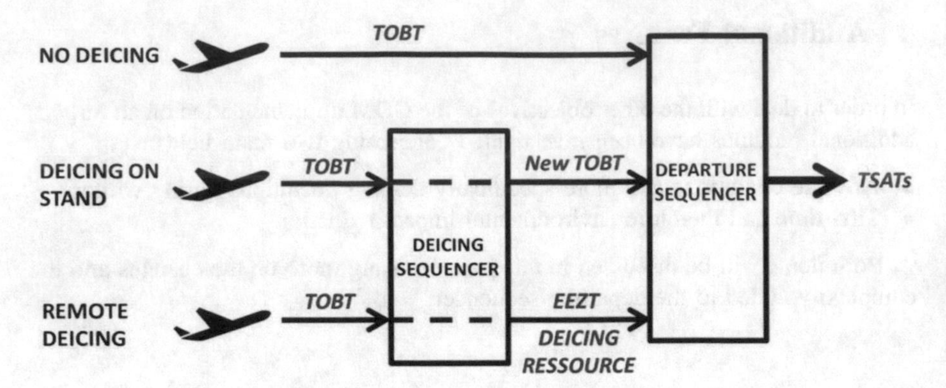

Fig. 5 Diagram of the deicing sequencing process

departure sequencer but to highlight the complexity of that particular problem. A diagram of the process is presented in Fig. 5.

4.2 Taxi Time Optimization

The objective function used in the algorithm and described previously aims at increase the overall capacity and reduce the total delay of the departing aircraft. But depending on the airport infrastructure and the environmental objectives, it might be relevant to investigate the possibility to have the taxi time in the objective function. This part concerns two different aspects:

- The optimization of the pressure parameter: this parameter described previously is the maximum waiting time at the runway of an aircraft. This value has a great importance. If it is set too low, there will not be enough queuing to ensure that the runways always have a pool of aircraft ready to depart and departure slots might therefore be wasted. If it is set too high, the waiting time might be too long for no reason.
- The choice of the departure runway: this aspect concerns airports with at least two departure runways in use simultaneously. Depending on the aircraft's TMA (Terminal Manoeuvring Area) exit point and the traffic load within the TMA, aircraft can use both or one departure runway. An example is given in Fig. 6, where some of the TMA exit points for the departures in CDG are shown. While aircraft leaving towards the north should preferably use the north departure runway to avoid crossings in the airspace, aircraft leaving for one of the three exit points located east of the airport can use either the north or the south departure runway. It is then possible to take into account the taxi time for the two runways in the objective function, and try to minimize it.

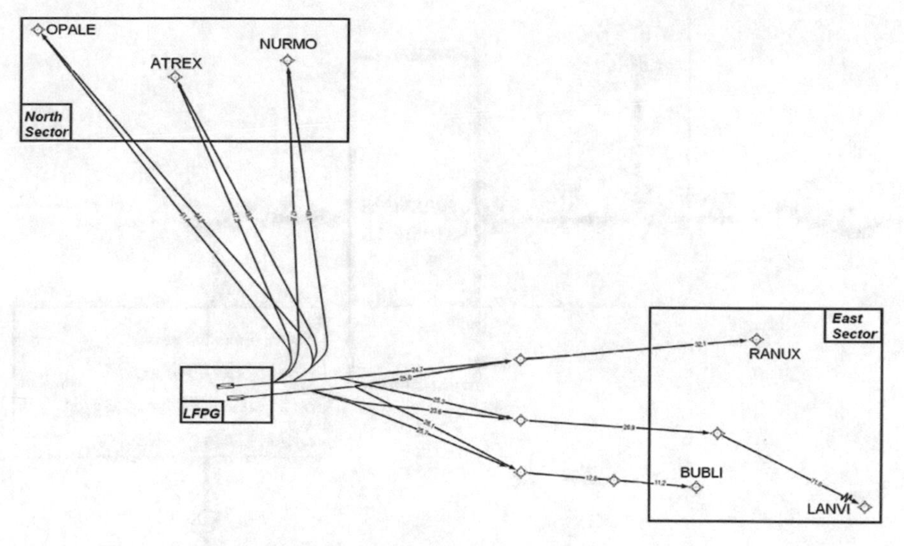

Fig. 6 Scheme of the Paris TMA east and north exit points

The taxi time optimization for a selection of eastern and western TMA exit point has been implemented in a simple way for testing purposes on the CDG case. The calculation is first made as described in the previous part. Once the calculation is done, a test is made on all aircraft which leave the TMA from the east or the west exit points: if the taxi time is improved by using the other runway and if changing the departure runway does not deteriorate the other objectives, the runway is switched. The last step is to ensure that a departure heading for the same exit point is not leaving at the same time from the other departure runway. The process is summarized in Fig. 7.

Switching runways only shows benefits during off-peak hours when slots availability at the runway is high. But as in the case of the deicing, the objective is again to highlight a new optimization horizon and not to provide a complete integration in the main sequencing algorithm.

5 Results and Benefits

The main sequencer and the additional features have been fine tuned for the case of Paris–Charles de Gaulle airport and were backtested using a sample of recorded data. The CDM has been deployed at CDG in 2010 and this airport was one of the first to get the label from EUROCONTROL. Its complex layout and traffic structure make it a well indicated case study. Due to the specificities of the traffic structure of CDG, various traffic samples had to be considered for the tests. Those traffic samples can be classified in three main categories: low traffic samples, medium traffic samples and high traffic samples.

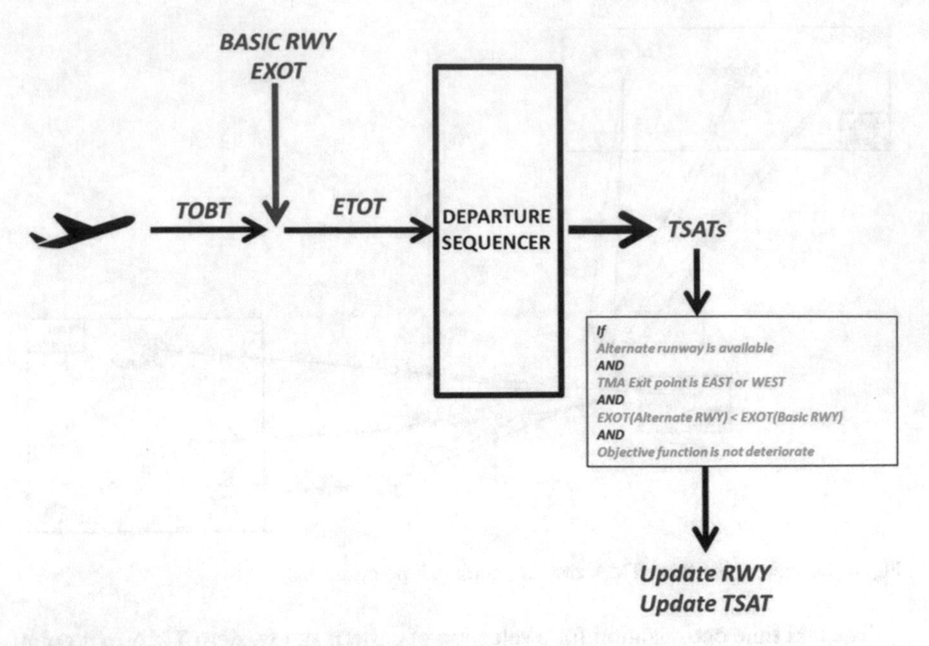

Fig. 7 Diagram of the taxi-time optimization process

These samples also differ in terms of aircraft type diversity, especially among the high traffic samples. This parameter is particularly critical because the more important this diversity is, the more benefits will be obtained by reordering the sequences to minimize the separations between successive aircraft will.

The algorithm was not tested in live operations (dynamically) but was tested using recorded data. The TSATs were computed with the new algorithm based on the information which was assumed to be available at the time at which the corresponding real world TSATs were calculated. These TSATs issued by the new algorithm were then compared to the TSATs computed by the current PDS. The indicator used to compare the results is the overall off-block delay (see Eq. (7)). The number of take-off within one hour could unfortunately not be calculated because the information (TTOT) was not available in the database used. Since the algorithm was only tested in a static way using recorded data, there could be some differences between the results shown below and the results which could be obtained after a live experiment.

5.1 Low Traffic Samples

During departure off-peak hours, the new sequencing algorithm does not improve the results already obtained by the current PDS concerning the overall delay. Indeed, an important number of departure slots at the runway are not used because the traffic is not high enough. Therefore, the modification of the sequence has no impact. Neither of the two systems creates off-block delays.

However the taxi time optimization module does provide a significant improvement over the current standard PDS. The numerous free departure slots allow runway switching without creating any delay for the other aircraft. Approximately 40% of the departing flights (data from 2009) leave the TMA from a west or east exit point. Depending on their stand on the airport, a significant proportion of these flights might be affected by a taxi time reduction. In every low traffic sample, several flights experienced a runway modification which led to a taxi time reduction ranging from 5 min (this value was set as the minimum value for runway modification) to 15 min.

5.2 Medium Traffic Samples

These medium traffic samples relate to the situations where all the departure slots are used, but the number of aircraft waiting at the runway is small. The sequence modification will typically have a more important impact in these situations. With the current PDS, two categories of off-block delay can be observed:

- Aircraft with a departure slot sent by the NMOC: the objective is here to have a departure as close as possible from the CTOT.
- Aircraft without a CTOT: the objective is here to sequence the off-block time as close as possible to the TOBT.

The results concerning the overall delays are here much more significant than for the low traffic periods. By taking into account the true wake vortex separation between two successive aircraft, the sequencer is able to reduce by 30–40% the overall off-block delays on all the medium traffic test samples (see Fig. 8). The average delay thus decreases from 94 to 56 s. Flights with a CTOT are much closer to their slots and there is almost no delay for the flights without CTOTs.

Regarding the taxi time, no relevant results were found as the number of free slots at both runways was close to zero. Therefore, any runway modification would have created delay for the following aircraft.

5.3 High Traffic Samples

The samples are classified in the high traffic category when all slots at the runway are used and the aircraft waiting time at the runway reaches the maximum value. These samples have a high overall off-block delay and are the most critical periods within the day. Most of the high traffic periods in CDG present a high diversity in terms of aircraft type.

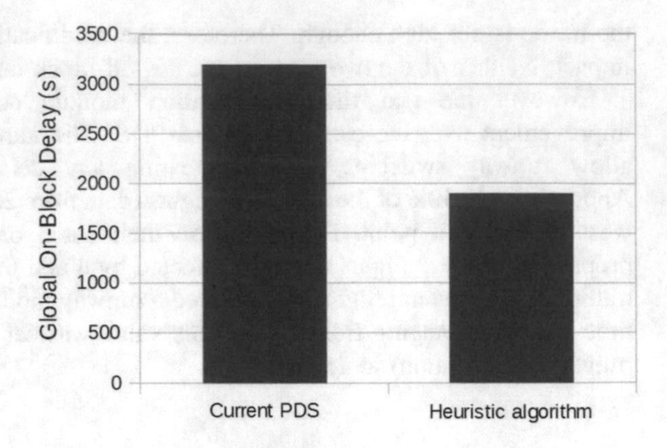

Fig. 8 Overall off-block delay reduction for medium traffic samples

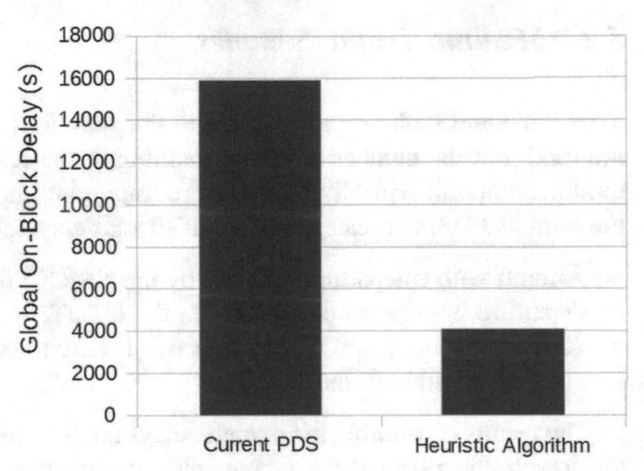

Fig. 9 Overall off-block delay reduction for high traffic samples

The current PDS is able to sequence all the flights with an average off-block delay of around 3 min per aircraft. The negative impact on a single flight is thus relatively limited. But due to the important number of aircraft movements during this departure peak, the overall off-block delay reaches 4–5 h. For an airline like Air France, which represents around 50% of all flights, it represents a cumulative delay of 2–3 h during a single departure peak.

The results of the heuristic algorithm are here again very promising; indeed the average delay has been almost divided by three, from 3 min to 65 s. The average overall delay on all samples decreases from 4h25 to 1h09 (see Fig. 9). The average delay for the high traffic periods is thus brought by the heuristic algorithm to a delay level similar to the one observed during medium traffic periods.

It is important to note that the traffic structure in CDG highlights particularly well the added value of the heuristic algorithm. The diversity of the aircraft fleets visiting this particular airport is important which means the true separation between aircraft also has an important diversity. On another airport such as Lyon airport where all aircraft are medium aircraft, the impact of the heuristic algorithm might be less significant.

6 Dynamic Testing Using Simulation

In addition to the static test made on traffic sample described in the previous part, the heuristic algorithm has also been tested using a simulation model. The tool used for this experimentation was CAST Aircraft [24], which was chosen for its ability to communicate with an external module using a specific interface. Using this interface, it was therefore possible to control, monitor or feed the software with dynamically modified data during a simulation run. A diagram of the process is shown on Fig. 10. CAST outputs milestones regarding the flight status at each time-step (the interval between two time steps being defined by the modeler):

- TSAT Locked: the current time in the simulation is too close from the TSAT and therefore it cannot be changed.
- TSAT Unlocked: the TSAT can still be modified by the PDS.
- TAKE-OFF: the flight took off and is therefore out of the PDS range.

The communication is done using "Methods" on simulation objects. "Methods" are commands that allow actions (display or modify values or milestones for example) on a defined property. The main method used here is "SetTSAT" which modifies the off-block time of an aircraft. With this architecture, the complexity of the simulation model is kept as low as possible and all the heuristic part is done by an external dedicated tool which could therefore offer optimized calculation.

The main drawback concerns the time required to perform the calculation inside the PDS module, from the analyst point of view. As it has been designed for real-time, it takes a few seconds to calculate all the new TSATs which fit the

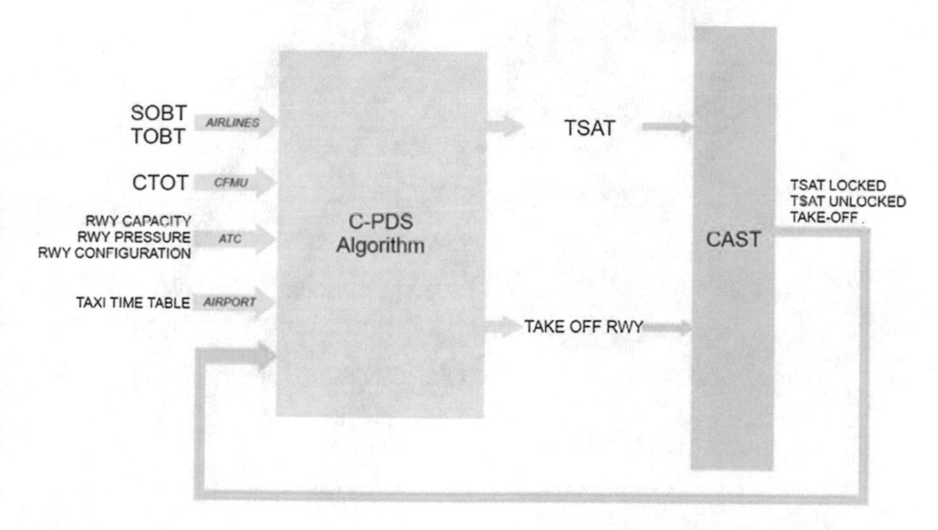

Fig. 10 Scheme of the communication process between the external module and CAST

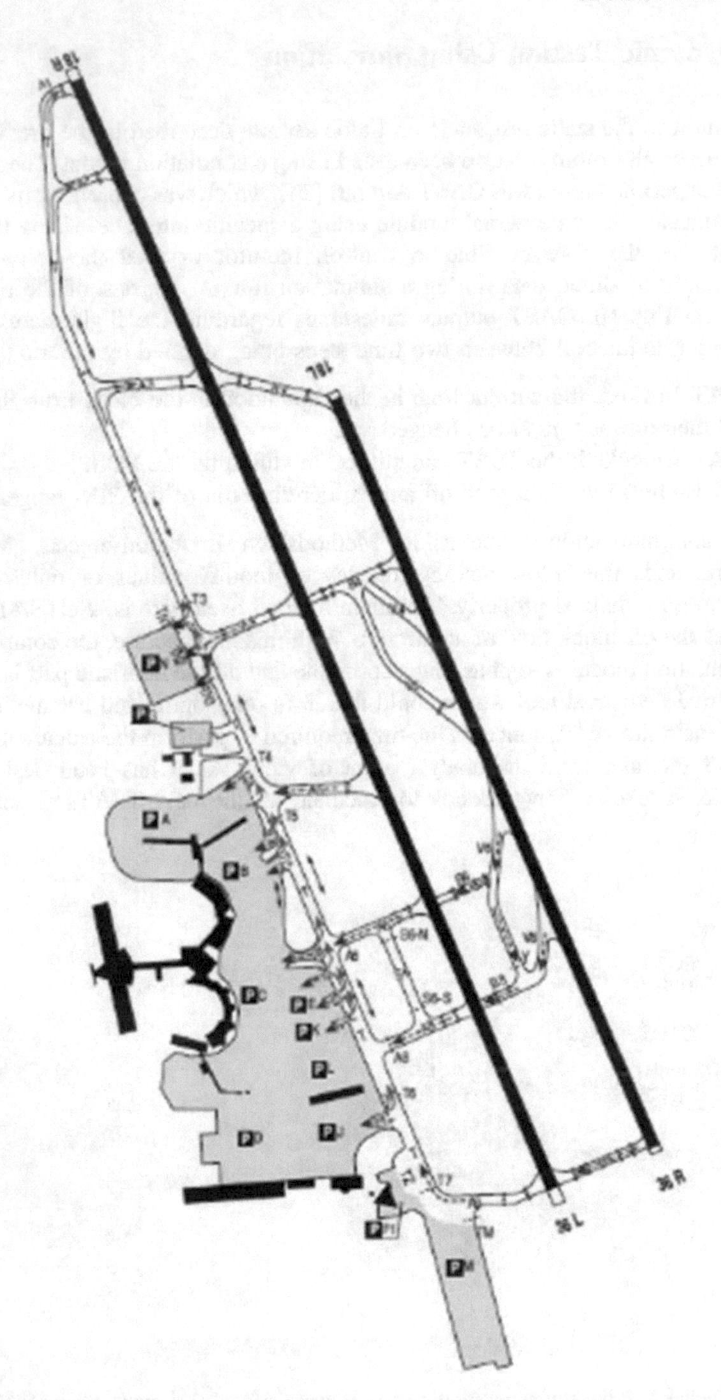

Fig. 11 Lyon Saint-Exupéry airport layout

requirements for a real-time process but it prevents the fast-time simulation tool to run as fast as it could. A solution to this particular problem would be to optimize the PDS code to allow faster calculation time and therefore meet the requirements for a fast-time usage.

Although it would have been technically possible to test the PDS algorithm using a model of Paris-CDG, the airport of Lyon Saint-Exupéry has been chosen for the first feasibility test to evaluate the potential of the PDS in a dynamic environment through simulation. Indeed, Lyon Saint-Exupéry airport has a simpler layout than Paris-CDG, as well as a lower traffic. It leads to less complexity within the simulation tool (simpler model due to the simpler layout) as well as within the PDS (less traffic leads to less sequences to calculate). Lyon Saint-Exupéry airport is still an interesting case as it is currently deploying the CDM concept and is therefore adding an actual PDS to its systems. A chart of the airport layout is shown on Fig. 11.

This first test was a great proof of concept and as such showed particularly promising results. The communication between the simulation tool and the external module done using Transmission Control Protocol (TCP) and a remote interface developed specifically for CAST has been working seamlessly although the time currently required to perform the sequencing phase within the external module prevents the simulation software to run as fast as possible as was expected. As a second phase, the current PDS algorithm behavior should be replicated in the external module as well to allow comparisons with the heuristic tool using the simulation software. A test on a bigger airport with more traffic diversity should also be performed to emphasize the performance of the heuristic algorithm.

7 Conclusions

The heuristic algorithm presented in this paper has proved to be a high quality tool to improve the overall operation on a platform such as Paris-Charles de Gaulle by reducing the overall delay and increasing runway performance. The method used fits well the time-frame constraint for the calculation and the results are very promising.

The overall delay, as well as the average delay per aircraft can be significantly reduced using this new sequencer. The additional features, especially the taxi time optimization part have shown their possibilities to improve the initial algorithm with multi-level optimization.

In a next phase, the sequencer has been tested using fast-time simulation to evaluate its performance on a dynamic level using Lyon Saint-Exupéry airport layout and traffic. This allowed testing on an unstable environment, and results concerning the capacity benefits and fuel consumption reduction could be measured using the results produced by the software. In order to go further, the current PDS algorithm must be optimized to reduce the calculation time to allow testing on a bigger airport with a higher traffic level.

The diversity of the aircraft fleets visiting an airport is also important. On an airport where the aircraft types show an important uniformity, the impact of the heuristic algorithm might be less significant.

Further tests will have to be made to evaluate more deeply the algorithm and the additional features on different airports with other traffic structures which were not done yet due to expected speed performance.

Acknowledgements The authors would like to thank Aéroports de Paris and the CDM team for the provided data used for the test phases.

References

1. EUROCONTROL. (2013). Challenge of growth: Summary report.
2. EUROCONTROL. (2016). Homepage, [Online].
3. EUROCONTROL. (2012). Airport CDM Implementation Manual v4.
4. Lawler, E., Lenstra, J., & Rinnoy Kan, A. (1982). Recent developments in deterministic sequencing and scheduling: A survey. In *NATO advanced study institutes series* (Vol. 84).
5. Lawler, E., Lenstra, J., Rinnooy Kan, A., & Shmoys, D. (1993). Sequencing and scheduling: Algorithms and complexity. In *Handbooks in operations research and management science* (Vol. 4).
6. Graham, R., Lawler, E., Lenstra, J., & Rinnooy Kan, A. (1979). Optimization and approximation in deterministic sequencing and scheduling: A survey. *Annals of Discrete Mathematics, 5*.
7. Bennel, J., Mesgarpour, M., & Potts, C. (2011). Airport runway scheduling. *4 OR: A Quarterly Journal of Operations Research, 9*.
8. Atkin, J., Burke, E., Greenwood, J., & Reeson, D. (2004). *Departure runway scheduling at London Heathrow airport*. University of Nottingham.
9. Atkin, J. (2008). On-line decision support for take-off runway scheduling at London Heathrow Airport, Ph.D. Thesis, University of Nottingham.
10. Atkin, J., Burke, E., Greenwood, J., & Reeson, D. (2007). Hybrid metaheuristics to aid runway scheduling at London Heathrow airport. *Transportation Science, 41*.
11. Hesselink, H., & Basjes, N. (1998). *MANTEA departure sequencer: Increasing airport capacity by planning optimal sequences*. National Aerospace Laboratory, NLR.
12. Van Leeuwen, P., Hessenlink, H., & Rohling, J. (2002). *Scheduling aircraft using constraint satisfaction*. Grado, Italy.
13. Hesselink, H., & Paul, S. (1998). *Planning aicraft movements on airports with constraint satisfaction*. Cardiff, UK.
14. Deau, R. (2010). Optmisation des séquences de pistes et des mouvements au sol sur les grands aéroports, Ph.D. Thesis, Université de Toulouse.
15. Gotteland, J. (2004). Optimisation du trafic au sol sur les grands aéroports, Ph.D. Thesis, Université de Toulouse.
16. Balakrishnan, H., & Chandran, B. (2010). Algorithms for scheduling runway operations under constrained position shifting. *Operations Research, 58*.
17. Trivizas, D. (1998). Optimal scheduling with maximum position shift (MPS) constraints: A runway scheduling application. *Journal of Navigation, 51*.
18. Beasley, J., Sonander, J., & Havelock, P. (2001). Scheduling aircraft landings at London Heathrow. *Journal of the Operational Research Society, 52*.
19. Beasley, J., Krishnamoorthy, M., Sharaiha, Y., & Abramson, D. (2000). Scheduling aircraft landings—The static case. *Transportation Science, 34*.

20. Ernst, A., Krishnamoorthy, M., & Storer, R. (1998). *Heuristic and exact algorithms for scheduling aircraft landings*. Wiley.
21. SESAR Joint Undertaking. (2011). Basic DMAN Operational Service and Environment Definition (OSED) (2nd ed.).
22. Aarts, E., & Lenstra, J. (1997). *Local search in combinatorial optimization*. Wiley.
23. Mao, X., Mors, A., Roos, N., & Witteveen, C. (2006). Agent based scheduling for aircraft deicing. Netherlands Organisation for Scientific Research, Project No. CSI4006.
24. Airport Research Center. (2016). Homepage. [Online].

Simulation and Optimization Applied to Power Flow in Hybrid Vehicles

Guillermo Becerra, Luis Alvarez-Icaza, Idalia Flores De La Mota and Jose Luis Mendoza-Soto

Abstract This chapter describes the application of optimization to power flow in hybrid electric vehicles, first using a strategy based on bang-bang optimal control and then comparing it with Pontryagin's alternative. The first strategy, known as the planetary gears system (PGS), focuses on satisfying the kinematic and dynamic constraints of the gears system, starting from the allocation of the electric machine power. The second uses Pontryagins minimum principle (PMP) to solve the energy management problem and decide the amount of power that the electric machine and combustion engine should provide. The approach of the PMP strategy entails three basic elements, namely: first of all, getting the demanded power to be supplemented by the drive machines; secondly, maintaining the state of charge at a level in and around a reference so as to avoid discharging and overloading the batteries and thirdly, saving on fuel. By using the above considerations, a cost function is set out that considers the power from both machines to be inputs. The simulations were performed in Matlab's Simulink using detailed models of the elements of a hybrid diesel-electric city bus in parallel configuration. The demands are represented by driving cycles while the combustion engine and electric machine are coupled using a planetary gears system.

G. Becerra (✉)
CONACYT - Universidad de Quintana Roo, Boulevard Bahia s/n esq. Ignacio Comonfort, Del Bosque, 77019 Chetumal, Q Roo, Mexico
e-mail: gbecerra@conacyt.mx

L. Alvarez-Icaza · I. Flores De La Mota
Universidad Nacional Autónoma de México,
Av. Universidad No. 3000, 04510 Coyoacan, Mexico
e-mail: alvar@pumas.ii.unam.mx

I. Flores De La Mota
e-mail: idalia@unam.mx

L. Alvarez-Icaza · I. Flores De La Mota
Ciudad Universitaria, Ciudad de México, Mexico

J.L. Mendoza-Soto
CINVESTAV, Av. Instituto Politecnico Nacional 2508, San Pedro Zacatenco,
07360 Gustavo A. Madero, Ciudad de Mexico, Mexico
e-mail: eemsj03@yahoo.com.mx

M. Mujica Mota and I. Flores De La Mota (eds.), *Applied Simulation and Optimization 2*, DOI 10.1007/978-3-319-55810-3_7

1 Introduction

Because of our natural need to move around, humanity has developed different methods of transporting people, animals, foodstuffs, things, etc. Since their invention, engine vehicles have been particularly popular among the majority of users, owing to their versatility and wide range of applications. However, nowadays, large cities have many transport-related problems, such as poor mobility, road congestion and the deterioration of air quality, as well as the problems associated with the politics surrounding fuels.

According to Schaefer and Victor [1], the demand for passenger transport and the number of vehicles per inhabitant rises alongside in direct proportion to a societys economic possibilities. In certain populations (such as Japan, Europe, the United States, etc.), the density of car use was approximately 400–800 vehicles per 1000 inhabitants. Guzzella and Sciarretta [2] mention that up until 2007 countries such as China (with 1,300 million inhabitants) or India (with 1,100 million inhabitants) had a car density of around 30 vehicles per 1000 people; however, the trend was for car density to increase.

The International Energy Agency (IEA) was created in 1974 mainly in response to problems with fossil fuels [3]. It is principally responsible for organizing energy policies to ensure oil supplies for its member countries. Who are seeking to avoid a repetition of phenomena such as the silver cycle that brought about the ruin of the city of Potosí, Bolivia [4], a place that was, in its day, a thriving city with the main silver mine in the world but, like everything, this was not permanent. Thus, many people are worried about oil being a non-renewable fuel and its imminent end, while our need to move from one place to another is seemingly growing every day. Figure 1 illustrates some of the problems caused by our transport needs.

Electric vehicles (EV) are an alternative to counteract some of the problems mentioned above. For example, they lower the level of pollution in the city, as EVs use electric power that has been produced elsewhere. However, such energy may not be clean: in Mexico, only 18.3% of the energy produced is clean, as shown in Fig. 2 [5]. It is important to remember that electric vehicles need electric power produced by

Fig. 1 Transport problems

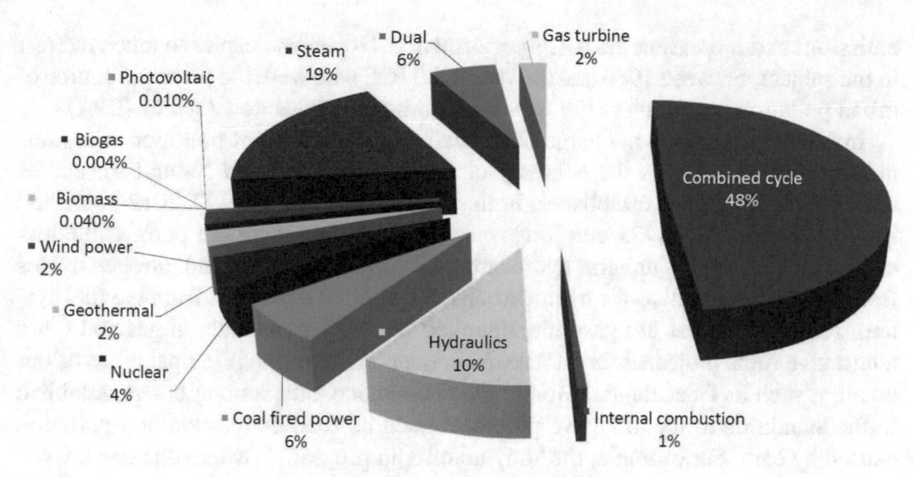

Fig. 2 Electric power production in Mexico, 2013 (275, 522 GWh)

a power plant and only the power produced by hydroelectric plants, wind turbines, solar power, etc. that are renewable do not generate emissions (as they do not consume any type of fossil fuel) are considered to be clean energy. However, electric vehicles have not been very successful for a variety of reasons of which the main one might be the fact that the energy density of gasoline is much greater than the electrochemical density that a battery can offer.

Electric drive vehicles have been around for a long time; for example, until the end of the nineteenth century most of the first cars on the road were electric or steam driven. Furthermore the first electric vehicles had the same problems as electric vehicles have today. However, the internal combustion engine (ICE) vehicle grew in popularity because it could travel long distances, reached a high speed and was much cheaper to buy.

At the start of the twentieth century thousands of hybrid and electric vehicles were being designed and manufactured; in fact, they were the peoples choice. In 1900, 38% of the cars sold were electric and the rest were steam or gasoline powered; just as a reference, more steam vehicles were sold than gasoline-powered ones [6]. Electric vehicles did not have the vibration, smell and noise of gasoline-powered cars. Moreover, they did not need a crank to start them and there was no need for a transmission or gear change. These were the main reasons of that the electric unit was preferred to ICE vehicles. However, in 1904 Henry Ford overcame some of the common objections to gasoline-powered cars (noise, vibrations and smells) and thanks to assembly-line production was able to offer gasoline-powered vehicles at very low prices.

In the car market, technological improvement is not only interested in the design of (aerodynamic and convenience) bodywork or interiors. Owing to the predominance of ICE vehicles, the transport technology that has more power density (volume/power ratio), many research and development efforts are aimed at lowering emissions and increasing the ICE's performance. However, despite so much interest in the subject, between 1980 and the year 2000 ICE cars were the principal source of urban pollution according to the U.S. Environmental Protection Agency (EPA) [7].

In Mexico, the body in charge of regulating the quantity of pollution emissions produced by vehicles is the Ministry of the Environment and Natural Resources (SEMARNAT), which establishes, in the NOM-076-SEMARNAT-2012 published in the Federal Official Gazette in November 2012 the maximum permissible levels for the emission of unburnt hydrocarbons, carbon monoxide and nitrogen oxides from exhausts, as well as for hydrocarbons evaporative emissions from the fuel system, for vehicles that use gasoline, liquefied petroleum gas, natural gas and other alternative fuels projected for car use. Moreover, some of the principal cities of our country, such as Guadalajara, Monterrey and Mexico City, among others, establish traffic standards to try too solve problems such as road congestion and pollution caused by cars. For example, the "hoy no circula program" (where one day a week every car of a certain age has to stay off the roads).

In general, hybrid electric vehicles (HEV) are a suitable compromise between fuel economy and autonomy. The purpose of the design of hybrid technologies is to combine two or more energy sources in such a way as to get the best qualities of each of them, while, at the same time, seeking to maximize the economic benefits of hybrid systems against their manufacturing cost. To illustrate the advantages of combining power in hybrid vehicles, in this chapter we are employing two power sources: an internal combustion engine and an electric machine; in particular a diesel ICE, the most efficient of its type, is used.

Hybrid electric vehicles offer some important advantages in comparison to conventional vehicles (the ones only with internal combustion engines), despite having additional components, greater complexity and costing more: hybridization can significantly cut down on fuel consumption as well as helping to significantly reduce the pollution emissions. Hodkinson and Fenton [8] mention that hybrid technology could reduce fuel consumption by up to two thirds and emissions by a third in comparison to conventional vehicles. Some of the advantages of hybrid vehicles derive from the fact that the total power is divided between the fuel power and electric power: this fact poses interesting challenges from the point of view of control; which some of them we aim to describe in this chapter. The main reason for this chapter is to do with the problems generated by the large number of vehicles now in circulation around the world, causing environmental pollution and very high levels of fuel consumption. In the search for solutions, new alternative fuels have been developed and important research is being carried out into finding ways to substantially lower its consumption. This chapter focuses on the second point.

One of the problems detected in power flow distribution strategies is their sensitivity to changes of driving cycle, for example, for the ECMS strategy, which optimizes fuel consumption for a specific driving cycle and whose parameters change drastically in another driving cycle.

Another significant problem is the large number of calculations involved in dynamic programming strategies that make it unviable for implementation in real time. Therefore, they are generally used to compare the other strategies in simulation.

In model predictive controls, the prediction horizons tend to be very small, so the solutions are similar to those in a regulation problem and sometimes it is necessary to add information about the road that requires other instruments, such as GPS in the case of the A-ECMS strategy, to adapt the parameters of the strategy to the new management profile.

We believe there are opportunities for improving the strategies the literature provides for power flow control in hybrid vehicles. We are mainly aiming to achieve a simple and effective strategy for distributing the power in hybrid drive trains that can be implemented online.

A further opportunity that we have taken from an analysis of the papers available in the literature is that the vast majority do not assess them using detailed models of the elements involved in the main function of the vehicle. If the ones that have the biggest influence on the end result are used we will have fulfilled at least one of the objectives of this chapter.

In view of the above, the purpose of this chapter is to use optimal control theory to design a strategy for the power distribution in the hybrid drive train for un hybrid electric vehicle in parallel configuration, on condition that the strategy can be implemented in real time.

This chapter is divided into five sections as follows: in section two hybrid electric vehicles and their configurations are described. In Sect. 3 the control of the internal combustion engine and electric machines is described, Sect. 4 is about energy management in the hybrid drive train, and finally in Sect. 5 we present the simulations obtained from coupling the models of each element for a hybrid electric vehicle.

2 Hybrid Electric Vehicles

A hybrid vehicle is one that combines two or more energy sources that can, directly or indirectly, provide drive power. Ideally, each of the sources of drive works to improve the efficiency and the performance of the rest of them and thus reduce the disadvantages to a minimum. The hybrid electric vehicles in current use have a conventional internal combustion engine and a battery-powered electric machine—bigger than the DC starter motor-, and some have more than one electric machine.

The logic behind the use of two power sources is simple: for a conventional vehicle, the combustion engine has more power than is needed for most driving situations. Only 20–40% of the power of the ICE is needed to maintain a cruising speed. The rest is only needed for acceleration and to overcome loads, such as going up a hill. These high-power engines use more fuel when they are asked to accelerate. An electric motor does not consume fuel and can supply energy almost instantly.

Hybrid electric vehicles typically use a smaller combustion engine and an electric motor to provide the power needed for acceleration and to overcome loads.

Hybrid vehicles use much less fuel in the city than conventional ones. This is because they do not have to provide all the power required for starting and stopping in the traffic. The electric motors power supplements the power of the combustion engine. There is also an improvement in fuel efficiency on the highway, owing to the use of smaller and more efficient engines. In most cases, these advanced engines cannot produce the power required for strong surges in acceleration without the help of the electric motor.

Many countries have locomotives that are diesel-electric hybrids and in some cities diesel-electric buses are used that can run on electric power provided by the electric charge from high voltage cables, when they do not have fuel, or even when they do not have a source of electric power, the drive is by means of the ICE.

Just some of the main advantages of HEVs in respect of conventional vehicles are the fact that it is sometimes possible to get up to twice the efficiency, which in turn lowers the pollution emissions, as mentioned by Zhao and Wang [9]; moreover, the regenerative braking system offers the possibility of recuperating kinetic energy for its later use, something that cannot be done in a conventional vehicle.

At the present time several car manufacturers are investing in these types of technologies, seeking to lower fuel consumption and, consequently, air pollution, while a variety of types of hybrid vehicles have been created for different functions and uses, with some very serious research going into the development of hybrid systems that combine an internal combustion engine and an electric machine (EM) being the hybrid systems as can be seen in Toyotas vehicle production since 1997 [10].

There are three main architectures for combining power in hybrid vehicles: series, parallel and series-parallel. The main advantages have already been analyzed in different papers such as Tim et al. [11], Wirasingha et al. [12], Miller [13], Ehsani et al. [14] and others are given below.

2.1 *Configuration of Series HEV*

The series configuration (see Fig. 3) was developed by adding a small combustion engine-generator set to a pure electric vehicle, in order to offset the power discharge in the batteries. In Fig. 3, ICE is the combustion engine; EG the generator; EM the electric machine that in most cases operates as an engine for the drive, and Dif., which is the differential for sending the drive to the wheels. This configuration is used in locomotives [15]. The first HEV with a combustion engine was designed by Dr. Jacob Ferdinand Porsche in 1899 for the company Lohner in Austria [16]. The vehicle had a series configuration with electric motor on the front wheels, an internal combustion engine and a generator that supplied the electricity to the motor. It is worth mentioning that Porsche called it a vehicle with a mixed drive system.

The main advantages of the series configuration are: Mechanical decoupling of the combustion engine from the drive wheels, which allows the combustion engine

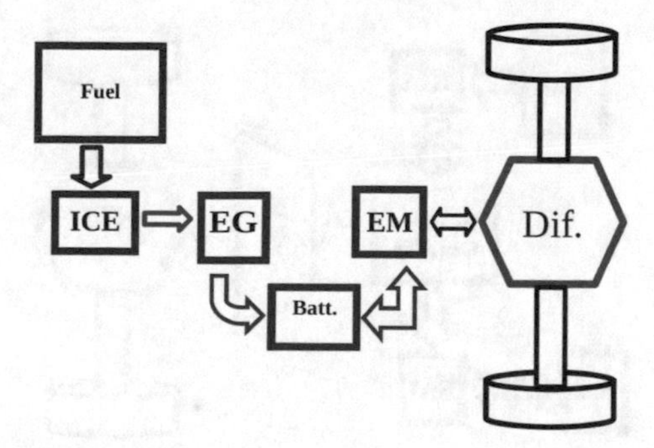

Fig. 3 Series configuration of HEV

to operate in its optimum region. The only source of torque for the drive wheels is an electric motor that simplifies the speed control (similar to the accelerator pedal control). The electric motors almost ideal torque-speed control renders a multi-gear transmission unnecessary and avoids the mechanical components of axles, transmission, gears, clutch, etc. Simple structure, drive control and easy handling (the combustion engine—generator, batteries and the drive motor are only connected by electric cables).

However, the series hybrid drive does have some disadvantages that are listed below:

Dual energy conversion (from the combustion engines mechanical energy into electrical energy through the generator and then into mechanical energy again through the drive motor) that causes more energy losses Two electric machines (a generator and motor) are required. A large electric engine is needed as it is the only source of torque for the drive wheels.

Taking advantage of its structure and the simple control, the series hybrid drive is used in heavy vehicles. The main reason is that large vehicles have enough space for the voluminous engine and generator system.

2.2 Configuration of Parallel HEVs

In the parallel configuration for the hybrid drive system, the internal combustion engine (ICE) and the electric machine (EM) can supply its torque directly to the drive wheels through a mechanical coupling, as shown in Fig. 4, and this chapter has been developed around this concept.

For the parallel configuration, the power flow from the ICE can go to drive only or to drive in combination with the electric machine to recharge the batteries. The power from the ME can go from the batteries to the drive axle in hybrid drive mode, from the ICE to the batteries in hybrid battery recharge mode or from the drive axle to the batteries in regenerative braking mode, when the ICE is freed by using a clutch and brake to decouple it.

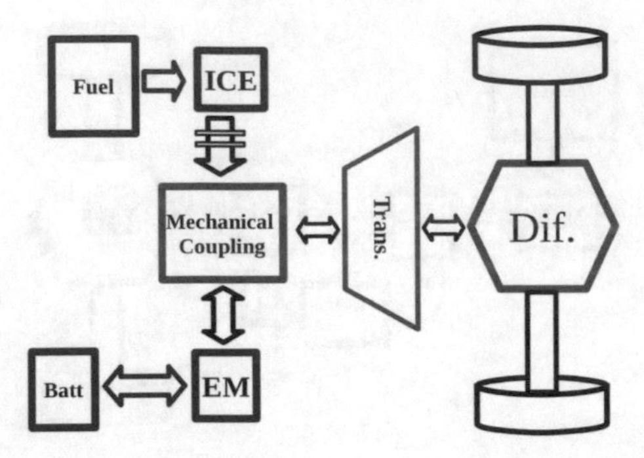

Fig. 4 Parallel configuration of HEVs

The mechanical coupling could be a gearbox, a unit of belts and pulleys, a unit of gears and chains, a planetary gear system (PGS), or even a single axle.

The advantages of the parallel hybrid drive are:

The combustion engine and the electric engine can directly provide the torque to the drive wheels and the dual energy conversion is not produced, so there is less loss of energy. It is more compact, does not need the generator and drive motor, a single electric machine that can perform both functions.

However, it also has disadvantages such as:

The mechanical connection between the combustion engine and the drive wheels, means that the engines operating points are not always in the optimum speed region. The structure and control are more complex.

Owing to its compact characteristics, the parallel configuration is used more in small vehicles, recently being used in the Honda Civic Hybrid 2013 [17], Volkswagen Jetta Hybrid 2013, however, it also is employed in the Volvo 7700 Hybrid Bus, being tested in Mexico City.

2.3 *Series-Parallel HEV Configuration*

This configuration decouples the speed of the combustion engine from the speed of the wheels. It combines the advantages of the series and parallel drives, shown in Fig. 5, however, it needs an additional electric machine (two electric machines EM1 and EM2) and a planetary unit that is the mechanical coupling in Fig. 5, which makes the control of the drive more complicated. Another alternative series-parallel drive is an electric machine with a floating stator (called a transmotor). In this configuration, the stator is connected to the combustion engine and the rotor is connected to the drivetrain of the wheels by means of gears. The motor speed and the relative speed between the stator and the rotor can be controlled to adjust the speed of the combustion engine to any given vehicle speed. This drive has similar operational characteristics to those of a planetary gear drive.

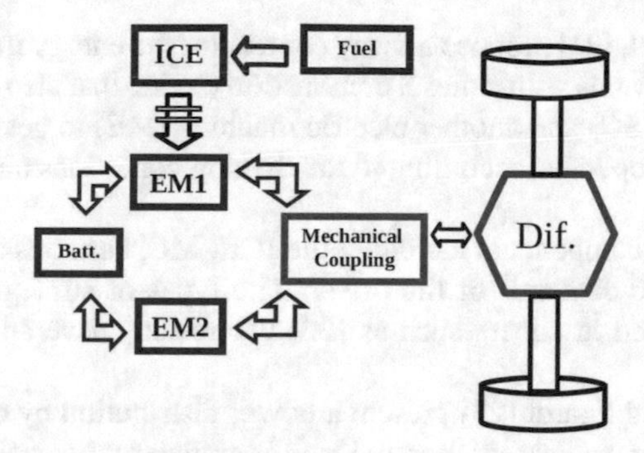

Fig. 5 Series-parallel HEV configuration

The combustion engine, the planetary gear unit and the electric machine constitute the power flow paths. When the electric machine speed is negative (opposite direction to the torque), the machine operates in generator mode. The power of the combustion engine is divided into two parts: one is transferred to the power train and the other to the generator. When the speed of the electric machine is positive, the machine operates in motor mode and adds its power to the drive wheels. This way, the engine speed can be adjusted to its optimum region by means of the electric machines speed control.

The engine generator can be removed from the drive, blocking the engine generators stator and rotor and de-energizing it. This way, the planetary gear unit is converted into a simple gearbox with a fixed gear ratio. Another source of power (torque) is the drive motor EM2 (Fig. 5) that directly adds torque to the wheels [14].

In Mexico the Toyota Prius 2014 is now being marketed with this configuration.

2.4 *Power Distribution in HEV*

Different categories of strategies for power distribution in hybrid electric vehicles have been identified: in the first place, rule-based strategies [18], that may use heuristic, fuzzy logic, neural networks, etc. Another category uses the optimal control theory, where a cost function based on the fuel consumption is minimized. Of these, the dynamic programming (PD) approach, which is generally used for comparing how the strategies perform particularly, stands out. There are strategies that minimize a cost function and that include, apart from fuel consumption, the consumption of electric power.

Since Delprat et al. [19] the aim has been to minimize fuel consumption in hybrid vehicles by power distribution for at least one particular driving cycle, using heuristic methods. Tzeng et al. [20] apply a fuzzy control to a parallel hybrid vehicle with a continuous variable transmission (CVT) of belts and pulleys, the speed/torque ratio is selected for the combustion engine and electric engine by means of servo engines. Xiong et al. [21] propose a fuzzy control for the energy management, for an application in a city bus with a diesel combustion engine that also uses an integrated starter generator (ISG) and another electric machine (ME) to get the series-parallel configuration that operates according to the driving conditions and the needs of the driver.

The size of the components including the ICE, ME, batteries and gear ratio was chosen to meet the demands of the driver. The types of strategies mentioned are still being employed in papers such as [22] where they have an application in an excavator.

Johannesson and Egardt [23] present a power distribution by dynamic programming for a hybrid vehicle of parallel configuration, where simple models are employed for the elements of the vehicle, such as the ICE, battery and electric motor. Using the simplified models of the subsystems that make up the vehicle, they seek to shorten the computational time for the solution as well as for the performance of the iterations.

Koot et al. [24] compare two dynamic programming and quadratic programming strategies that minimize a function of the fuel and the pollution emissions, and find the quadratic programming strategy to be better at lowering pollution emissions. For the simulations, they employ static maps of the engines and engine that link torque, speed, fuel consumption and emissions as well as the dynamics in the state of charge and the vehicle in movement. Paganelli et al. [25] compare control strategies for the power mix in the parallel hybrid electric vehicle; the algorithms are designed to minimize fuel consumption by using the drive torque when changing the gear ratio to work in more efficient regions. Paganelli et al. [26] describe a formulation for the power distribution control problem in hybrid vehicles, which is based on distributing the power between the ICE and the ME as well as properly choosing the gear ratio.

Paganelli et al. [27] and Sciarretta et al. [28] describe an algorithm that distributes the power between the ICE and the ME for the purpose of minimizing fuel consumption. An optimizing criterion is set out that minimizes fuel consumption and proposes an equivalency between fuel and energy, hence its name, Equivalent Consumption Minimization Strategy (ECMS). One interesting detail for the above strategy is that, in order to minimize the desired performance, it is necessary to calibrate some parameters that depend on the driving cycle, on leaving them set and changing the cycle, the battery is discharged or recharged. Moreover, the equivalent consumption factor that is used for evaluating the difference between the initial and final state of charge does not always correspond to the efficiency in the ICEs fuel use.

Delprat et al. [29] propose a strategy for energy management in the parallel hybrid drive for distributing the power between the ICE and the ME, in such a way that it is an efficient tool for evaluating the minimum fuel consumption that can be achieved in simulation. This proposal is based on the classic optimal control theory, using the Lagrange multipliers method. The performance is assessed by means of static maps for the ICE, and the model or static map used is not mentioned for the ME. The battery is taken as a dynamic element, then the analysis and optimization is carried out. Moreover, there is no mention of how the difference between the initial and final state of charge is offset to determine net fuel consumption.

Musardo et al. [30] propose the adaptable version of the algorithm for energy management in hybrid vehicles (A-ECMS), an enlargement on the work of Paganelli et al. [27] and Sciarretta et al. [28]. The A-ECMS algorithm adaptable results from adding a new variable equivalence factor in accord with the driving conditions to the ECMS strategy. In this case the idea is to use outside information, such as a global positioning system (GPS). This strategy is simulated with the parallel hybrid drive, where the idea is to economize on fuel and equivalent fuel consumption to that of electric power is obtained. However, one important problem that is similar to the one in the earlier paper is the relationship between the difference in the state of charge and the fuel consumption where the difference in efficiencies between the ICE and the ME-battery are not taken into account. Moreover, Pisu and Rizzoni [31] compare a rules-based control, the A-ECMS strategy and a H_{∞} control, to get better performance out of the adaptable strategy.

Borhan et al. [32] present a strategy for distributing the power flow in HEVs based on a model predictive control (MPC) for the series-parallel configuration. In order to use the MPC, it is necessary to linearize the model around each operating point that depends on the torque demand and the state of charge. The strategy is solved by means of quadratic programming for the linear system, where the parameters are adjusted during the prediction horizons. Yan et al. [33] present a MPC strategy for a parallel HEV that incorporates the qualities of the diesel engine to mix the efficiencies together with the ME. The strategy is compared to a PI control that only depends on the tracking error, while the MPC has a minimization criterion, so the comparison seems unfair.

Ngô et al. [34] use an MPC algorithm to select the best gear ratio in the transmission to economize on fuel. They mention that the proposed algorithm can function in real time and the comparison of the MPC is against an optimum algorithm that mixes dynamic programming with Pontryagins minimum principle (PMP). For the proposed driving cycles, the optimum strategy produces a saving in comparison to a conventional vehicle of 35.9 to 43.5.

Serrao et al. [35] analyze three main optimum strategies for the power distribution in hybrid vehicles (HEV): dynamic programming, Pontryagins minimum principle and the equivalent consumption strategy ECMS. The last of these three stands out because of the speed of solution, however, as in other papers; they also mention that the ECMS strategy has a problem in the tuning of the parameters for different driving cycles.

Kim et al. [36] report a strategy for series/parallel hybrid electric vehicles using Pontryagins minimum principle. They compare the strategy with dynamic programming and the ECMS. They only include dynamics in state of charge in the batteries. The strategy is related to ECMS. Fixing the fuel-electric energy equivalence they find a 1% difference between ECMS and their PMP formulation.

Yuan et al. [37] present a similar paper to the above, but only compare Pontryagins minimum principle strategy, which was developed in the aforementioned paper, with dynamic programming and it is to be noted that the simulation time is significantly less in PMP than in PD. In the objective function they only include the state of charge and the fuel flow, apart from the fact that the dynamics shown only pertain to the vehicle and batteries. A variation in the efficiency of the electric machine and the total efficiency through the coupling of the components is also considered, however, no mention is made of the efficiency of the combustion engine, which is extremely important. The power is mixed by means of an automated manual transmission that replaces the usual automatic transmission.

2.5 *Modeling the Vehicle*

A hybrid electric vehicle consists of a variety of elements and mathematical models are employed in this research to describe their dynamic response using ordinary differential equations. As the emphasis is on the design of the drive train, the elements being modeled are:

The diesel internal combustion engine, considered to be of primary interest as its performance makes it the most popular combustion engine in freight transport and passenger vehicles The clutch employed to couple and decouples the ICE with the drive system. This mechanical device is still found useful in mechanical systems for joining and separating moving parts. The battery bank, as more and more care is needed in the cars electric power storage, owing to the growing presence of electronic components. The electric machine that has a significant influence on the end performance of a hybrid vehicle. The vehicle in movement, that has an influence on the users acceleration and braking behavior, as well as on the driving cycles employed to test the performance, demand and efficiency of the car. Another system that integrates the hybrid vehicle and is a central part of the chapter is the planetary gear system used to couple the power sources to the drive, which is also modeled with its respective dynamics. The mechanical power that should be contributed by the ICE or the electric machine, as appropriate, is allocated through the planetary gear system to deliver the power demanded by the driver, or the speed and torque setting allocated by the driving cycle in this case, in the drive.

3 Control of the Internal Combustion Engine and Electric Machine

For the control of the main elements that supply power to the vehicle for its movement, techniques used in control theory are employed such as: sliding modes, passivity, feedback linearization, to mention just a few. However, they are only mentioned in this section to give a panorama and have an idea of the variables we are dealing with.

The combustion engine gets its energy from fuel, while the electric machine gets its energy from the battery. Both machines have the mechanical variables of speed and torque in common and this is just how they are coupled for the hybrid electric vehicle. The main idea is for the speed the user desires to be represented by a driving cycle. The goal is to design the control strategies in such a way that, when all the dynamics of the subsystems are integrated, the output speed is equal to the speed of the driving cycle.

By using the control models that have speed as a common variable, this section describes the controllers that make the diesel internal combustion engine and the electric machine keep to the speed that has been allocated to satisfy the demands of the driver, providing the corresponding torque to each one of them. The aim is for the power sources to incorporate their own controllers, first of all so that they obey a specific speed setting and a separate torque setting, and then for them to be coupled to the hybrid vehicles drive system.

For the internal combustion engine, the controller defines the fuel flow that enters the combustion chamber and the speed reference is related to the air flow. Moreover, the load torque must also be offset by this controller. The fuel injector is in charge of dosing the diesel that has to enter the combustion chamber in accordance with the compression /air-fuel mixture of ratio. The comparison and performance of some controllers for the model of a diesel engine are described in Guzman et al. [38], which we used in this chapter for the hybrid vehicle.

The electric machine is controlled by the voltage and current in the stator. When it operates as an engine, it is supplied voltage and current in order for the machine to provide the necessary speed and torque. When it operates as a generator, it is supplied torque and speed in order for it to provide the voltage and current needed to recharge the batteries.

4 Power Flow Control in the Hybrid Drive Train

For practical effects or at least in the context of vehicle systems, the terms power flow control and energy management are interchangeable [39]. However, power is an instant value and energy involves a period of time when the power is applied. Therefore, when we say that power is controlled or distributed, in general we are talking about energy management. Owing to the costs associated with the production and use of energy, more and more research focuses on the production and management of the elements involved. As has already been analyzed in different papers such as Nersesian [40]; Black and Flarend [41], the production, use and management of energy tends to employ renewable sources such as sun, wind, sea, among others, for their conversion into electric power. Whereas the energy obtained by extraction or conversion from biomass, such as ethanol or methane, is similar to gasoline or natural gas. Furthermore, the aim is to always find more efficient ways to use or manage the equipment and systems employed.

All types of hybrid vehicles need a power flow control that determines how to operate each power source to satisfy the demands of the driver, which in this case means the monitoring of the driving cycle with the loads present. Furthermore, one of the main objectives of power distribution is to lower the use of power and, in

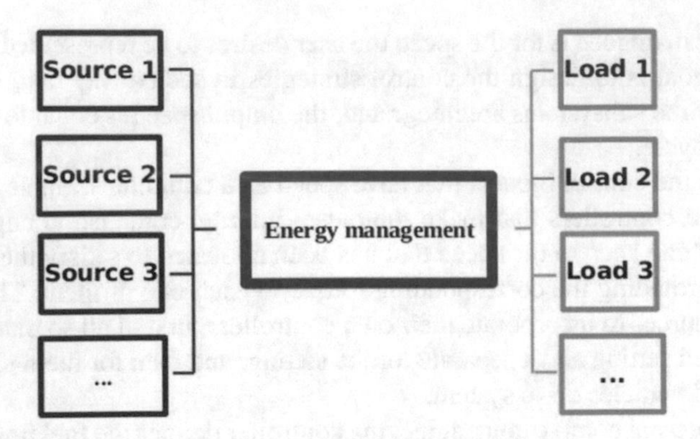

Fig. 6 Power sources

the case of combustion engines, pollution emissions. A general approach is given in Fig. 6 that reproduces a generic system that involves different power/energy sources and loads as well as including the distribution of power to the loads.

As the general idea of a hybrid vehicle is to combine different energy sources, Fig. 6 illustrates the fact that each source can be of a particular type. For example, the sources can be, starting with a combustion engine, that transforms the chemical energy from the fuel into mechanical energy, source 2 could be an inertia drive that transforms kinetic energy into mechanical energy, another source can be a solar cell, that converts solar power into electric power, a battery, (a device commonly used for storing electric power), a capacitor or fuel cell that transforms chemical energy into electric power. In general different types of sources can be used, as they can also be hydraulic, pneumatic, among others.

The power sources can be very similar to each other. For example, sources 1, 2 and 3, battery, fuel cell, super capacitor or even using two combustion engines, one diesel and the other gasoline, to mention just a few.

The loads in Fig. 6 as is usual in vehicle systems, also come in different forms, where the main load is mechanical to meet the need to move a certain mass of vehicle. The electrical loads are lights, electric motors for different applications such as cooling, and the rest of the elements. Other loads, which are usually to be found in the most recent vehicles for the convenience of the users, are mechanical loads, such as an air-conditioning compressor, hydraulic pumps for steering support or electric loads such as a heater, etc.

The power flow distribution and control problem mainly centers on satisfying energy demands (loads) using the power sources, however, from the point of view of energy policy in respect of energy saving, and the control of gas emissions, great care should be taken with the functionality and operation of the power sources in order to choose the best combination for each situation. For example, if we have an electrical load and two sources to supply it, a battery and a supercapacitor, the best way to supply the demanded power is for the battery to operate when the demand is constant and the supercapacitor to operate when dealing with sudden surges (peaks) of power, as a result of the proper performance and operation of each device [42].

Another example occurs when mechanical loads have to be supplemented and sources can be two combustion engines, a diesel one and the other a gasoline one, which have a different efficiency map (in general, a diesel engine is more efficient than a gasoline engine). However, as the efficiency characteristics can vary greatly, in certain torque-speed operating points, a gasoline engine may be more inefficient.

This section presents the chapters proposal and, in that sense, is subdivided into three sections. The first section describes an energy management strategy that is inspired by optimal control, which is a "bang-bang" controller variant that was presented in Becerra and Alvarez-Icaza [43, 44]. It is worth mentioning that this strategy is implementable in real time. It also describes the tuning methodology for getting it to perform well. In the second section we propose a power flow control strategy derived from optimal control that involves Pontryagins principle and gives a more general result than the earlier one which has been the subject of previous papers also presented in Becerra et al. [45, 46]. Now the tuning methodology is also added. In the third section we describe the allocation of speed and torque in the aforementioned strategies, after the power distribution.

4.1 Strategy Using the Planetary Gears System (PGS Strategy)

The PGS strategy is based on distributing the power demanded by the driver through the constraints of the planetary gears system to each power source in addition to the battery state of charge and the improved efficiency of the internal combustion engine. The main idea is derived from the following observations.

1. The most important requirement in the HEVs power flow control is the ability to satisfy the total power demanded by the driver or the driving cycle.
2. All the optimum solutions for the power flow control must preserve the battery state of charge, on average, over a sufficiently long period of time. If the tests are carried out during a driving cycle, the cycle must start and end with the same state of charge in order to corroborate the performance.
3. To minimize fuel consumption, the ICE must operate in highly efficient regions, as when efficiency is maximized, fuel consumption is minimized.

Observation 2, which is key in the strategy, points out that all the optimum solutions that are based on driving cycles preserve the batterys initial state of charge at the end of the cycle, otherwise the vehicle cannot sustain a repetition of the same driving cycle.

A similar observation is made in Musardo et al. [30], when the fine tuning of the adaptable equivalent consumption minimization strategy (A-ECMS) is discussed. Observation 3 can be verified, for example, in Ehsani et al. [14], this is mainly the reason why HEVs are more efficient than conventional vehicles.

The control problem to be solved is how to distribute the power required in the PGS between the two power sources in order to economize on fuel. This problem has multiple solutions, as the combination of torque and speed in each power source may be arbitrary while the demanded power is being obtained. The above observations result in the approach that involves the equations to be solved, that acknowledges that, in order to find the solution, we should resort to the PGS model, given by Eqs. (1), (2)

$$P_p = T_{em}\omega_{em} + T_{ice}\omega_{ice} \tag{1}$$

$$\omega_p = \frac{1}{(k+1)}\omega_{em} + \frac{k}{(k+1)}\omega_{ice} \tag{2}$$

where P is the power; T, the torque; ω, the speed; k, the PGSs gear ratio constant and the subscripts p, *ice* and *em* represent the planetary holder, ICE and EM, respectively. Considering that P_d is the power required by the vehicle's drive to meet the driver's requirements (demanded power) and P_p is the power that both machines contribute between them.

Two strategies are employed to lower fuel consumption: using the ME as much as possible and operating the ICE at the maximum possible efficiency. Assuming that the state of charge in the batteries can be taken as a reference value, in the case of drive, $P_p \geq 0$, the cost criterion to be used derives from the bang-bang optimal control [47], and is as follows

$$J_1 = max \int_0^{T_c} (sign(P_p)sign(soc - soc_{ref}))P_{em}dt \tag{3}$$

where T_c is the length of time of the driving cycle, soc_{ref} is the soc reference and P_{me} the power of the ME. This expression is used in the case of drive and the battery-recharging drive. In other words, any time the user demands power.

When it is necessary for the vehicle to slow down, in other words, in the case of braking, $Pp < 0$, the criterion employed must change slightly while conserving the original idea, using the function that involves the regenerative braking, with which the idea of employing the electric machine is conserved, so now the criterion is as follows:

$$J_2 = max \int_0^{T_c} (sign(P_p))P_{em}dt \tag{4}$$

Using the criteria to be maximized, Eqs. (3) and (4) as well as the kinematic power constraint because of the coupling (1), the power that each source of energy should contribute is decided using the methodology described below.

4.1.1 Power Allocation

The power distribution strategy consists of two main parts: first the power from each machine is allocated according to the battery state of charge, to avoid discharging or overloading them. Initially the power corresponding to the electric machine is allocated, taking into account the maximum power it can contribute. Then the power from the ICE supplements the total demanded power. The second part consists of allocating the angular speeds: in this case, starting with the ICE and ending with the EM.

Electric Power Allocation P_{em}

In the first place, the power corresponding to the electric machine is allocated, therefore the value of Eqs. (3)–(4) is maximized when,

$$P_{em} = min\left\{sign(P_p)P_p, sign(P_p)P_{em}^{max}\right\}$$

where P_{em}^{max} is the maximum power of the EM (assumed to be equal in the case of the motor and generator). However, on employing a control of this type, also known as a maximum strain control, the vehicle and its components, in general, are submitted to severe mechanical stress that can cause, apart from damage or failure in the components, serious accidents. To avoid the abrupt change induced by the function $sign(P_p)$ a soft function is used that depends on the state of charge in the batteries. The above is aimed at avoiding the aforementioned problems and the new constraint is as follows:

$$P_{em} = P_{em}(soc) = \alpha_i(soc)P_{em}^{max}, \tag{5}$$

where subscript i in Eq. (5) is 1 when $Pp \geq 0$ and 2 when $Pp < 0$, clearly $\alpha_i \in [-1, 1]$, as we must not exceed the maximum strain that the electric machine can provide. Note that the criteria in Eqs. (3) and (4) are represented by Eq. (5).

We take as known the power P_p and speed ω_p demanded by the driver that, in a conventional vehicle with the sensors being used nowadays, are known variables or come from a driving cycle. The proposed solution for the power flow control problem starts by substituting Eq. (5) in Eq. (1), in other words

$$P_p = \alpha_i P_{em}^{max} + P_{ice} \tag{6}$$

The value of α_i depends on the drive or braking power in planetary holder P_p, as well as the value of the state of charge in the batteries. The form of $\alpha_i(soc)$ determines how much electric power is absorbed or delivered at a given power point. One possible form for $\alpha_i(soc)$ is given in Fig. 9 that is described by,

$$\alpha_1 = \tanh(A_1(soc - soc_{ref})) \qquad Pp \geq 0 \tag{7}$$

$$\alpha_2 = 0.5 - 0.5(\tanh(A_2(soc - soc_{fin}))) \qquad Pp < 0 \tag{8}$$

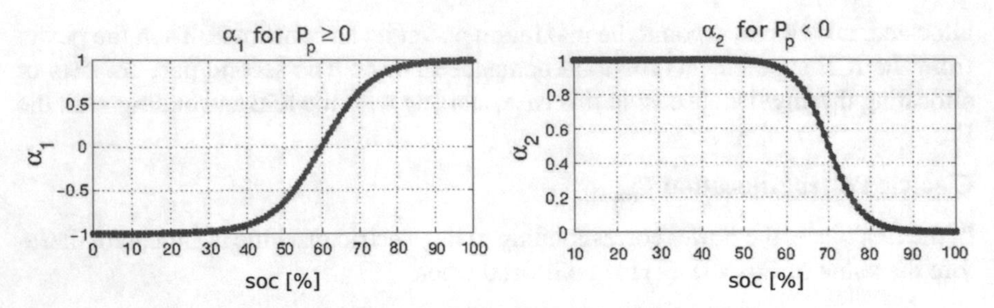

Fig. 7 Distribution of electric power versus battery state of charge (*soc*)

where soc_{ref} is a reference value for the battery state of charge for this status to stay around that point and soc_{fin} is a reference value to avoid the battery from being overloaded during regenerative braking.

Figure 7 reveals that when $Pp \geq 0$, $\alpha_i \in [-1, 1]$, depending on the soc; if 1 is positive, the electric machine operates as a motor, otherwise it operates as a generator. When the power in the planetary holder, $Pp < 0$ requires the vehicle to be braked, $\alpha_i \in [0, 1]$, regenerative braking is possible and the EM can only operate as a generator until the batteries are totally recharged. The rest of the power is dissipated through the friction brakes as described below.

Allocation of Mechanical Power (P_{ice}) and Friction Brakes

Using the function α_i, the electric power in the balance equation in the planetary holder, Eq. 1, 1 is set. The power of the combustion engine (P_{ice}) is obtained from the same balance, as follows:

$$P_{ice} = min(P_p - P_{em}, P_{ice}^{max}); \qquad P_p \geq 0 \tag{9}$$

where P_{ice}^{max} is the maximum power value that the combustion engine can contribute and in Eq. (9) it is used to indicate that the ICE is supplying the power up to its maximum capacity. The ICE acts when power is need to drive the vehicle or when the batteries are discharged. When it is necessary to brake the vehicle, the ICE's power is exchanged for the friction brakes, in other words, in the case of regenerative braking, $P_p < 0$, the generator has a limit on the power it can recover and therefore the rest of the power is dissipated through the brakes, which is described by the equation,

$$P_{Fr} = max\left\{0, P_p - \alpha_2 P_{em}^{max}\right\} \tag{10}$$

where P_{Fr} is the power dissipated through the friction brakes that only depend on the power that the generator can recover and how fast the vehicle stops.

4.1.2 Tuning the PGS Strategy

The tuning of the PGS strategy has been detailed since its presentation in Becerra and Alvarez-Icaza [43], where we presented a preliminary paper and proposed an initial form of curves α_i and the parameters used. However, in Becerra et al. [44], a series of proposals is discussed for the choice of parameters, the variation of the state of charge and fuel saving.

After various discussions and analysis of the simulations, we make a proposal for the election of the four parameters in the PGS strategy, that consists of first allocating curve α_1. The first parameter $soc_{ref} = 60\%$, was proposed as per Chaturvedi et al. [48] and Romero-Becerril and Alvarez-Icaza [49], who argue that the ideal operation for any battery must be around half the state of charge, however, a slightly higher level is proposed to prevent the battery being discharged as a consequence of some mishap.

Parameter A_1 determines how soft or abrupt curve α_1 is, which is directly related to the operation of the electric machine in generator or motor mode. In the case of the vehicle's drive, we proposed $A_1 = 0.4$ that permits a soft operating mode for the electric machine and avoids damaging the components through mechanical stress.

Alternatively, curve α_2 should be allocated for the braking mode in order to determine how much power can be recovered in the regenerative braking. Parameter A_2 now implies how steep the aforementioned curve is and soc_{fin} is the turning point, which is incorporated to avoid overloading the batteries. The parameters of the second curve α_2 are not very strict, as it will not be possible to recover more energy than the maximum the generator can operate or the exceed the maximum capacity of the batteries, so it may be steeper or softer and only be careful of the overload. $A_2 = 0.13$ and $soc_{fin} = 63.5$ were chosen in this chapter.

4.2 *Power Distribution by Optimization*

The PGS strategy that was mentioned in the above section, is an implementable strategy, however there is no way of comparing it with the best value that can be obtained when it is implemented. Therefore, in order to compare the proposed strategy with the optimal control approach, we look for a solution of the objective function with its respective constraints by using an optimum solution.

The criterion to be optimized is based on the observations made for the proposed strategy, where the biggest problem is setting out the problem and its constraints in terms of variables that permit its solution. We choose to use Pontryagins principle, mainly because out of all the optimization strategies, it is the one that enables us to get solutions that can be implemented in real time.

Calculation of variations and Pontryagin's minimum principle. As has already been described in Pantoja-Vazquez et al. [50] that recalls that most of the energy management strategies developed by means of optimal control use dynamic programming, and are mainly used for testing other strategies. This is due to their high computational load that increases with the complexity of the models employed for the vehicle.

Furthermore, implementable controls can be obtained using Pontryagin's minimum principle. This is why this chapter uses this, the usual formulation, with some important constraints that are described in this section. The development of the structure and the problem was inspired by the book of Kirk [51], which however uses another model and objective function. The problem to be solved is to find an admissible control $u^* \in U$ that makes a system like the following:

$$\dot{x}(t) = a(x(t), u(t), t) \tag{11}$$

It follows an admissible path $x^* \in X$ that minimizes a performance formulated using the structure

$$J(u) = h(x(t_f), t_f) + \int_{t_0}^{t_f} g(x(t), u(t), t)dt \tag{12}$$

where g and h are assumed to be soft functions, the initial conditions are specified, $x_{t_0} = x_0$, as is the initial time, t_0. As is usual in control systems, x is a vector of n state variables and u is a vector with m control inputs.

Assuming that h is a differentiable function, the aforementioned functional (12) can be expressed as follows:

$$J(u) = \int_{t_0}^{t_f} \left\{ g(x(t), u(t), t) + \left[\frac{\partial h}{\partial x}(x(t), t)\right]^T \dot{x} + \frac{\partial h}{\partial x}(x(t), t) \right\} dt \tag{13}$$

By introducing constraints for the systems differential equations, the objective function argument changes through the Co-states

$$p(t) = [p_1(t) \quad p_2(t) \quad p_3(t) \quad \ldots \quad p_n(t)]$$

For the augmented system to have the form

$$g_a(x(t), \dot{x}(t), p) = g(x(t), u(t), t) + p^T(t)[a(x, u, t) - \dot{x}(t)] \tag{14}$$

On adding the co-states as described in Eq. (14), the objective function is also affected and should incorporate the co-states, through the constraints, equal to end up as follows:

$$J_a(u) = \int_{t_0}^{t_f} \left\{ g(x(t), u(t), t) + \left[\frac{\partial h}{\partial x}(x(t), t)\right]^T \dot{x} + \frac{\partial h}{\partial x}(x(t), t) + p^T(t)\left[a(x(t), u(t), t) - \dot{x}(t)\right] \right\} dt \tag{15}$$

For systems with the aforementioned characteristics and objective function as obtained above, it is possible to define a Hamiltonian with the following structure,

$$H(x(t),u(t),p(t),t) = g(x(t),u(t),t) + p^T(t)\left[a(x(t),u(t),t)\right] \tag{16}$$

With the Hamiltonian of the function, the control that minimizes the objective depends on admissible states x^*, co-states p^* and time t. By using the Hamilton-Jacobi Bellman equation, it is possible to write the necessary conditions, to find states x^*, inputs u^* and co-states p^*, optimum that lead to the solution being sought.

They have a direct dependency with the Hamiltonian for finding the respective dynamics, as follows:

$$\dot{x}^*(t) = \frac{\partial \boldsymbol{H}}{\partial p}(x^*(t),u^*(t),p^*(t),t) \tag{17}$$

$$\dot{p}^*(t) = -\frac{\partial \boldsymbol{H}}{\partial x}(x^*(t),u^*(t),p^*(t),t) \tag{18}$$

$$0 = \frac{\partial \boldsymbol{H}}{\partial u}(x^*(t),u^*(t),p^*(t),t) \tag{19}$$

4.2.1 Approach to the Problem for the Optimum Allocation of Power

The main idea is to obtain a formulation like the one shown in the above section and find the conditions of the inputs, states and co-states that correspond to the constraints, using the dynamics of the complete system of the hybrid electric vehicle.

Of the power allocation priorities that we have already indicated, the power constraint of the planetary gears system that joins the hybrid vehicles power sources must be considered, Eq. (1). Complying with this constraint is one of the main requirements in the vehicle, as the power that the machines provide must meet the power demanded by the driver (in this case the driving cycle must be satisfied), in other words, Eq. (11) shall be employed further on to formulate one of the states of the system to be optimized.

The following parameter of interest is the battery state of charge *soc*, which should be preserved in order to avoid overloading or discharging the battery, also being a primary element in the hybrid vehicle. The soc is expressed in terms of power, so that it can also be used as a state of the complete system as well as to make sure that its contain similar units to the previous one. Therefore we have,

$$soc = soc_0 - \frac{1}{V_{bat}Q_{nom}}\int_0^t P_{bat}dt$$

$$\dot{soc} = -\frac{1}{V_{bat}Q_{nom}}P_{bat} \tag{20}$$

The power of the electric motor in respect of the battery is considered to be affected by the efficiency η_{em} of both components, in other words, the efficiency between the battery bank and the electric machine, i.e.

$$P_{em} = \eta_{em} P_{bat} \tag{21}$$

The third and no less important parameter to be considered is fuel, which is one of the main reasons for making the vehicle hybrid. Therefore, from the internal combustion engines output power, fuel flow m_f is found as follows:

$$\dot{m}_f = \frac{P_{ice}}{\eta_{ice} P_{th}} \tag{22}$$

Now then, by using the above Eqs. (20)–(22), the dynamics of the hybrid electric vehicles system are expressed using the following approach. The inputs vector is considered to be $x^T = [soc \quad m_f \quad P_p]$, while the power of each motor $u^T = [P_{bat} \quad P_{ice}]$ is taken for the inputs vector. Using the three states, it is possible to express the complete system in terms of the state errors $e_{P_p} = P_p - P_d$, $e_{soc} = soc - soc_{ref}$ and $e_{m_f} = m_f - m_{f_{ref}}$, where P_d is the power demanded by the driving cycle (driver of the vehicle), soc_{ref} and $m_{f_{ref}}$ are the soc and m_f references that are treated as fixed. Therefore, the dynamics of the system are expressed in the following way,

$$\begin{aligned} e_{P_p} &= \eta_{em} P_{bat} + P_{ice} - P_d \\ \dot{e}_{soc} &= -\frac{1}{V_{bat} Q_{nom}} P_{bat} - s\dot{o}c_{ref} \\ \dot{e}_{m_f} &= \frac{P_{ice}}{\eta_{ice} P_{th}} - \dot{m}_{f_{ref}} \end{aligned} \tag{23}$$

One important detail is that the efficiency of the ICE depends on the region of operation and the fuel consumption, which can be minimized if the efficiency is maximized. This can be expressed in terms of the power or torque and the speed at which the combustion engine operates, in other words

$$\eta_{ice} = \eta_{ice}(P_{ice}, \omega_{ice}) \tag{24}$$

Based on Eq. (24), the optimal control problem is defined if the efficiency of the combustion engine is maximized as a function of the speed

$$\bar{\eta}_{ice}(P_{ice}) = \max\nolimits_{\forall \omega_{ice}} \{\eta_{ice}(P_{ice}, \omega_{ice})\} \tag{25}$$

That corresponds to finding the maximum possible efficiency for some P_{ice}, assuming that the angular velocity ω_{ice} can be adjusted to that maximum.

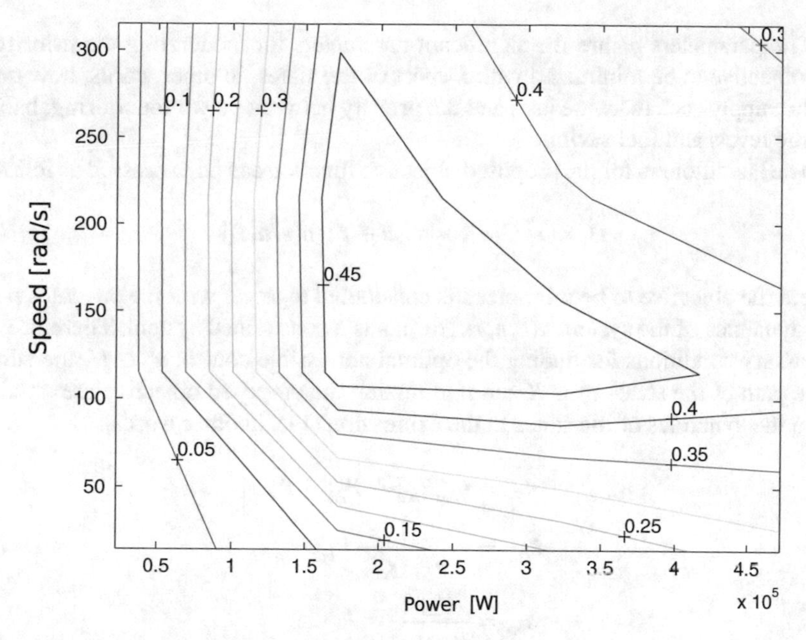

Fig. 8 Power/speed/efficiency curves

One example of the dependence of speed on power is given in Fig. 8, where the maximum efficiency can be chosen for each power. This Power/speed/efficiency map corresponds to the combustion engine model that we presented in the second section.

$$\min J = \int (e^T G_1 e + u^T G_2 u) \quad dt \tag{26}$$

For the system being approached in terms of error, see Eq. (23), we propose a quadratic cost function in terms of the states $e^T = [e_{P_p} \quad e_{soc} \quad e_{m_f}]$ and the inputs $u^T = [P_{bat} \quad P_{ice}]$, which are the energy sources. Therefore, the following objective function is minimized. The cost function in Eq. (12) can be written as

$$\min J = \int (e^T G_1 e + u^T G_2 u) \quad dt \tag{27}$$

where matrices G_1 and G_2 are chosen as follows:

$$G_1 = \begin{bmatrix} g_{11} & 0 & 0 \\ 0 & g_{12} & 0 \\ 0 & 0 & g_{13} \end{bmatrix} \qquad G_2 = \begin{bmatrix} g_{21} & 0 \\ 0 & g_{22} \end{bmatrix}$$

The parameters g_{ij} are the significant parameters for moderating the priority of the objective to be minimized with respect of the states, in other words, how much of the supply each machine uses and the priority between power monitoring, battery charge levels and fuel saving.

The Hamiltonian for the proposed objective function can be expressed as follows:

$$H_e = e^T G_1 e + u^T G_2 u + p^T[a(e, u, t)] \tag{28}$$

where the objective to be minimized is considered together with the co-states p and the dynamics of the system $a(e, u, t)$, for this is a constrained system. Therefore, the necessary conditions for finding the optimal admissible control $u^* \in U$, the admissible path of the states $e^* \in X$ and minimizing the proposed objective are obtained from the dynamics of the states in the expression (17), in other words,

$$\begin{aligned} e_{P_p} &= \eta_{em} P_{bat} + P_{ice} - P_d \\ \dot{e}_{soc} &= -\frac{1}{V_{bat} Q_{nom}} P_{bat} \\ \dot{e}_{m_f} &= \frac{P_{ice}}{\bar{\eta}_{ice} p_{th}} \end{aligned} \tag{29}$$

where the ones derived from the soc and m_f references are made equal to zero by assuming that they are constants.

Equation (18) is used for the co-states and their dynamics turn out to be

$$\begin{aligned} \dot{p}_1 &= -2g_{11} e_{P_p} \\ \dot{p}_2 &= -2g_{12} e_{soc} \\ \dot{p}_3 &= -2g_{13} e_{m_f} \end{aligned} \tag{30}$$

Expression (18) is used to find the inputs and the constraints end up as

$$\begin{aligned} 0 &= 2g_{21} u_1 + p_1 \eta_{em} - p_2 \frac{1}{V_{bat} Q_{nom}} \\ 0 &= 2g_{22} u_2 + p_1 + p_3 \frac{1}{\bar{\eta}_{ice} p_{th}} \end{aligned} \tag{31}$$

The inputs are found for Eq. (30) to obtain control u that produces the optimum admissible path. In other words:

$$u_1^* = -\frac{p_1\eta_{em} - p_2\frac{1}{V_{bat}Q_{nom}}}{2g_{21}} \qquad (32)$$
$$u_2^* = -\frac{p_1 + p_3\frac{1}{\bar{\eta}_{ice}P_{th}}}{2g_{22}}$$

If Eq. (31) are substituted for the optimal control $[P_{bat} \quad P_{ice}]^T = [u_1^* \quad u_2^*]^T$ in the states of the Eq. (28), the following simultaneous equations are obtained, three for the states and three for the co-states,

$$\begin{aligned} e_1^* &= \eta_{em}u_1^*(p_1,p_2) + u_2^*(p_1,p_3) - P_d \\ \dot{e}_2^* &= -\frac{1}{V_{bat}Q_{nom}}u_1^*(p_1,p_2) \\ \dot{e}_3^* &= \frac{1}{\eta_{ice}P_{th}}u_2^*(p_1,p_3) \\ \dot{p}_1^* &= -2g_{11}e_1^* \\ \dot{p}_2^* &= -2g_{12}e_2^* \\ \dot{p}_3^* &= -2g_{13}e_3^* \end{aligned} \qquad (33)$$

In Eq. (32) the system distributes the power between the sources EM and ICE optimally in accordance with the chosen parameters in the objective function. In a similar way to the PGS strategy, in the case of braking, the energy in excess of the maximum possible from the regenerative braking should be dissipated by using Eq. (10). Moreover, the speed and torque in each machine is distributed in a similar fashion to the distribution in the PGS strategy and is described further on.

4.2.2 Tuning of PMP Strategy

One important problem in the solution of power distribution in a hybrid vehicle is the selection of parameters for adjusting its operation during the route chosen by the driver.

There are five parameters for the Pontryagin solution strategy, of which the first three correspond to power monitoring, state of charge and fuel consumption errors, while the two remaining parameters directly correspond to the considered inputs of power from the machines.

The parameters g_{ij} of cost function (26) are tuned by trial and error taking into account the fact that the intention is to save fuel, preserve the state of charge and monitor the driving cycle. The final choice is given in Table 1.

Table 1 Parameters in an optimum PMP strategy

g_{11}	g_{12}	g_{13}	g_{21}	g_{22}
20	2	1	0.5	0.0613

4.3 *Allocation of Speed and Torque*

Once the power from each machine is allocated, we look for their speed and torque. Owing to the fact that, in general, a combustion engine is less efficient than an electric motor, the first thing to be considered is the allocation of the ICEs speed and torque in order to get the best advantage from it.

Given the power from the least efficient source allocated in the above section is, in this case, the power of the internal combustion engine P_{ice}, the angular velocity at which the ICE should operate, ω_{ice}, is obtained from the efficiency map in the power-speed graph. This curve has a similar shape to the most highlighted superimposed line that is shown in Fig. 9.

For the chosen combustion engine, with the model and the parameters that have already been presented in section two, by obtaining the efficiency curves for each power value required, it is possible to find the curve that stands out in Fig. 9 that corresponds to the speed value that is more efficient for a given power value and

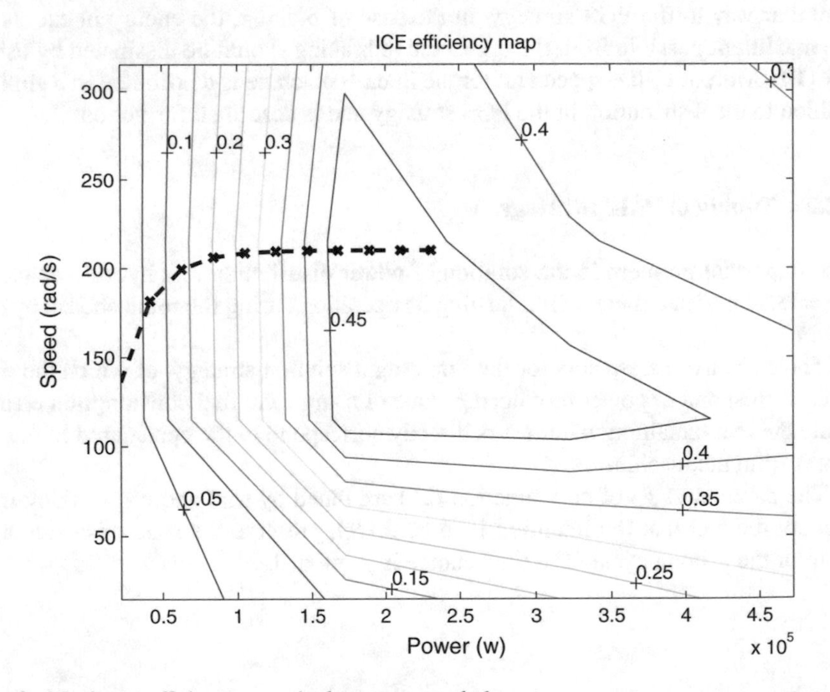

Fig. 9 Maximum efficiency curve in the power-speed plane

corresponds to the solution of Eq. (25). The curve was approximated by using the function

$$\omega_{ice} = 210(1 - e^{-(1/21000)*P_{ice}}) \quad (34)$$

For the combustion engine that is employed in this chapter, the maximum power at which it can operate is $205[kW]$, for this reason the curve is truncated when it reaches this power. Once the engine speed ω_{ice} is obtained by means of the high efficiency curve, the next step is to determine the angular velocity of the electric machine for Eq. (34)

$$\omega_p = \frac{1}{(k+1)}\omega_{em} + \frac{k}{(k+1)}\omega_{ice} \quad (35)$$

The Eq. (34) represents the sum of the velocities for each machine, for the mechanical coupling; therefore, ω_{em} is determined by using the following Eq. (35)

$$\omega_{em} = \frac{(k+1)}{k}(\omega_p - \frac{1}{(k+1)}\omega_{ice}) \quad (36)$$

Once the speeds of the mechanical power sources are allocated, the torque that the ICE should provide is obtained from the equivalence of mechanical power in a rotational system, by using Eq. (36)

$$T_{ice} = \frac{P_{ice}}{\omega_{ice}} \quad for \quad \omega > 0 \qquad and \qquad 0 \quad for \quad \omega = 0 \quad (37)$$

Using a similar technique to the above, the torque T_{em} of the electric machine is determined that, in the case of drive, can operate as a motor or generator, as follows:

$$T_{em} = \frac{P_{em}}{\omega_{em}} \quad (38)$$

Moreover, in the case of regenerative braking, when braking is required, in other words, $P_p < 0$, the power of the combustion engine is not involved, $P_{ice} = 0$; and the electric power is directly recovered. The speed and torque that need to be dissipated are obtained by similar means to the method shown above,

$$\omega_{Fr} = (k+1)(\omega_p - \frac{k}{(k+1)}\omega_{em})$$

$$T_{Fr} = \frac{P_{Fr}}{\omega_{Fr}}$$

5 Simulation Results

We present the simulations obtained from coupling the models of each element for a hybrid electric vehicle. The development of this chapter has focused on city buses, as their manufacturing process is still manual, unlike compact vehicles, such as Toyota, Honda, Ford, Audi, to name but a few, where the manufacturing process is highly automated and it is hard to imagine being able to compete with such firms. Moreover, their technologies are so protected it is hard to find any divulgation in the literature. Whereas in the case of buses, there has not been much distribution yet and the first prototypes are in the testing stage.

The idea is for the city bus to have driving cycle monitoring in the city and for this to be used to analyze the performance of the strategies being put forward for the power distribution in the hybrid electric vehicle. We are mainly proposing driving cycles for Mexico City that have already been already in Sect. 2 for the speeds: slow, moderate and high.

The proposed city bus has a total mass of 15,000 kg, where the mass in components is considered to be 7,000 kg and the variable load up to 8,000 kg. The internal combustion engine is diesel with compression ignition, having 205 kW capacity, 45% maximum efficiency, that is coupled to a planetary gears system, with a gear ratio of $k = 5$, by means of a clutch system to the sun gear. The electric machine is an brushless DC motor with 93 kW capacity, 92% maximum efficiency, powered by the battery bank with a capacity of 25 Ah, at 288 V.

First of all, the simulations of the machines (power sources) were carried out separately. As the speed monitoring is done through the vehicle, tests were carried out with speed references for the machines using the respective controllers presented in Sect. 3, where each machine model was checked to make sure that the speed and torque settings that had been set were being followed. In this section the respective signals are set that they should follow and that altogether result in the power demanded by the driving cycle. Then a comparison of the PMP and PGS strategies is considered. First of all we analyze the most important requirement, the demand of power, which is checked using the tracking of the driving cycles. Later the results of the behavior of the state of charge in the batteries and the fuel consumption are analyzed.

5.1 *Tracking of the Hybrid Vehicles Routes*

Simplified and detailed models of the machines are used to test the power flow control strategies. The tests are performed as follows: first, the driving cycle, which is considered to be equivalent to the demand of the driver, determines the angular velocity and torque needed in the wheel (affected by its radius). These variables are borne, through the differential ratio, to the transmission and the planetary gears

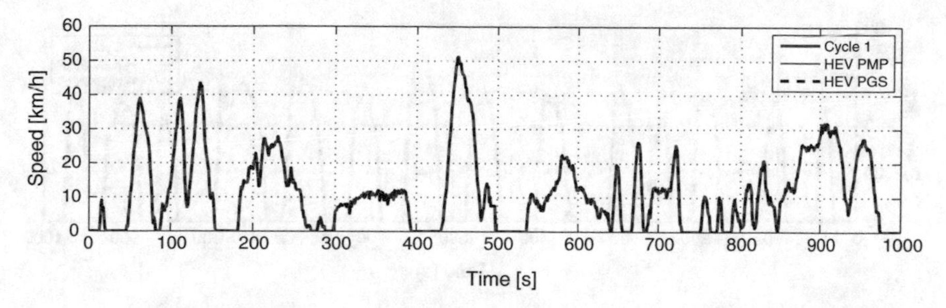

Fig. 10 Slow speed driving cycle

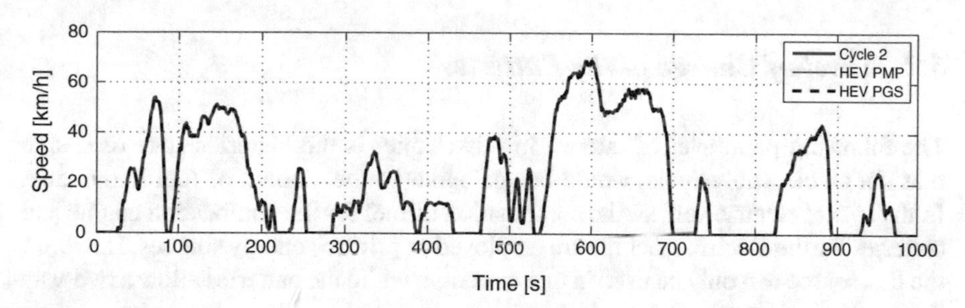

Fig. 11 Moderate speed driving cycle

system, to the axles of each machine, which then receive the required power value (speed-torque settings) to comply with the driving cycle.

Figures 10 and 11 show the tracking of the hybrid electric vehicle at slow and moderate speed driving cycles, called cycle 1 and 2, respectively, in Mexico City that correspond to the buses that drive through the City when there is a considerable and moderate amount of traffic. Both the power flow control strategies, the PGS strategy and the PMP strategy, are employed. We can see that there is no difference between them and that by using the two strategies the hybrid vehicle manages to follow the desired route.

The first route, called cycle 1, has an approximate maximum speed of 50 km/h and an average of under 20 km/h, as well as a very irregular speed because of traffic conditions. The second route to be followed corresponds to cycle 2 with an approximate maximum speed of 70 km/h, faster than the first, and an average speed of 25 km/h, which is also faster.

Figure 12 shows the tracking of the hybrid vehicle on the route posed for the bus driving cycle along bus lanes that corresponds to high-speed cycle 3 in Mexico City. In this case the maximum speed reached is 70 km/h and the average speed is higher than the above as are the levels of acceleration and demand of power that will be shown in the following sections.

As can be verified in the simulations, the two proposed strategies comply with the first demand, as the power mix in the machines manages to move the city bus at the desired speed and acceleration, without any problem whatsoever.

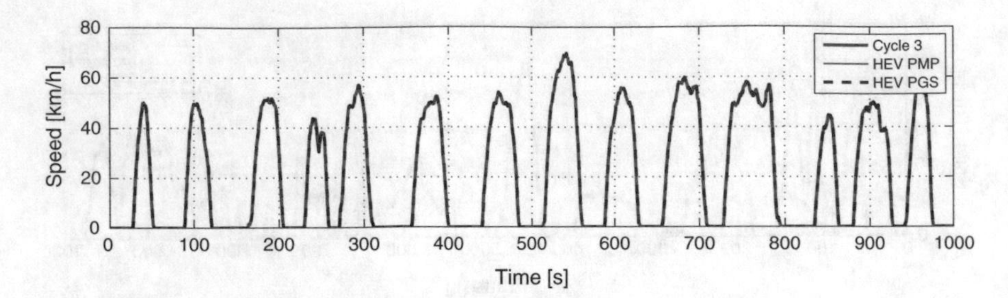

Fig. 12 High speed cycle

5.2 *Level of Charge in the Batteries*

The following parameter of interest in this chapter is the batteries state of charge that, for an electric vehicle, would be analogous to the amount of fuel in the tank. In the hybrid electric vehicle the combination of fuel for the combustion engine and batteries for the electric machine are employed as primary energy sources. However, the first source can only be used in one direction, while the batteries allow a two-way flow.

Numerous simulations were carried out for different conditions in each driving cycle, only keeping the results where the end state of charge coincides with the initial state of charge, while discarding the rest of the simulations. This is because the vehicles real fuel consumption is shown and we thus avoid the need to adjust a (different initial from the final) state of charge, by means of the difference and a factor of equivalence to chemical fuel that can slant the end result.

Figures 13 and 14 show the dynamics of the state of charge in the battery bank for the bus in the tests of the driving cycles at slow and moderate speed, respectively, called cycle 1 and 2. The main differences are because of the forms of the corresponding driving cycle and it is worth mentioning that the behavior of the two PGS and PMP strategies is very similar, under the proper tuning.

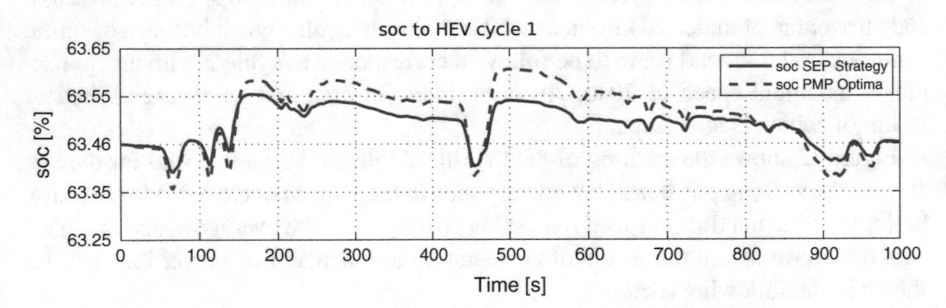

Fig. 13 State of charge in the battery bank for the bus in the tests of the driving cycles at slow speed

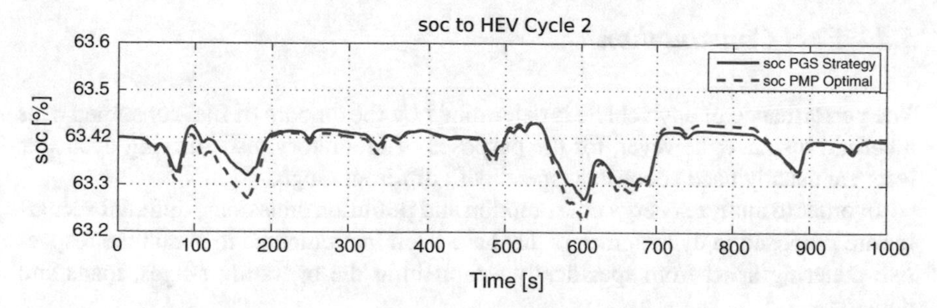

Fig. 14 State of charge in the battery bank for the bus in the tests of the driving cycles at moderate speed

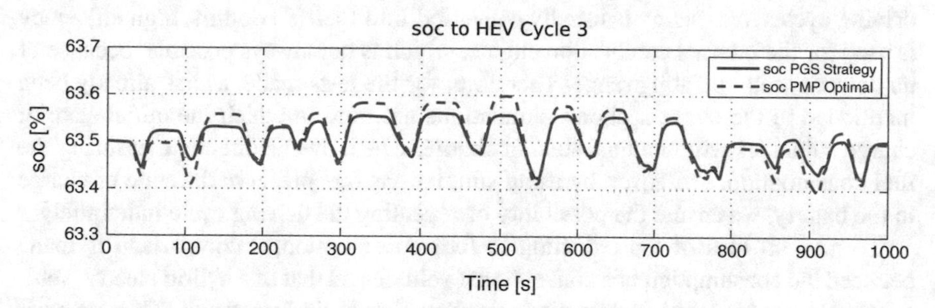

Fig. 15 Dynamics of the soc for driving cycle 3, for the bus when the route is along bus lanes in Mexico City

It was necessary in the two driving cycles to start at a much lower value than the reference value for the driving profile and energy recovery, to fulfill the constraint of preserving the state of charge, in other words to start and end with the same value. In cycle 2, Fig. 16, we can better appreciate how the regenerative braking profile directly shown in the *soc* is less because of the protracted slowing down.

Figure 15 shows the dynamics of the soc for driving cycle 3, corresponding to the bus when the route is along bus lanes in Mexico City. For this driving cycle, the initial and final status are equal to the parameter $soc_{ref} = 63.5\%$.

One interesting point in the state of charge simulations that is possibly the most visible one for comparing the results between the PMP strategies and the PGS strategy is the similarity of dynamics, on top of what has already been shown, i.e. compliance with the driving cycle. The behavior of the soc is practically the same, with the difference lying in some abrupt changes.

The strategies simulations use the same tuning and the change in driving cycle can be seen in the final value of coincidence of *soc*. This is obtained when the cycle tests are repeated over and over again, using the final state of charge from the previous test as the initial state of charge and so on until the desired coincidence is achieved.

5.3 Fuel Consumption

The performance of any vehicle is determined by the amount of fuel consumed over a certain distance; however, for the purposes of laboratory tests, driving cycles or tests are usually used at constant speed with different ranges.

In order to analyze energy consumption and pollution emissions, different scenarios are proposed in dynamometers that have been instrumented to obtain the respective metering apart from specifically establishing the operating ranges, loads and demands.

One difference that this chapter has in respect of other published material in the literature is that in other papers the battery does discharge or recharge during the driving cycles and energy is usually converted into fuel. To do this, high efficiency is used for the internal combustion engine, which is not always possible, because of its power-speed operating range. Therefore, for the tests made, as has already been mentioned in the above sections, simulations are used in which the initial state of charge coincides with the final state of charge, which means directly measuring the fuel consumption. Moreover, by using simulations that preserve the state of charge in the battery, we ensure the possibility of repeating the driving cycle indefinitely.

To give an idea of the reduction in fuel consumption, a comparison is made between the consumption of a conventional vehicle and that of a hybrid electric vehicle operated under the aforementioned power distribution strategies. The comparison is given in Table 2, whose first column gives the consumption corresponding to the conventional vehicle, with only a combustion engine. The second column corresponds to the heuristic PGS strategy. The third column is the result of using the simplified solution from the optimum PMP strategy. In the fourth column the PMP strategy is assessed using simulations of the detailed models of the electric machine and engine.

In every case, the tuning parameters presented in Table 2 are used. The simulation in fuel consumption for PGS and PMP shows that the difference between them is very small, between 1% and 2.4%.

The distance between the simulated solution PMP (PMP-simp.), compared with the detailed PGS strategy solution is even shorter as it is now between 0.4 and 1.8%.

Table 2 Fuel consumption with $soc_0 = soc_f$

		Only ICE	PGS	PMP	PMP-sim.
Cycle	kg	2.158	1.803	1.752	1.762
MX1	%	100	83.54	81.18	81.65
Cycle	kg	2.803	2.222	2.194	2.205
MX2	%	100	79.26	78.27	78.67
Cycle	kg	4.549	2.734	2.692	2.716
MX3	%	100	60.10	59.17	59.71

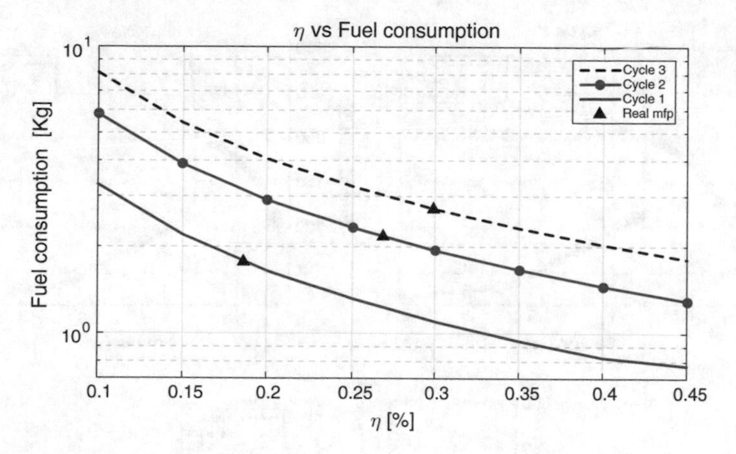

Fig. 16 ICE variation versus fuel consumption

The difference is minimal for cycles 2 and 3 and increases for cycles 1 and 4 that are the slowest speed.

Apart from the above, a comparative analysis was made with the results shown in the literature, by using the constant efficiency of the internal combustion engine $\bar{\eta}_{ice}$, in the formulation of a simplified PMP. Paganelli et al. [27] and Kim et al. [36] give an argument for the validity of the results obtained by using the ECMS formulation, where a constant efficiency value is employed to show the fuel saving or use equivalent to the state of charge in the battery. The results obtained in the literature mention that these are very similar when the variable or constant efficiency is employed.

In this work we first proceed to identify the average efficiency value corresponding to each driving cycle, said value is found when the same fuel consumption is obtained in the PMP solution using constant efficiency as the consumption obtained when the detailed models are used.

Figure 16 illustrates the variation in the hybrid electric vehicles fuel consumption using the strategy of Pontryagins minimum principle by considering the efficiency of the ICE to be constant for a range of power values. The points marked with a triangle in Fig. 16 correspond to the fuel consumption obtained by the PMP strategy and provide a average constant efficiency $\bar{\eta}_{ice}$ for each driving cycle.

Once the corresponding average efficiency is obtained, it is used in the detailed models. It is worth mentioning that it is not always possible to use the internal combustion engine at constant efficiency (for the chosen parallel configuration and the different tests in the driving cycles), therefore, we propose two options for trying to get the ICE to operate with constant efficiency.

The first proposal, that does not give a good result because of the capacity of the machines, is for the power of the ICE that cannot be delivered at the chosen the constant efficiency to be supplemented by the electric machine, in order thus to try

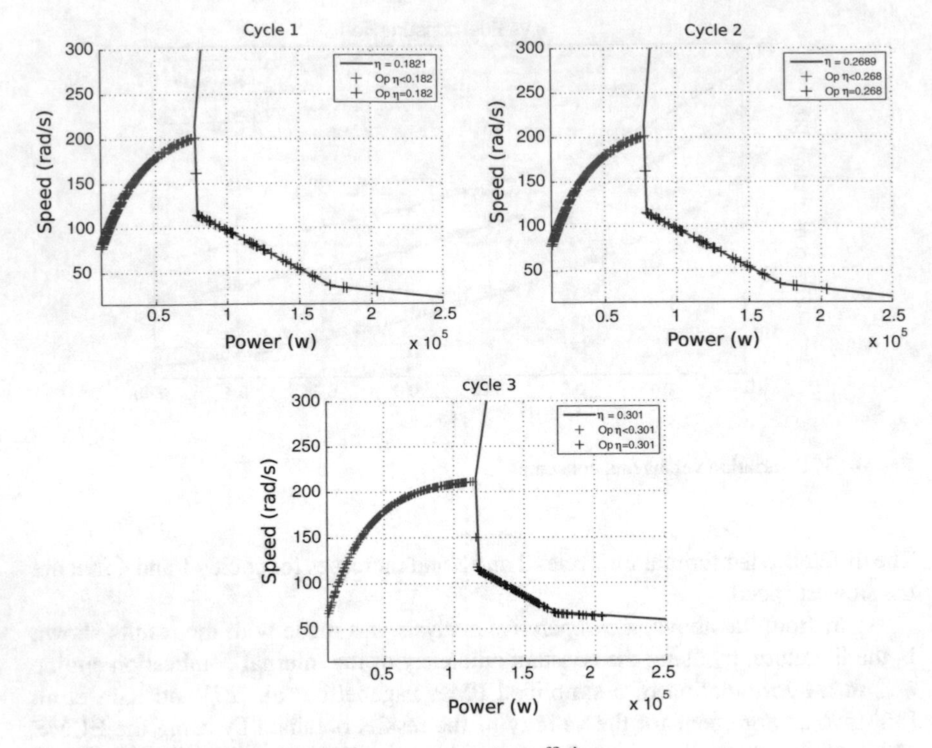

Fig. 17 ICE Operation points on the average constant efficiency

to carry out the driving cycle. This option is not viable as it is beyond the capacity of the EM, so the cycle cannot be carried out.

The second option considered is to use the ICE at an efficiency of less than the constant $\bar{\eta}_{ice}$ that was previous selected, when it is not possible to maintain the constant efficiency to carry out the driving cycle.

Figure 17 illustrates the operating points of the combustion engine on the average constant efficiency curve corresponding to each driving cycle (points on the right), as well as the ICE's operation when the power is less than the minimum that can be delivered with the corresponding constant efficiency (points on the left) for the four driving cycles.

Table 3 Comparison of fuel consumption (PMP), η_{ice} variable versus $\bar{\eta}_{ice}$ constant, $soc_0 = soc_f$

Cycle	$\bar{\eta}_{ice}$ (%)	fuel (Kg)	(%)
MX1	18.21	2.025	93.85
MX2	26.89	2.585	92.22
MX3	30.11	3.611	79.38

Table 3 shows equivalent fuel consumptions when using the ICEs average constant efficiency $\bar{\eta}_{ice}$ in the PMP formulation for each driving cycle, when the ICE operates on the points marked in Fig. 17 when using this solution in the detailed models. This result also makes it possible to check that the strategy proposed in this chapter has a better performance than the ECMS strategy in the literature, at least when constant equivalence is used.

As we can see, consumption rises significantly between 12 and 30% more than in the solution proposed by the PMP and PGS strategies.

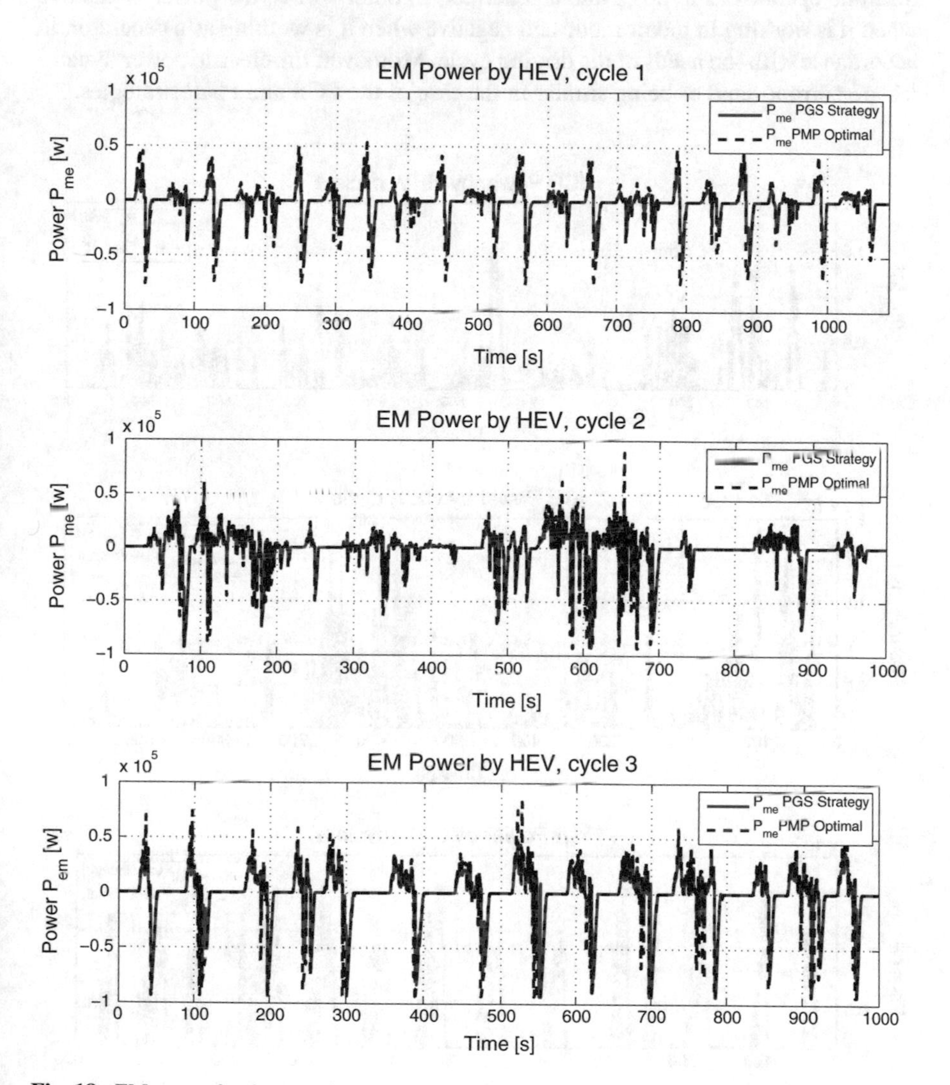

Fig. 18 EM power for three cycles

5.4 Engine Power

This section compares the powers used by the engines, using the two strategies contemplated in this proposal. Because of the way they behave in the state of charge, we would expect the powers and the combustion engines operating points to have very similar behavior in the dynamics.

The power in the electric machine is given in Fig. 18, which illustrates that this machine operates as a motor and a generator, in other words, the power is positive when it is working in motor mode and negative when it is working as a generator, in accordance with the needs of the driving cycle. Moreover, the electric power dynamics are corroborated as being similar in the case of the PGS and PMP strategies.

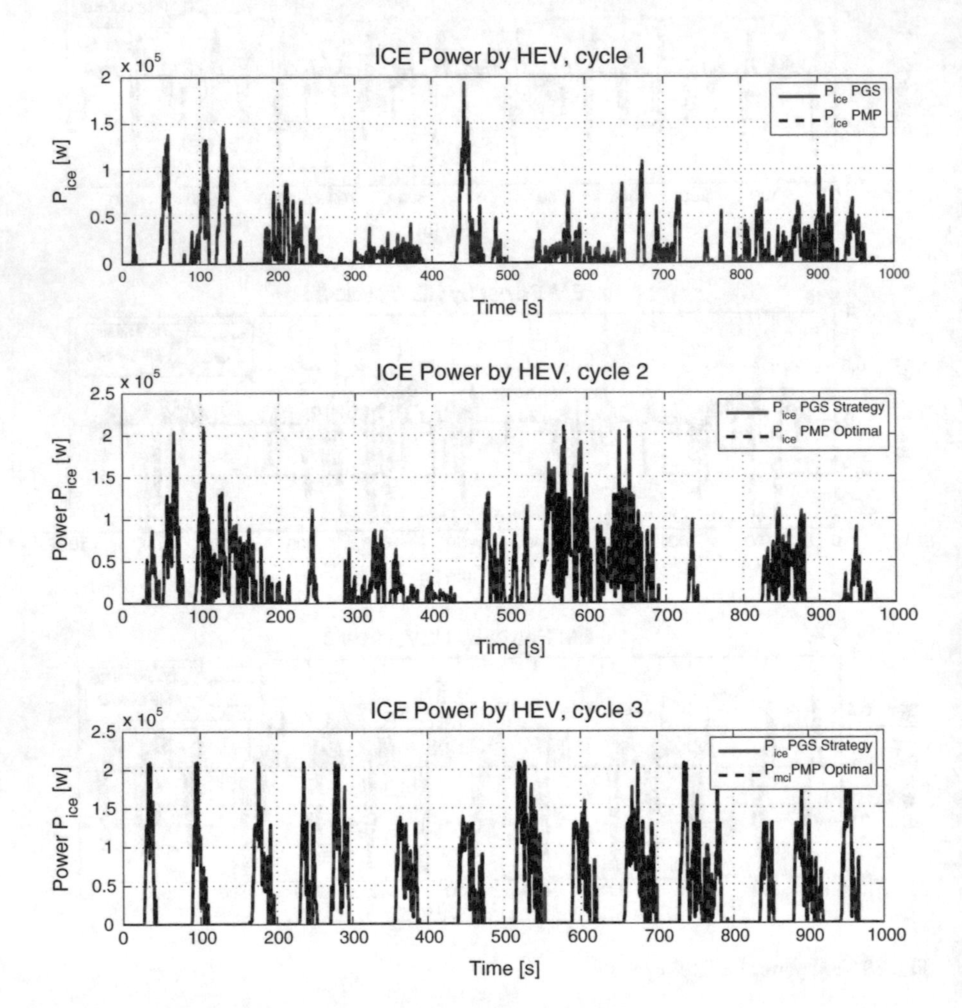

Fig. 19 ICE power for the three cycles

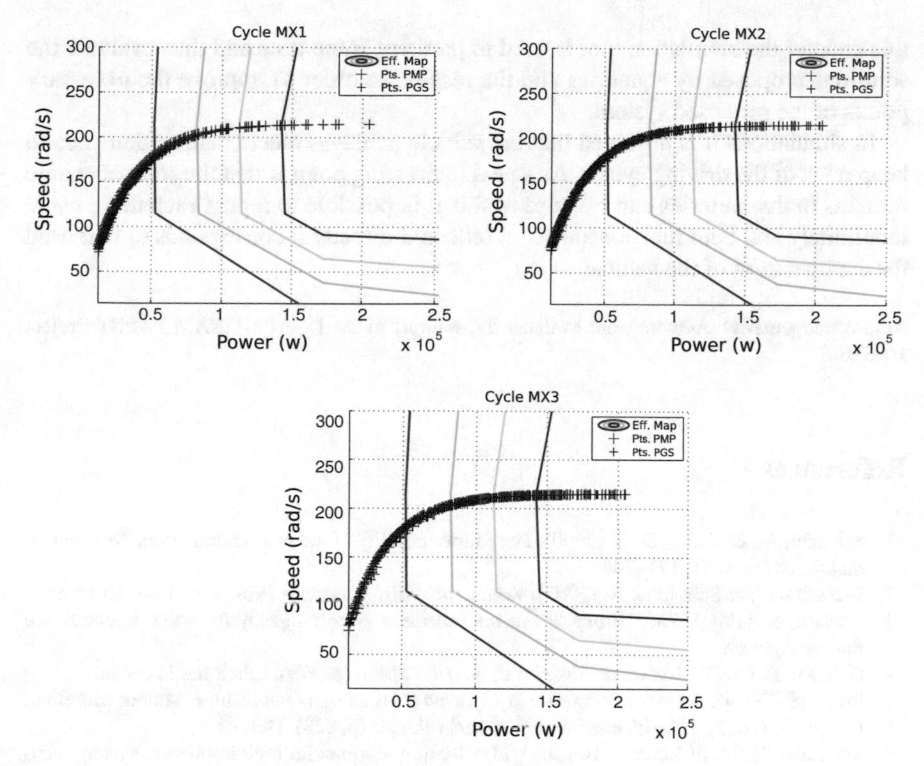

Fig. 20 ICE operation points for PMP and SEP strategies and three cycles

In cycle 3, that has the highest demand and recovery of energy for its range of speed, both strategies enable the highest amount of energy to be recovered in the regenerative braking.

The comparison of the power employed by the diesel internal combustion engine is given in Fig. 19 for the four driving cycles used in the tests. In cycles 2 and 3, the ICE at some instants of time is used at maximum capacity, while in cycle 1 this is not necessary. Moreover Fig. 20 illustrates the ICEs operating points during every driving cycle, which indicates the priority given to the speed setting imposed for operating as efficiently as possible in respect of the efficiency map of Fig. 8.

6 Conclusions

On the one hand optimization helps to solve problems in a good way and in this chapter is used as an engineering tool for the distribution of power in hybrid electric vehicles. Hybrid vehicles help in some part to solve the problems related to transport, which as in Mexico City, exist in different cities on the planet. On the other hand, in this chapter the simulation tool is used to perform some tests and thus evaluate the solutions proposed by equations and the results, in order to improve the necessary points of the proposed system.

In simulations it is observed that the vehicle achieves meet operator demands to keep track of the driving cycles. As a next interesting point is that the state of charge remains in the batteries and because of this it is possible to repeat each drive cycle indefinitely and both fuel consumption reflected directly it corresponds to that used for the operation of the vehicle.

Acknowledgements Authors want to thank the support to the DGAPA-UNAM PAPIIT Project 109316.

References

1. Schaefer, A., & Victor, D. G. (2000). The future mobility of the world population. *Transportation Research A, 34*, 171–205.
2. Guzzella, L., & Sciarretta, A. (2007). *Vehicle propulsion systems* (Vol. 2, p. 344). Springer.
3. Richard, S. (1994). *The history of the international energy agency* (p. 453). International Energy Agency.
4. Galeano, E. (1971). *Las venas abiertas de Amrica latina* (p. 379). Ediciones la cueva.
5. Romo, S. (2014). *Sector elctrico ante la Reforma y las energas renovables*. Mxico: Morelos.
6. Erjavec, J. (2013). *Hybrid, electric & fuel-cell vehicles* (p. 428). Delmar.
7. MacLean, H. L., & Lave, L. B. (2003). Evaluating automobile fuel/propulsion system technologies. *Progress in Energy and Combustion Science., 29*, 1–69.
8. Hodkinson, R., & Fenton, J. (2001). *Lightweight electric/hybrid vehicle design* (p. 280). Plant a tree.
9. Zhao, J., & Wang, J. (2014). Model predictive control of integrated hybrid electric powertrains coupled with aftertreament systems. In *Dynamic systems and control conference*.
10. Becerra, G. (2010). *Modelado y control del acoplamiento entre fuentes de potencia de vehculos hbridos*. Available via DIALOG. http://www.ptolomeo.unam.mx:8080/xmlui/bitstream/handle/132.248.52.100/3993/becerranu%C3%B1ez.pdf?sequence=1. Retrieved May 15, 2016.
11. Tim, H., Stuart, B., & Guido, H. (2014). Current hybrid-electric powertrain architectures: Applying empirical design data to life cycle assessment and whole-life cost analysis. *Applied Energy*. **119**, 314–329.
12. Wirasingha, S. G., Gremban, R., & Emadi, A. (2012). Source-to-wheel (STW) analysis of plug-in hybrid electric vehicles. *IEEE Transactionson Smart Grid, 3*, 316–331.
13. Miller, J. M. (2010). *Propulsion systems for hybrid vehicles* (p. 567). IET Renewable Energy Series.
14. Ehsani, M., Gao, Y., & Miller, J. M. (2007). Hybrid electric vehicles: architecture and motor drives. *IEEE, 95*, 719–728.
15. American-Rails: American-Rails. (2014). The Montreal locomotive works. http://www.american-rails.com. Retrieved May 06, 2016.
16. Gmotors: American-Rails. (2014). http://www.gmotors.co.uk. Retrieved May 06, 2016
17. Honda, Mxico: Civic Hybrid. (2013). http://www.honda.mx/civic-hybrid/. Retrieved May 06, 2016.
18. Luk, P. C. K., & Rosario, L. C. (2006). Power and energy management of a dual-energy source electric vehicle policy implementation issues. In *Power electronics and motion control conference*.

19. Delprat, S., Paganelli, G., Guerra, T. M., Santin, J. J., Delhom, M., & Combes, E. (1999). Algorithmic optimization tool for the evaluation of HEV control strategies. In *Proceeding electric vehicle symposium*
20. Tzeng, S., Huang, K. D., & Chen, C. C. (2005). Optimization of the dual energy-integration mechanism in a parallel-type hybrid vehicle. *Applied Energy, 80*, 225–245.
21. Xiong, W., Zhang, Y., & Yin, C. (2009). Optimal energy management for a series-parallel hybrid electric bus. *Energy Conversion and Management, 50*, 1730–1738.
22. Yao, H., & Wang, Q. (2015). The control strategy for improving the stability of a powertrain for a compound hybrid power excavator. *Journal of Automobile Engineering*.
23. Johannesson, L., & Egardt, B. (2008). Approximate dynamic programming applied to parallel hybrid powertrains. In *Proceeding of the 17th IFAC, World Congress*.
24. Koot, M., Kessels, J. T. B. A., de Jager, B., Heemels, W. P. M. H., van den Bosch, P. P. J., & Steinbuch, M. (2005). Energy management strategies for vehicular electric power systems. *Transactions on Vehicular Technology*.
25. Paganelli, G., Guerra, T. M., Delprat, S., Santin, J-J., Delhom, M., & Combes, E. (2000). Simulation and assessment of power control strategies for a parallel hybrid car. *Journal of Automobile Engineering*.
26. Paganelli, G., Ercole, G., Brahma, A., Guezennec, Y., Rizzoni, G. (2001). General supervisory control policy for the energy optimization of charge-sustaining hybrid electric vehicles.
27. Paganelli, G., Delprat, S., Guerra, T. M., Rimaux, J., & Santin, J. J. (2002). Equivalent consumption minimization strategy for parallel hybrid powertrains. In *Proceeding of the 55th IEEE vehicular technology*.
28. Sciarretta, A., Back, M., & Guzzella, L. (2004). Optimal control of parallel hybrid electric vehicles. *Transactions on Control Systems Technology*.
29. Delprat, S., Lauber, J., Guerra, T. M., & Rimaux, J. (2004). Control of a parallel pybrid powertrain: optimal control. *Transactions on Vehicular Technology*.
30. Musardo, C., Rizzoni, G., Sataccia, B. (2005). A-ECMS: an adaptive algoritm for hybrid electric vehicle energy management. In *Proceeding of the Conference on Decision and Control*.
31. Pisu, P., & Rizzoni, G. (2007). A comparative study of supervisory control strategies for hybrid electric vehicles. *Transactions on Control Systems Technology*.
32. Borhan, H. A., Vahidi, A., Phillips, A. M., Kuang, M. L., & Kolmanovsky, I. V. (2009). Predictive energy management of a power-split hybrid electric vehicle. In *Proceeding of the American Control Conference*
33. Yan, F., Wang, J., & Huang, K. (2012). Hybrid electric vehicle model predictive control torque-split strategy incorporating engine transient characteristics. *Transactions on Vehicular Technology*.
34. Ngo, V., Hofman, T., Steinbuch, M., & Serrarens, A. (2011). Predictive gear shift control for a parallel hybrid electric vehicle. In *Proceeding of the Vehicle Power and Propulsion Conference*.
35. Serrao, L., Onori, S., & Rizzoni, G. (2011). A comparative analysis of energy management strategies for hybrid electric vehicles. *Journal of Dynamic Systems, Measurement and Control*.
36. Kim, N., Cha, S., & Peng, H. (2011). Optimal control of hybrid electric vehicles based on Pontryagin's principle. *Transactions on Control Systems Technology*.
37. Yuan, Z., Teng, L., Fengchun, S. & Peng, H. (2013). Comparative study of dynamic programming and Pontryagin's minimum principle on energy management for a parallel hybrid electric vehicle. *Energies*.
38. Guzman, E., Becerra, G., Moreno, J. A., & Alvarez-Icaza, L. (2014). Controladores para motores diésel con incertidumbres paramétricas. In *Proceeding of the XVI Congreso Latinoamericano de Control Automático*
39. Mi, C., Masrur, M. A., Gao, D. W. (2011). *Hybrid electric vehicles* (p. 435). Wiley.
40. Nersesian, R. L. (2007). *Energy for the 21st century: a comprehensive guide to conventional and alternative sources* (p. 425). M.E. Sharpe, Inc.
41. Black, B. C., & Flarend, R. (2010). *Alternative energy* (p. 223). Greenwood.
42. Romero-Becerril, A. (2015). Reduccin y validacion de modelos electroquimicos de celdas de iones de litio y supercapacitores. Available via DIALOG. http://www.ptolomeo.unam.mx:8080/xmlui/handle/132.248.52.100/7307. Retrieved May 15 2016.

43. Becerra, G., & Alvarez-Icaza, L. (2010). Modelado y control del acoplamiento de fuentes de potencia en vehículos híbridos. In *Proceeding of the reunión de otono de potencia* (Electrónica y Computación). ISBN: 978-607-95476-1-5.
44. Becerra, G., & Alvarez-Icaza, L. (2011). Control del flujo de potencia en vehículos híbridos. In *Proceeding of the asociación de México de control automático A. C. (AMCA)*. ISBN: 978-607-95508-1-3.
45. Becerra, G., Pantoja-Vazquez, A., Alvarez-Icaza, L., & Flores, I. (2013). Simulation and optimal control of hybrid electric vehicles. In *Proceeding of the 25th European modeling & simulation symposium (EMSS)*.
46. Becerra, G., Alvarez-Icaza, L., & Pantoja-Vazquez, A. (2016). Power flow control strategies in parallel hybrid electric vehicles. *Journal of Automobile Engineering*.
47. Bryson, A. E., & Ho, Y. C. (1975). *Applied optimal control* (p. 428). Taylor & Francis.
48. Chaturvedi, N. A., Klein, R., Christensen, J., Ahmed, J., & Kojic, A. (2010). Algorithms for advanced battery management systems. *IEEE Control Systems Magazine*, 49–68.
49. Romero-Becerril, A., & Alvarez-Icaza, L. (2011). Comparison of discretization methods applied to the single-particle model of lithium-ion batteries. *Journal of Power Sources*, 267–269.
50. Pantoja-Vazquez, A., Alvarez-Icaza, L., & Becerra, G. (2015). Virtual serial strategy for parallel hybrid electric. *Journal of Automobile Engineering*.
51. Kirk, D. E. (2004). *Optimal control theory: An introduction*. Mineola, New York: Dover Publications.

A Simulation-Based Optimization Analysis of Retail Outlet Ordering Policies and Vendor Minimum Purchase Requirements in a Distribution System

Gerald W. Evans, Gail W. DePuy and Aman Gupta

Abstract This chapter presents an approach involving both discrete event simulation (DES) and optimization to address operational problems faced by a distribution system. In the system modeled, vendors may require minimum purchase requirements for each order. The model can be used to determine whether retail outlets should order product directly from the vendors, or through a centralized warehouse, as well as whether each retail outlet should violate its pre-specified inventory policy in order to meet vendor-minimum requirements. In addition, the model can be of use as an aid in negotiation with vendors with respect to minimum purchase requirements. The work is based on a project performed for an actual company with a centralized warehouse, located in Louisville, Kentucky, and 19 retail outlets, located throughout the United States.

Keywords Inventory policy · Simulation · Optimization · Distribution

1 Introduction

The operation of a typical three-echelon distribution system involves the shipping of stock keeping units (SKUs) from vendors to distribution center(s), from distribution center(s) to retail outlets and directly from vendors to retail outlets (see Fig. 1). Often, the SKUs are organized into product lines, where a specific product line corresponds to a particular vendor. A vendor may specify that purchases within a particular product line meet a minimum dollar or a minimum weight requirement;

G.W. Evans · G.W. DePuy
Department of Industrial Engineering, University of Louisville,
Louisville, KY 40292, USA

A. Gupta (✉)
Department of Decision Sciences, Embry-Riddle Aeronautical University - Worldwide,
Daytona Beach, FL 32114, USA
e-mail: aman.gupta@erau.edu

M. Mujica Mota and I. Flores De La Mota (eds.), *Applied Simulation and Optimization 2*, DOI 10.1007/978-3-319-55810-3_8

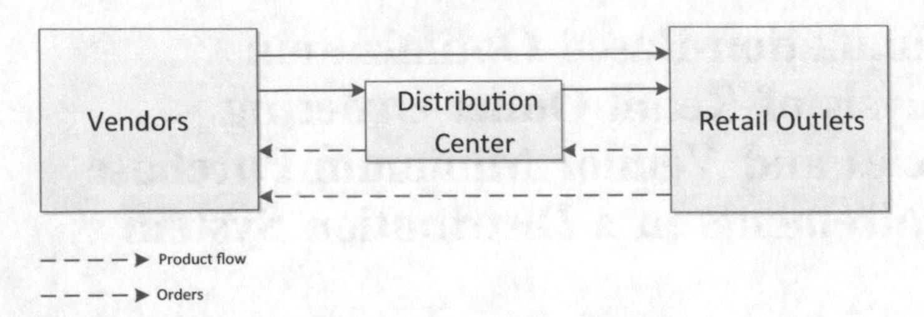

Fig. 1 A three-echelon distribution system

however, if a retail outlet orders the SKUs associated with a product line from a distribution center, there typically is no minimum size order requirement.

In these three-echelon distribution systems, operating policies must account for many inter-dependent decisions, including, for various combinations of SKUs, product lines, retail outlets, and distribution centers, policies corresponding to the answers to the following questions:

1. Should a retail outlet order a product line directly from the vendor, or from the distribution center?
2. What reorder point and order quantity should each retail outlet or distribution center use for each SKU?
3. If an order is to be placed with a vendor (as opposed to a distribution center) by a retail outlet, and the minimum size order requirement is not met, which SKUs (within the product line), if any, and corresponding order quantities should be added to the order to meet the requirement?

Note that with respect to question 3 above, one must consider tradeoffs between lost sales/backorders and inventory carrying charges; for example, a particular retail outlet may add specific SKUs to an order in an effort to meet the minimum order requirement specified by a vendor, thereby incurring fewer lost sales/backorders, but also increasing inventory carrying charges. The answers to these questions will depend on a number of factors, such as (1) customer demands at the various retail outlets, (2) shipping charges, (3) inventory carrying charges, (4) purchase costs, (5) selling prices, etc. In this chapter, we will describe, and illustrate the use of, a simulation-optimization approach that can be used to answer the questions posed above. Such a simulation-optimization approach allows for an accurate representation of the relationship between policy variables and performance measures (as opposed to an analytical modeling approach); in addition, the optimization procedure allows for the implicit consideration of a large number of alternative policies.

This chapter is organized as follows. In the next section, we review some of the extensive literature in this area. In the third section, we describe a three-echelon distribution system of the type described above for which a simulation model was constructed for analysis. This model was developed for an actual organization facing the problems described in this paper. Section four contains a description of the simulation model that was developed to model the system. In the fifth section of the chapter, we provide the formulation for an optimization model, to be interfaced with the simulation model. In the sixth section, we discuss the results of experimentation with the simulation-optimization model for answering the questions posed above. Data used for the simulation-optimization experiments described in this section was actual data for a particular product line of the organization. Finally, the last section of the paper contains a summary and conclusions.

2 Literature Review

The published research related to this paper can be categorized according to several different characteristics, including (1) the type of system studied (e.g. one distribution center vs. multiple distribution centers, one product vs. multiple products, etc.), (2) the decisions addressed (e.g., facility location decisions, inventory policy decisions, vendor decisions), (3) assumptions made (e.g., Poisson distributed demands, etc.), and (4) modeling and optimization methodologies employed.

DES models, analytical models, and optimization techniques have been employed both separately and in a joint fashion for several decades as an aid in the design and operation of distribution systems. A major value of DES models is that they can represent the time dynamic, probabilistic aspects of real world distribution systems. Optimization techniques allow the identification of a "best" policy without having to enumerate every possible policy; this feature is obviously important when the number of possible policies is large, such as in situations when there is a combination of integer decision variables (e.g., reorder points for several retail outlets in an organization) or continuous decision variables. Even though it makes sense to do so, these methodologies have only been rarely applied jointly as described in this paper.

In spite of their drawbacks, a variety of publications describing the use of analytical models have appeared in the literature. These analytical models are typically easier to employ with optimization techniques because of their closed form nature and the fact that one does not obtain just *estimates* for outputs, as is the case with stochastic DES models. However, these models are often greatly restricted with respect to the types of demand patterns that can be represented and the fact that they are often static in nature.

As an example of these analytical models, Abdul-Jalbar et al. [1] formulated a mixed integer, nonlinear optimization model for determining the inventory policy variable values for a system consisting of one warehouse and multiple retailers. Their work differs from many of the other analytical models for this type of problem in that backorders are allowed and that, instead of a customer demand at each of the retailers being uniformly distributed, the customer demands are modeled as power patterns.

Boute et al. [6] developed an analytical model for analyzing a two-echelon supply chain (one retailer and one manufacturer). Their model considered only one type of product with independent, identically distributed demands and employed the variability of the order rate from the retailer as the primary decision variable; in addition, their model allowed two types of strategies by the manufacturer: one involving a flexible capacity which would ensure constant lead times with low inventory levels, and the other an inflexible capacity resulting in stochastic lead times and increased retailer inventory level.

Miranda and Garrido [25] addressed both the location and service level problem for a two-stage (or three-echelon) distribution network involving a plant/central warehouse, regional warehouses, and customer demand zones. Specifically, they developed two optimization models, solved iteratively and in sequence. The first model determines optimal regional warehouse locations and customer assignments for a given service level, while the second model addresses the inventory service-level optimization problem for the given locations.

Additional research in this area involving the use of analytic models has been presented in [8, 9, 23, 26, 27, 32, 38].

Several researchers have developed metamodels of distribution systems/supply chain simulation models. For example, Hayya et al. [16] developed regression equations from fractional factorial experiments performed on a simulation model. Specifically, they considered "order crossover" (defined as the situation where orders are not received in the same sequence as placed) for the problem of determining optimal reorder quantity and reorder point for a situation with a single product type.

Tee and Rossetti [36] explored the validity of Axsater's [5] analytical models for a two-echelon system with one warehouse and multiple retailers for a scenario of nonstationary demand (a major assumption of Axsater's models). In particular, they developed a simulation model of the same two-echelon system to determine optimal reorder points and reorder quantities. Tee and Rosetti determined that Axsater's models worked well in certain situations with low demand and large batch size orders, but not so well under other conditions.

As noted above, each methodology (analytical modeling and simulation modeling) has its own advantages and disadvantages when compared to the other methodology. For example, Persson and Araldi [29] noted that, although many of the supply chain models from the literature are of a closed form (an analytical model) and involve the use of optimization techniques, these models are not able to represent the dynamic perspective allowed by simulation models. Their work involved using the Supply Chain Operations Reference (SCOR) model [34] as a basis for the development of an Arena [22] DES model of a supply chain. Two case studies are illustrated in their paper; the results of these case studies indicate where the SCOR template could be improved, and therefore become an even more useful tool.

Hung et al. [18] also noted that analytical models, although useful in some cases, are too simplistic to be of practical use for complex supply chains. They developed a general simulation model in which generic nodes were used to represent plants, warehouses, and retailers, respectively. The model allows the investigation of various replenishment control policies, including two continuous review policies and two periodic review policies. Latin Supercube Sampling is used as an experimental procedure in the application of the model. A case study of a system with two markets and four products was employed to demonstrate the model.

Jain [19] used a simulation model to represent a large supply chain consisting of customers located in all 50 states and 27 countries, customer service centers, distribution centers, and suppliers. Through sampling of a relatively small number of part types, Jain was able to develop insights into various relationships between procurement times and administrative business process times (ABPT), between ABPT and service levels, etc., with an ultimate aim of analyzing the tradeoffs between service levels and low inventories.

Albino et al. [2] studied cooperation among supply chain members within the context of industrial districts (defined as "specific production systems characterized by a high level of fragmentation of the production process into several phases"). The cooperation studied involves assignment of orders so that utilization of production capacities is balanced among the various firms and unserved customer demand is minimized. The cooperation is modeled through the use of an agent-based simulation model.

Crnkovic et al. [11] presented a simulation-based decision support framework for exploring tradeoffs in supply chains. In particular, they addressed the problem of determining the amount of an item to produce in a volatile external environment in which the production-sales interval is small under a variety of supply chain configurations.

Son and Sheu [33] addressed a problem similar to the one addressed in this paper, involving deviations from replenishment policies in a decentralized supply chain. In particular, they employed Sterman's Beer Game simulation as a hypothetical case study to simulate the effects of deviations from a coordinated order replenishment policy.

Tsai and Zheng [37] developed a simulation-based optimization approach for a two echelon system similar to the one studied in this paper. More specifically, the system studied consisted of a central warehouse and multiple "field depots", each of which supplies parts to customers who require those parts for machines which fail. The basic problem addressed is to set the various stocking levels so that the total inventory cost is minimized subject to constraints on the expected response time at the depots. Tsai and Zheng noted that the use of a simulation model as opposed to an analytical model avoids the assumption of independence between the depots. They employed a sample average approximation technique, a cutting plane method, and a ranking and selection procedure for solving their problem.

Chen et al. [7] addressed the management of a multi-echelon-production-distribution supply chain for clinical trials of new drugs via a simulation-optimization approach. Specifically, a set of mixed integer linear programs were solved to determine the manufacturing and shipping plans for the system. The effectiveness of these plans was then assessed through the use of Monte Carlo simulation models for various demand scenarios.

Fang and Li [14] developed a multiobjective hybrid simulation-optimization approach for the optimization of inventory policies in a complex distribution network. The policies involved determining a reorder point and an order quantity at each of the inventory locations. The two conflicting performance measures of fill rate and inventory cost were addressed through the use of a multiobjective genetic algorithm (based on the work in [12]) interfaced with a simulation model of the system. The approach was demonstrated on a system with a central distribution center and 14 regional distribution centers.

Eruguz et al. [13] provided a comprehensive review of guaranteed service models for the allocation of safety stocks (to minimize cost while meeting target service levels) in multi-echelon distribution systems. These guaranteed service models assume that a deterministic service time can always be satisfied at each stage of the distribution system.

Chu et al. [10] developed a simulation-based optimization framework for multi-echelon inventory problems with the objective of minimizing the inventory cost while maintaining the service levels. The authors proposed an agent-based simulation approach followed by a Monte-Carlo method. Lastly, the optimization problem is solved by a cutting plane algorithm.

Güller et al. [15] presented a multiobjective optimization approach to determine inventory control parameters with the objectives of minimizing total inventory cost and maximizing the service level. To assess the control parameters an object-oriented framework to develop the simulation model is also presented.

A simulation-optimization approach for the optimal operation of pull control systems was proposed by Pedrielli et al. [28]. The proposed approach has the ability to perform parameter optimization and performance evaluation in the same model.

Kochel and Nielander [24] also noted the difficulty of modeling realistic inventory systems with an analytical modeling approach. They investigated the use of simulation and optimization methodologies for determining continuous review order point and order quantity policies; the hypothetical system studied involved a single product which flowed through multiple locations in a multi-echelon inventory system.

Rosen et al. [31] described a new method for interfacing simulation and optimization that explicitly considers the user's preference structure over risk and uncertainty over multiple performance measures. They evaluated their approach against two simulation-optimization methods that employ deterministic multicriteria strategies, and show that their approach yields significantly better results under a variety of experimental settings.

Finally, the model described in this paper can be considered as a "data-driven simulation model" since the daily demands at each of the retail outlets are read by the model from an external file. Tannock et al. [35] used a data driven simulation model from the civil aerospace sector to show that this type of model can be useful in the improvement of supply chains. In particular, users can apply such a model to different scenarios just by changing the input data.

Additional papers of interest in the area of applications of simulation for distribution system design and operation include [4, 17, 20, 21, 30].

As seen from the above review, most of the analytical models cannot be used for purposes other than "high-level" decision-making in a distribution system/supply chain since they are for the most part static in nature and require overly restrictive assumptions. Even many of the simulation models described in the literature have been developed to represent only hypothetical systems involving, for example, only a single part/product type, inflexible delivery schedules, etc. Even when actual systems with multiple part/product types are modeled, relationships that exist when products belong to the same product line are not considered. The simulation model described in this paper addresses these various issues in an accurate fashion, and hence allows for improved decision making.

3 Description of the System

A company operates a three-echelon physical distribution system for thousands of different SKUs and is interested in analyzing two aspects associated with the operation of its system: (1) the policies employed by its retail outlets with respect to whether the SKUs are ordered directly from the vendor or through the company-owned distribution center, and (2) the vendor minimum purchase requirements.

The company's distribution system is composed of a single distribution center and 19 retail outlets which distribute approximately 5000 different SKUs from approximately 100 different vendors. These 5000 different SKUs are categorized by product line—typically, each vendor has one product line. Customers can purchase items at the retail outlets as well as the distribution center (i.e., the distribution center also acts as a retail outlet in that it can sell product directly to customers). The retail outlets can purchase SKUs directly from the vendor for that product line, or through the company-owned distribution center. The distribution center purchases SKUs directly from each vendor. The distribution center follows a regular shipping schedule to each of the retail outlets.

Currently, separate methods are employed by the company to determine their ordering policies at the retail outlets, depending on whether the SKU is ordered directly from the vendor or through the distribution center.

3.1 *Orders Placed Directly with the Vendor*

When the on-hand inventory for a particular SKU at a retail outlet falls below the reorder point for that SKU-outlet combination, a possible order to the vendor is triggered. More specifically, all SKUs (with on-hand inventory levels less than their respective reorder points) in the product line associated with the triggering SKU are used to form a possible order to the vendor. If the resulting total order meets the minimum purchase amount (in cost or weight) for that vendor-product line combination, then the order is placed with the vendor. In some cases, the manager of the retail outlet arbitrarily places an order with a vendor for the minimum purchase amount even if the total order as originally computed did not meet the vendor's requirement, thus trading off higher inventory holding costs against the lower value of lost sales as a result of not being able to satisfy customer demand. This is accomplished by adding SKUs to the order which had inventory levels above their respective reorder points. This decision process associated with ordering SKUs for a product line directly from a vendor is illustrated in Fig. 2.

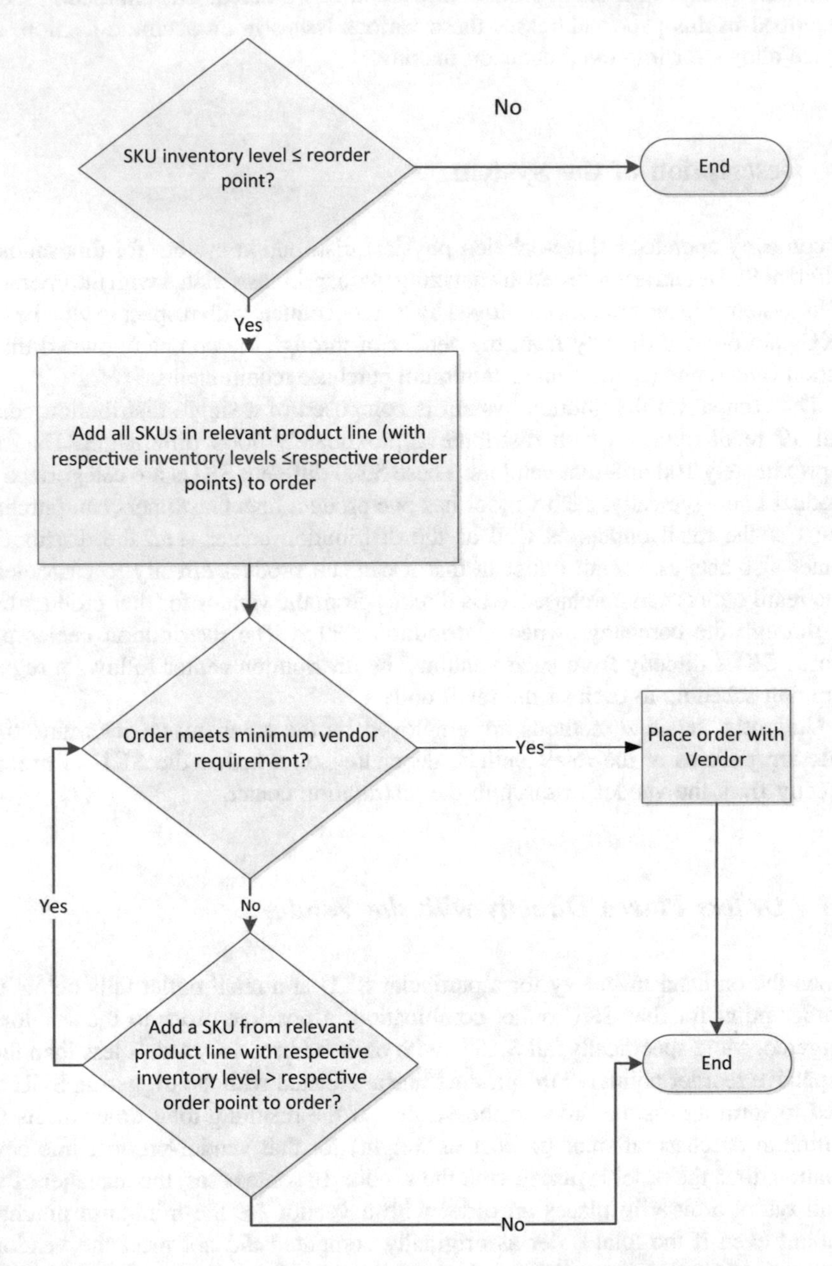

Fig. 2 Decision process associated with ordering product directly from a vendor

The reorder points and order quantities for each respective SKU are computed as functions of several parameter values, including customer demand, projected lead time for delivery from the vendor, sales price, and a safety factor value. These reorder points and order quantities can vary by specific SKU-retail outlet combination. Note that in computing the values for the reorder points and order quantities for the distribution center, the customer demands for the SKUs at all retail outlets supplied through the distribution center must be considered, as well as the customer demand at the distribution center.

3.2 *Orders Placed with the Distribution Center*

One of the main advantages associated with ordering through the distribution center for a retail outlet is that there is no minimum-sized order requirement. The decision for a retail outlet of whether to order a product line from the distribution center or from the vendor is determined as a function of several "static" parameter values, including average customer demand, lead time, and prices for the items in the product line. When the ordering policy for a particular product line dictates that every SKU in that product line is to be ordered through the distribution center, a min-max (i.e., an (s, S)) inventory policy is followed. In other words, whenever the inventory level for any particular SKU that is ordered from the distribution center falls below a pre-specified minimum value, an order is placed to the distribution center for that SKU and for all other SKUs in the product line with inventory levels below their pre-specified minimums. The order quantity for each SKU in the order is set equal to a pre-specified maximum minus the current inventory level. Note that these pre-specified maximum and minimum values for a SKU could vary among the retail outlets.

The values associated with these pre-specified maximums and minimums are computed as functions of several parameter values, including customer demand, delivery time from the distribution center, sales price, and a safety factor value. As stated previously, the distribution center follows a set schedule with respect to delivery to the retail outlets.

4 The Simulation Model

The simulation model of the distribution system was developed with the Arena Software Package [22]. This section of the chapter provides an overview of the model's operation; the inputs to the model, categorized as control variables or parameters; and the output from the model.

The model is composed of several sub-models, entitled:

1. Read Input Data,
2. Compute Initial Variable Values,
3. Day of Week and Day of Simulation Update,
4. Decrease Inventory Levels by Customer Demand and Modify Order to DC and Order to Vendor Variables,
5. Generate Order Entities to DC and Vendors,
6. Schedule Shipments to Depart DC and Arrive at Retail Outlets,
7. Schedule Shipments to Depart Vendors and Arrive at Retail Outlets, and
8. Output Variable Computation.

Figure 3 illustrates the relationships between these submodels.

The sub-model **Read Input Data** reads data for the system from several different Excel input files. This data includes the simulation duration, the warm-up period for the simulation, the number of retail outlets, the fixed ordering costs, the number of product lines, the specific SKUs in each product line, the product lines supplied by each vendor, the minimum purchase requirement (in terms of weight or dollars) for each product line, the vendor lead times to each retail outlet, the shipping schedule

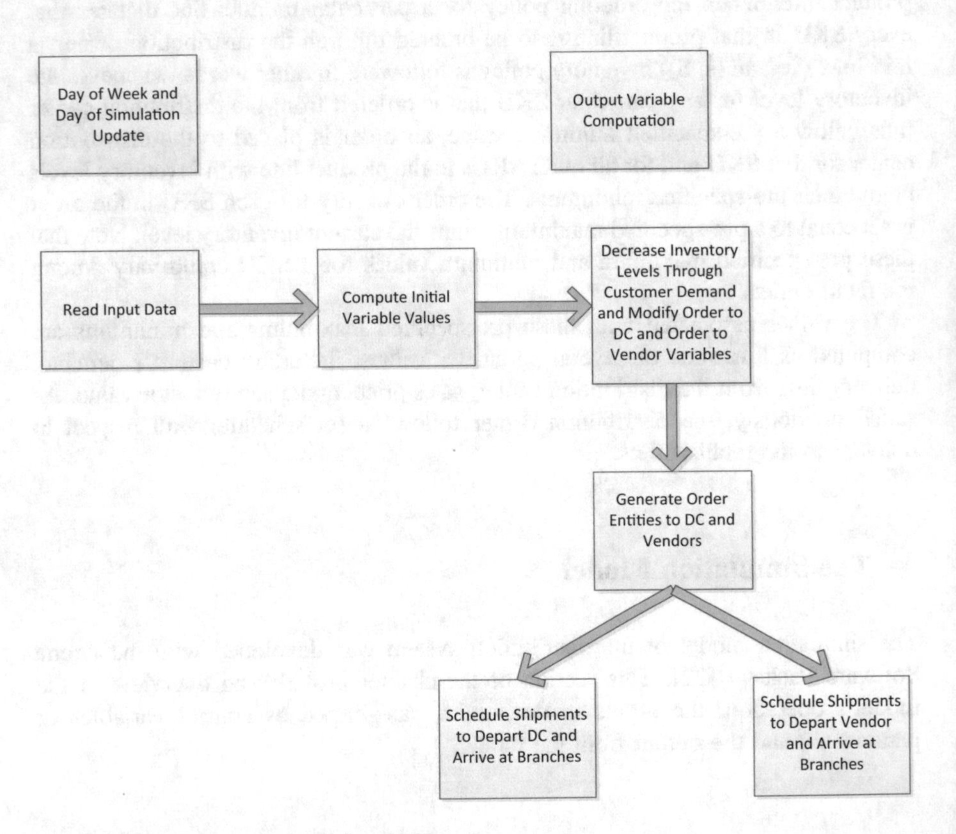

Fig. 3 Submodels of the simulation model

from the distribution center to each retail outlet, the shipping costs (from vendors to the retail outlets and from the distribution center to the retail outlets), the cost and selling price for each SKU, the inventory carrying charges (expressed as a percent of the value of the inventory), and the cost associated with a "lost sale" (expressed as a percent of the selling price).

The sub-model **Compute Initial Inventory Values** computes the values associated with several different model parameters, with the main ones being the inventory policy values: reorder points, order quantities, minimums, and maximums for each SKU at each retail outlet. Note that the values for reorder points and order quantities are computed only for those SKUs that are ordered directly from the vendor by the retail outlet, and that the values for maximums and minimums are computed only for the SKUs that are ordered from the distribution center. The values for the control variables associated with the decisions associated with ordering directly from the vendor, or through the distribution center, discussed in more detail below, are input through initialization of the Variables Modules of Arena. Both of the sub-models **Read Input Data** and **Compute Initial Inventory Values** are executed at simulated time 0.

The sub-model **Day of Week and Day of Simulation Update** is used to update the values for the variables (1) Day of Week and (2) Day of Simulation as the simulation is running. Day of Week is defined as having a value of 1 to represent Monday and a value of 7 to represent a Sunday. The variable, Day of Week, is important in representing the shipping schedule for the distribution center to the retail outlets.

The sub-model **Decrease Inventory Levels by Customer Demand and Modify Order to DC and Order to Vendor Variables** reads an Excel data file which contains the customer demand for each SKU at each retail outlet for each day of the simulation run. These are the actual customer demands, as the firm authorizing the study was interested in determining an optimal policy with respect to the actual demands for a specific year. Inventory levels are appropriately decreased and the variable values associated with lost sales are also incremented when there is a demand for a SKU with a zero level of on-hand inventory. In addition, a set of variables which are used to form orders to the distribution center and to the vendors are updated. This sub-model is executed once each simulated day.

The sub-model **Generate Order Entities to DC and Vendors** is executed each simulated day immediately after execution of the sub-model **Decrease Inventory Levels by Customer Demand and Modify Order to DC and Order to Vendor Variables**. It uses the set of variables referred to in the previous paragraph in order to form order entities representing orders to the distribution center and the vendors, depending on which ordering policy is being followed (as determined by the product line). This sub-model checks for orders to be placed by each retail outlet each day of the simulation run. The order entities generated by this sub-model contain attributes which represent the number of SKUs in each order, the specific SKUs in the order, the quantity ordered for each SKU, and the specific retail outlet making the order.

When the policy indicates that a particular product line is ordered through the distribution center, only those SKUs which have inventory positions (on-hand inventory level plus number of units on order) less than the SKU-retail outlet "minimum" are placed in the order.

Conversely, when the policy indicates that a particular product line is to be ordered directly from the vendor for a particular retail outlet, this sub-model checks to see whether the vendor minimum requirement is met by forming an order of all SKUs with inventory positions below respective reorder points. The order quantity for each SKU in the order is set by the computation performed in the sub-model **Compute Initial Variable Values**. If the vendor minimum requirement is not met, than SKUs are added to the order in a sequential manner, according to a control variable: Fraction Extra Order (b, p); this control variable represents the maximum fraction over the reorder point at retail outlet b for product line p for which a SKU will be added to the vendor order.

For example, if the reorder point for a particular SKU in product line 3 at retail outlet number 6 is 40, and the current inventory position is 43, then this position is $100 \times (43 - 40) = 7.5\%$ over its reorder point. Therefore, this SKU would be eligible for addition to an order if the value for Fraction Extra Order (6, p) were 0.075 or higher, but not eligible otherwise.

Each time a SKU is added to a vendor order, a check is made to see if the vendor minimum requirement is made. If the vendor minimum requirement is met, then an order is placed; if not, then a check is made for the next SKU in the product line to determine if it can be added to the order. If, after checking every SKU in the product line, the vendor minimum requirement is not met, an order is not placed for this product line.

The sub-model **Schedule Shipments to Depart DC and Arrive at Retail Outlets** is executed based on entities which represent orders from the retail outlets to the distribution center. This sub-model forms shipment entities which have as attributes the specific SKUs in the shipment and appropriate shipment quantities. The sub-model first finds the first available day (following the current day) for which the distribution center is scheduled to make a shipment to this retail outlet based on the shipping schedule input, and schedules an appropriate delay until the shipment is to leave the distribution center. At that simulated time, the program cycles through each SKU in the order and determines if the distribution center has enough inventory to completely satisfy the order. If so, the shipment entity is formed accordingly. If not, the shipment quantity for the relevant SKU is based on the amount available in the distribution center inventory.

If there is not enough inventory at the distribution center to satisfy the order for a particular SKU for at a retail outlet, then the variable: Number of Insufficient Shipments from the DC is increased by 1, and the attribute values associated with the shipment quantities are appropriately decreased.

Appropriate changes are made to variables representing inventory levels, volume of inventory, and value of inventory at the distribution center, as well as inventory in transit from the distribution center to the retail outlets. Variables representing shipping charges are also updated. In addition, this sub-model revises variables which will set up potential orders from the distribution center to the vendors, based on the revised inventory levels at the distribution center.

Following a simulated delay associated with the lead time corresponding to the shipping schedule, the shipment arrives at the retail outlet. At this time, the variables representing the on-hand inventory levels, inventory on order, value of inventory, and volume of inventory at the retail outlets are updated.

The sub-model **Schedule Shipments to Depart Vendors and Arrive at Retail Outlets** is executed based on entities which represent orders from the retail outlets to the vendors. This sub-model is somewhat simpler than the **Schedule Shipments to Depart DC and Arrive at Retail Outlets** sub-model, since inventory levels at the vendors do not have to be explicitly represented as they do at the distribution center. This sub-model first updates the values for the variables representing ordering costs and costs for purchase of stock. Following an appropriate delay corresponding to the lead time from the vendor to the retail outlet variables corresponding to inventory levels, values, and volumes are updated, along with shipping charges.

The final sub-model, **Output Variable Computation**, is executed at the end of the simulation run. As its name indicates, this sub-model computes and outputs the performance measure values for the simulation run.

The simulation model described above computes values for a variety of performance measures, including:

1. Inventory Holding Costs,
2. Shipping Charges from the Vendors,
3. Shipping Charges from the Distribution Center,
4. Fixed Ordering Costs,
5. Cost of Purchase of SKUs from the Vendor,
6. Cost associated with Lost Sales,
7. Gross profit from the sales of SKUs,
8. Value of all Inventory at the End of the Warm-Up Period of the Simulation Run.
9. Value of all Inventory at the End of the Simulation Run.

All costs and profits (performance measures 1 through 7) were calculated for each retail outlet, the distribution center and the entire system except for the shipping charges from the distribution center (performance measure 3) which was only calculated for each retail outlet and the entire system.

The inventory holding costs are computed as a percentage of the time average value of the inventory in stock over a 1-year period of time. The input data for the model allows for this annual inventory holding interest rate to vary by retail outlet, however the company uses a rate of 25% for each retail outlet in the system.

The shipping charges are also computed as a percentage of the value of each shipment. These percentages, which are inputs to the model, are also allowed to vary depending upon whether the shipment was being made directly from the vendor, or from the distribution center, and also depending on the retail outlet to which the shipment was being made.

The fixed ordering cost for an order is calculated as a constant multiplied by the number of different SKUs in an order. The constant is allowed to vary depending upon whether the order was being placed with the distribution center or directly with the vendor. Since the system's management is not very concerned about these fixed ordering costs, a constant $1 is used.

The cost associated with the purchase of SKUs from the vendor is computed as a constant (determined from the system's database) for a particular SKU multiplied by the number of SKUs of that type purchased.

The penalty cost associated with any lost sales is computed as a constant, which was allowed to vary by SKU, multiplied by the number of units of demand not met. For the experiments reported later in this chapter, a constant of 1% of the cost of the SKU per unit from the vendor was used. The company recommended this small penalty percentage as they believe the major cost associated with lost sales is lost profit. Hence, this lost profit cost is already implicitly considered through the decrease in sales.

Initial inventory levels (i.e., at simulated time 0) for each SKU at each retail outlet are set equal to the halfway point between the minimum and maximum inventory levels employed as policy variables when the retail outlet orders inventory from the distribution center (as discussed in Sect. 3.2). Initial inventory levels at the distribution center for each SKU are set at a value equal to 1.2 times the reorder point value for that SKU. The calculation of these reorder points considers the customer demand at the distribution center as well as half of the demand associated with the 19 retail outlets (implicitly assuming that approximately half of the retail outlets will order their product through the distribution center as opposed to directly from a supplier). Since setting inventory levels at these values is somewhat arbitrary, the model was allowed to warm up for a period of three months, before the start of data collection.

Following the 3-month warm up period, the value of all inventory in the system is computed (called **Value of all Inventory at the End of the Warm Up Period of the Simulation Run** in the list above). This variable value, as well as the value for the variable **Value of all Inventory at the End of the Simulation Run**, were very complex computations, because these values reflect the policy being simulated with respect to whether the various product lines are being ordered through the distribution center or directly from the vendor for each retail outlet. Without going into specific detail on these calculations, they consider the profit from selling the inventory minus the inventory holding cost minus the shipping charges for shipments from the DC to the retail outlets. The calculation assumes that during the future period for which the inventory levels are depleted, the demand rate is constant and equal to the average demand rate for the SKU at the retail outlet over the entire simulation run.

5 Optimization Model

An optimization model, to be interfaced with the simulation model described in Sect. 4, can be defined through the use of decision variables (control variables for a simulation model) and parameters, an objective function (or functions if there are multiple objectives to consider) of the control variables, and constraints.

In the discussion which follows in this section, we will assume that there is only **one** product line, consisting of a specified number of SKUs. This assumption of one product line will allow for a simplified model presentation, as well as simplified experimentation. Since the product lines operate independently, the extension of the optimization model to represent multiple product lines is straightforward.

We begin by providing notation for several parameters. Let

$$\text{B} = \text{the number of ``branches'', or retail outlets, and}$$
$$\text{S} = \text{the number of different SKU's in the product line.}$$

A wide variety of control variables could be considered for this simulation model. Corresponding to the desires of company management, we focused on two categories of control variables for our experimentation:

$$y_b = \begin{cases} 1, \textit{if the product line at retail outlet b is ordered through the distribution center}, \\ 0, \textit{otherwise (i.e., the product line is ordered directly from the vendor)}, \\ \textit{for } b = 1, \ldots, B \textit{ and} \end{cases}$$

$$x_b = \textit{Fraction Extra Order}\,(b, 1);\ \textit{for}\ b = 1, \ldots, B.$$

Note that the second category of control variable, x_b, is only employed for a retail outlet b and when $y_b = 1$. Also when $y_b = 0$, retail outlet b orders **all** SKUs in the product line directly from the vendor. The Fraction Extra Order variable was defined in Sect. 4; as noted, this control variable allows the adding of a SKU to an order from a retail outlet to a vendor even if the inventory level for that SKU is greater than its "small s" value. The purpose is of this variable is to allow for the composition of an order which meets the vendor minimum requirement.

Additional control variables which could have been considered include those related to the shipping schedule from the distribution center to the retail outlets and the values employed for the inventory policy variables. However, in keeping with the company's desires and also to keep the study relatively simple, we focused on the two categories defined above.

A variety of criterion models could have been developed to consider the various performance measures listed above. Any such criterion model would contain an objective function of these performance measures to be optimized, and a set of constraints. For example, one might want to maximize the gross profit in overall sales from the system, subject to constraints on inventory carrying charges at some of the retail outlets. One criterion model developed for experimentation was the maximization of net profit, defined as follows:

Net Profit = Gross profit from the sales of SKUs + (Value of all Inventory at the End of the Simulation Run – Value of all Inventory at the End of the Warm Up Period of the Simulation Run) – Inventory Holding Costs – Fixed Ordering Costs – Shipping Charges from the Vendor – Shipping Charges from the Distribution Center – Cost of Purchase of SKUs from the Vendor – Cost associated with Lost Sales.

A simpler model was also developed for experimentation. This model, which allows for the study of tradeoffs between inventory holding costs and lost sales, can be defined as follows:

$$\text{Minimize Inventory Holding Costs} \tag{1}$$

subject to:

$$x_b = b_1, \text{ for } b = 1, \ldots, B, \tag{2}$$

$$\text{Units of Lost Sales} \leq b_2, \tag{3}$$

$$\text{Minimum Purchase Order} = b_3, \tag{4}$$

$$y_b \text{ is } 0, 1 \text{ and } x_b \geq 0 \text{ for } b = 1, \ldots, B. \tag{5}$$

In the optimization model above, the b_1, b_2, b_3 represent constants which are the right hand sides of the constraints. The objective function, **Inventory Holding Costs**, as well as the left hand side of (3), **Units of Lost Sales**, are determined as output from the simulation model. The values for the left hand sides of (2) and (4), x_b and Minimum Purchase Order, respectively, are determined by just setting these values equal to b_1 and b_3, respectively.

The basic idea in solving (1)–(5) is to determine the values for $y_1, y_2, \ldots, y_B$ that will minimize Inventory Holding Costs while satisfying the constraints. Since a simulation is used, there is no closed-form function which represents the relationship between $y_1, y_2, \ldots, y_B$ and Inventory Holding Costs or Units of Lost Sales. Hence, a metaheuristic, as employed in the software package, OptQuest, is used.

Results from experimentation with this model are reported in Sect. 6 of the chapter.

6 Simulation-Optimization Results

Simulation optimization experiments are reported here for one product line containing 25 different SKUs. As with the actual system, there are 19 retail outlets. Allowing each of the retail outlets, for b = 1, …, 19, to order the product line either directly from the distribution center ($y_b = 1$) or from the vendor ($y_b = 0$), resulted in an optimization model with 19 zero-one variables, corresponding to the 19 retail outlets.

This optimization model has $2^{19} = 524{,}288$ feasible solutions when all combinations of zeros and ones are considered. Since it is not computationally feasible to enumerate all of these solutions, an optimization algorithm is needed. In particular, the model was solved through the use of the OptQuest (April et al. [3]) tool, associated with the Arena software package. OptQuest relies on the metaheuristic optimization tools of scatter search, tabu search, and artificial neural networks to search the feasible decision variable space.

Results from the simulation-optimization runs are shown in Tables 1, 2, 3, and 4 for values of b_3 (the value set for minimum purchase order) of \$4000, \$3000, \$2000, and \$1000, respectively. Note that \$4000 is the current value for minimum purchase amount from the vendor for this particular product line. In the tables, each row represents information concerning the best solution found by OptQuest for the problem given by (1)–(4) for specific values for the right hand sides of the constraints (b_1, b_2, b_3).

Table 1 Optimization results for b_3 (minimum purchase order) = \$4000

Value of fraction extra order (b_1)	Upper limit of units of lost sales (b_2)	Units of lost sales associated with solution	Inventory holding costs for solution (in dollars)	Number of RO's ordering from the DC for solution
1	1400	1395	39,288	7
1	1000	986	40,964	8
1	600	574	49,558	11
0.75	1400	1372	35,041	6
0.75	1000	942	37,240	5
0.75	600	605	55,530	19
0.5	1400	1302	31,272	5
0.5	1000	996	33,131	8
0.5	600	605	50,558	19
0.25	1400	1369	28,057	4
0.25	1000	854	31,266	7
0.25	600	610	48,209	19
0	1400	1307	26,982	6
0	1000	972	36,173	10
0	600	628	44,382	19

Table 2 Optimization results for b_3 (minimum purchase order) = \$3000

Value of fraction extra order (b_1)	Upper limit of units of lost sales (b_2)	Units of lost sales associated with solution	Inventory holding costs for solution (in dollars)	Number of RO's ordering from the DC for solution
1	1400	1204	35,772	3
1	1000	959	37,500	4
1	600	596	42,945	11
0.75	1400	1261	32,320	5
0.75	1000	977	34,669	6
0.75	600	566	41,927	15
0.5	1400	1293	29,404	3
0.5	1000	853	32,561	3
0.5	600	605	52,427	19
0.25	1400	1388	26,958	3
0.25	1000	987	30,925	9
0.25	600	610	48,324	19
0	1400	1381	25,406	4
0	1000	972	27,700	7
0	600	613	44,737	19

Table 3 Optimization results for b_3 (minimum purchase order) = \$2000

Value of fraction extra order (b_1)	Upper limit of units of lost sales (b_2)	Units of lost sales associated with solution	Inventory holding costs for solution (in dollars)	Number of RO's ordering from the DC for solution
1	1400	755	37,146	2
1	1000	755	37,146	2
1	600	591	41,633	10
0.75	1400	784	34,050	3
0.75	1000	784	34,050	3
0.75	600	597	37,945	10
0.5	1400	842	31,599	2
0.5	1000	842	31,599	2
0.5	600	589	36,410	13
0.25	1400	922	28,630	2
0.25	1000	922	28,630	2
0.25	600	606	48,089	18
0	1400	1101	25,559	5
0	1000	982	26,697	4
0	600	613	45,217	19

Table 4 Optimization results for b_3 (minimum purchase order) = $1000

Value of fraction extra order (b_1)	Upper limit of units of lost sales (b_2)	Units of lost sales associated with solution	Inventory holding costs for solution (in dollars)	Number of RO's ordering from the DC for solution
1	1400	633	36,420	3
1	1000	633	36,420	3
1	600	579	37,396	7
0.75	1400	687	32,868	1
0.75	1000	687	32,868	1
0.75	600	599	35,597	7
0.5	1400	698	30,760	2
0.5	1000	698	30,760	2
0.5	600	592	32,659	8
0.25	1400	774	29,326	2
0.25	1000	774	29,326	2
0.25	600	605	48,965	19
0	1400	865	27,120	4
0	1000	865	27,120	4
0	600	609	46,262	19

The third column of each table shows the left hand side value for the constraint (3) for the solution found. The purpose of showing these values was basically to show whether or not that constraint (specifically the right-hand-side constant, b_2) was close to binding or not; note that feasible solutions were obtained in most, but not all cases; e.g., in the last line in Table 1, there were 628 units of lost sales, but $b_2 = 600$, indicating that the lost sales constraint was violated for the solution found.

The values found for the inventory holding costs for the solutions are shown in the fourth column of the tables, entitled Inventory Holding Costs for Solution (in dollars). Finally, the numbers of retail outlets ordering product from the distribution center (instead of directly from the vendor) for the near optimal solutions found are shown in the last column.

In order to more easily view the tabular results, polynomial functions were fit to the outputs, showing the relationships between lost sales and inventory holding costs for the various values of minimum purchase requirements from the vendor ($4000, $3000, $2000, and $1000, respectively). These graphs are shown in Fig. 4.

One should first note that the results are not always "nice", and in some cases are counterintuitive. This could be the result of several things associated with the study. First, not all possible control variables were considered. For example, the reorder points and reorder quantities employed in the study were computed according to the prespecified policy of the firm, according to the firm's desires; it can be noted however that this policy did account for whether an item was ordered through the distribution center or directly from the vendor for each retail outlet. Second, the

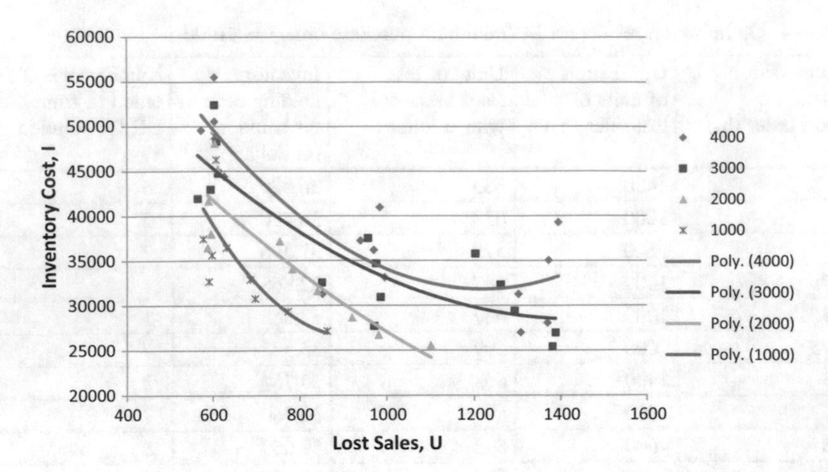

Fig. 4 Relationships between lost sales and inventory holding costs for various values of minimum purchase amounts from vendor

actual demands at the retail outlets were used in the simulation, again according to the firm's desires; unusual demand patterns could have resulted in counterintuitive results in some cases. Third, the optimization procedure is heuristic in nature, working better in some cases than in others. Exact algorithms are not appropriate for this problem since the functional relationships between the control variables and the responses are not of a closed-form nature. The OptQuest tool, which incorporates elements of scatter search, tabu search, and artificial neural networks, was convenient for use in this case since it is an "add-on" for the Arena software package.

As noted, each record in each of the tables represents a near optimal solution to the optimization problem (1)–(4) for specific values of b_1, b_2, and b_3, thereby giving the manager of the system much flexibility in choosing a solution.

One should also note that increasing the value of b_1 (the value of the fraction extra order parameter) is not the same thing as increasing reorder point values, as a value of b_1 other than zero only takes affect when the minimum purchase requirement by the vendor is not met.

In viewing the results as illustrated in Fig. 4, the tradeoff between lost sales and inventory holding costs is obvious. Of particular interest is the effect of lower minimum purchase requirement of the vendor. That is lower values for both lost sales and inventory holding costs can be achieved simultaneously as the minimum purchase requirement of the vendor is decreased. The effect is especially pronounced in a decrease from $3000 to $2000. This type of information can be especially useful in negotiations with the vendor.

The importance of the distribution center to the system is noted as the right-hand-side for the lost sales constraint is decreased. For example, as shown in Table 1, at the current value for minimum purchase requirement of $4000, and $b_1 = 0$, the number of retail outlets that order product through the distribution center (as opposed to directly from the vendor) increases from 6, to 10, to 19 (i.e., all retail outlets) as the right-hand-side for the lost sales constraint decreases from 1400, to 1000, to 600, respectively.

7 Summary and Conclusions

A detailed simulation model, representing the operation of a three-echelon distribution system consisting of vendors, a centralized warehouse, and multiple retail outlets, is described in this paper. The model represents shipments from the vendors to the centralized warehouse, from the vendors to the retail outlets, and from the centralized warehouse to the retail outlets, as well as orders from the retail outlets to the warehouse and vendors, and from the warehouse to the vendors. A wide variety of inputs and outputs are represented by the model, including daily demands for each SKU at each retail outlet, and inventory levels for each SKU at each retail outlet and at the warehouse. A key feature of the model is the use of control variables to represent the policy decisions of whether a retail outlet orders a product line directly from its vendor, or from the centralized warehouse, and the "amount of violation" by each of the retail outlets with respect to ordering a SKU when that SKU's inventory level is greater than its pre-specified reorder point (in order to meet a vendor minimum order requirement).

An optimization model was interfaced with the simulation model with control variables (as defined above) as decision variables. Several examples were run to illustrate the optimization. In some cases counterintuitive results were achieved, possibly as a result of not considering all possible control variables (such as reorder points and order quantities) in the optimization. This illustrates the importance of a system-wide perspective in systems as complex as supply chains and distribution systems, as well as the importance of detailed simulation models for the study of such systems.

Finally, a surprising aspect of the project itself was the complexity of the simulation model that was required to model the system. One would think that modeling a system in which products, organized into product lines, are depleted from some locations, and just moved from one location to another, would be relatively easy to model. However, tracking the various movements, inventory levels for various SKUs, and various variables representing different types of costs and profits, resulted in a very complex model.

Acknowledgements This work was funded from a contract received through the Center for Engineering Logistics and Distribution (CELDi), a multi-university, multi-disciplinary National Science Foundation sponsored Industry/University Cooperative Research Center (I/UCRC). The authors also acknowledge (1) the aid of two former graduate students from the Department of Industrial Engineering at the University of Louisville: Maria Chiodi and Elizabeth Forney, and (2) the helpful suggestions of anonymous referees.

References

1. Abdul-Jalbar, B., Gutierrez, J. M., & Sicilia, J. (2009). A two-echelon inventory/distribution system with power demand pattern and backorders. *International Journal of Production Economics, 122,* 519–524.
2. Albino, V., Carbonara, N., & Giannoccaro, I. (2007). Supply chain cooperation in industrial districts: A simulation analysis. *European Journal of Operational Research, 177*(1), 261–280.
3. April, J., Glover, F., Kelly, J. P., & Laguna, M. (2003). Practical introduction to simulation optimization. In S. Chick, P. J. Sánchez, D. Ferrin & D. J. Morrice (Eds.), *Proceedings of the 2003 Winter Simulation Conference* (pp. 71–78). Piscataway, New Jersey: Institute of Electrical and Electronics Engineers.
4. Arns, M., Fischer, M., Kemper, P., & Tepper, C. (2002). Supply chain modeling and its analytical evaluation. *Journal of the Operational Research Society, 53*(8), 885–894.
5. Axsater, S. (2000). Simple solution procedure for a class of two-echelon inventory problems. *Operations Research, 38*(1), 64–69.
6. Boute, R. N., Disney, S. M., Lambrecht, M. R., & Van Houdt, B. (2009). Designing replenishment rules in a two-echelon supply chain with a flexible or an inflexible capacity strategy. *International Journal of Production Economics, 119*(1), 187–198.
7. Chen, Y., Mockus, L., Orcun, S., & Reklaitis, G. V. (2012). Simulation-optimization approach to clinical trial supply chain management with demand scenario forecast. *Computers and Chemical Engineering, 40,* 82–96.
8. Chen, L.-H., & Kang, F. S. (2007). Integrated vendor-buyer cooperative inventory models with variant permissible delay in payments. *European Journal of Operational Research, 183* (2), 658–673.
9. Cheng, T. C. E., & Wu, Y. N. (2005). The impact of information sharing in a two-level supply chain with multiple retailers. *Journal of the Operational Research Society, 56*(10), 1159–1165.
10. Chu, Y., You, F., Wassick, J. M., & Agarwal, A. (2015). Simulation-based optimization framework for multi-echelon inventory systems under uncertainty. *Computers and Chemical Engineering, 73,* 1–16.
11. Crnkovic, J., Tayi, G. K., & Ballou, D. P. (2008). A decision support framework for exploring supply chain tradeoffs. *International Journal of Production Economics, 115,* 28–38.
12. Deb, K., Pratap, A., Agarwal, S., & Meyarivan, T. (2002). A fast and elitist multiobjective genetic algorithm: NSGA-II. *IEEE Transactions on Evolutionary Computation, 6,* 182–197.
13. Eruguz, A. S., Sahin, E., Jemai, Z., & Dallery, Y. (2016). A comprehensive survey of guaranteed-service models for multi-echelon inventory optimization. *International Journal of Production Economics, 172,* 110–125.
14. Fang, D. J., & Li, C. (2014). Simulation-based hybrid approach to robust multi-echelon inventory policies for complex distribution networks. *International Journal of Simulation Modelling, 13*(3), 377–387.
15. Güller, M., Uygun, Y., & Noche, B. (2015). Simulation-based optimization for a capacitated multi-echelon production-inventory system. *Journal of Simulation, 9*(4), 325–336.
16. Hayya, J. C., Harrison, T. P., & Chatfield, D. C. (2009). A solution for the intractable inventory model when both demand and lead time are stochastic. *International Journal of Production Economics, 122,* 595–605.
17. Holweg, M., & Bicheno, J. (2002). Supply chain simulation-a tool for education, enhancement and endeavor. *International Journal of Production Economics, 78*(2), 163–175.
18. Hung, W. Y., Kucherenko, S., Samsatli, N. J., & Shah, N. (2004). A flexible and generic approach to dynamic modeling of supply chains. *Journal of the Operational Research Society, 55*(8), 801–813.
19. Jain, S. (2004). Supply chain management tradeoffs analysis. In R. G. Ingalls, M. D. Rosetti, J. S. Smith & B. A. Peters (Eds.), *Proceedings of the 2004 Winter Simulation Conference* (pp. 1358–1364). Piscataway, New Jersey: Institute of Electrical and Electronics Engineers.

20. Jansen, D. R., van Weeen, A., Beulens, A. J. M., & Huirne, R. B. M. (2001). Simulation model of multi-compartment distribution in the catering supply chain. *European Journal of Operational Research, 133*(1), 210–224.
21. Katsaliaki, K., & Brailsford, S. C. (2007). Using simulation to improve the blood supply chain. *Journal of the Operational Research Society, 58*(2), 219–227.
22. Kelton, W. D., Sadowski, R. P., & Swets, N. B. (2010). *Simulation with arena* (5th ed.). Boston: McGraw-Hill.
23. Khouja, M. (2003). Synchronization in supply chains: Implications for design and management. *Journal of the Operational Research Society, 54*(9), 984–994.
24. Kochel, P., & Nielander, U. (2005). Simulation-based optimisation of multi-echelon inventory systems. *International Journal of Production Economics, 93–94,* 505–513.
25. Miranda, P. A., & Garrido, R. A. (2009). Inventory service–level optimization within distribution network design problem. *International Journal of Production Economics, 122,* 276–285.
26. Neale, J. J., & Willems, S. P. (2009). Managing inventory in supply chains with nonstationary demand. *Interfaces, 39*(5), 388–399.
27. Ng, C. T., Li, L. Y. O., & Chakhlevitch, K. (2001). Coordinated replinishments with alternative supply sources in two-level supply chains. *International Journal of Production Economics, 73,* 227–240.
28. Pedrielli, G., Alfieri, A., & Matta, A. (2015). Integrated simulation–optimisation of pull control systems. *International Journal of Production Research, 53*(14), 4317–4336.
29. Persson, F., & Araldi, M. (2009). The development of a dynamic supply chain analysis tool-integration of SCOR and discrete event simulation. *International Journal of Production Economics, 121,* 574–583.
30. Persson, F., & Olhager, J. (2002). Performance simulation of supply chain designs. *International Journal of Production Economics, 77,* 231–245.
31. Rosen, S. L., Harmonosky, C. M., & Traband, M. T. (2007). A simulation optimization method that considers uncertainty and multiple performance measures. *European Journal of Operational Research, 181*(1), 315–330.
32. Shin, H., & Benton, W. C. (2007). A quantity discount approach to supply chain coordination. *European Journal of Operational Research, 180*(2), 601–616.
33. Son, J. Y., & Sheu, C. (2008). The impact of replenishment policy deviations in a decentralized supply chain. *International Journal of Production Economics, 113,* 785–804.
34. Supply Chain Council (2011). Retrieved July 8, 2011, from http://www.supply-chain.org.
35. Tannock, J., Cao, B., Farr, R., & Byrne, M. (2007). Data-driven simulation of the supply-chain—Insights from the aerospace sector. *International Journal of Production Economics, 110*(1), 70–84.
36. Tee, Y. S., & Rossetti, M. D. (2002). A robustness study of a multi-echelon inventory model via simulation. *International Journal of Production Economics, 80,* 265–277.
37. Tsai, S. C., & Zheng, Y. X. (2013). A simulation optimization approach for a two-echelon inventory system with service level constraints. *European Journal of Operational Research, 229*(2), 364–374.
38. Wagner, S. M., & Friedl, G. (2007). Supplier switching decisions. *European Journal of Operational Research, 183*(2), 700–717.

Optimization and Simulation of Fuel Distribution. Case Study: Mexico City

Ann Wellens, Esther Segura Pérez, Daniel Tello Gaete and Wulfrano Gómez Gallardo

Abstract In this chapter, the combined use of optimization and simulation in the design of a distribution network for hazardous materials in the northern region of Mexico City is assessed. A mathematical programming model was developed to allow for fuel dispatch truck allocation, minimizing the total distribution cost. Heuristics were used to solve the model and different simulation scenarios were applied to do what-if analysis to be able to decide on different managerial situations. Reviewing simulation and optimization results, an appropriate estimate of the fuel quantity to order (EOQ), the best type of truck to carry out the supply, as well as the ordering schedule that minimizes the associated costs of distribution and inventory, is provided. This real-life Mexican case study shows how a combined optimization-simulation approach, specifically taking advantage of heuristic methods to diminish computing time, can provide a practical, efficient and flexible tool for optimization assessment in operational research.

1 Introduction

Fuel supply has been studied since 1959 when Dantzig and Ramser publish *The Truck Dispatching Problem*, assessing the optimization of the routing of vehicles transporting gasoline from a terminal to different service stations. Since then, a

A. Wellens (✉) · E. Segura Pérez · W. Gómez Gallardo
Facultad de Ingeniería, Universidad Nacional Autónoma de México,
Mexico City, Mexico
e-mail: wann@unam.mx

E. Segura Pérez
e-mail: esegurap@iingen.unam.mx

W. Gómez Gallardo
e-mail: wulfrano.gomez@comunidad.unam.mx

D. Tello Gaete
Facultad de Ciencias de la Ingeniería, Universidad Austral de Chile, Valdivia, Chile
e-mail: daniel.tello@alumnos.uach.cl

M. Mujica Mota and I. Flores De La Mota (eds.), *Applied Simulation and Optimization 2*, DOI 10.1007/978-3-319-55810-3_9

variety of bibliographical material on optimization and simulation of fuel supply has been published, most of them on optimization of the distribution from production sites to refineries, as well as from refineries to mayor storage terminals, mostly by pipelines.

For fuel distribution from a minor deposit or distribution terminal, as Dantzig and Ramser, most of the authors consider trucks that can dispatch part of its load at different service stations. However, due to ruling standards, in México only trucks without compartments and with only one valve are allowed, changing the nature of the fuel distribution problem.

A Mexican company that distributes gasoline in the north of Mexico City using C3 type trucks [63] having a 20 (exactly 20.108) m^3 capacity, wants to know if the inclusion of T3-S3 and T3-S2-R4 type trucks [63] with capacities of 45 (46.149) m^3 and 60 (61.504) m^3 respectively, will minimize the distribution costs given a constant demand. The previous problem corresponds to a Designing Distribution Networks (DDN) problem, where the main goal is to distribute the fuel in the cheapest possible way. Figure 1 represents the problem graphically.

This chapter is organized in the following way: Sect. 2 addresses the theoretical background on designing distribution networks for hazardous materials and inventory optimization and management, Sect. 3 presents the used methodology and Sect. 4 shows the observed results, including data collection, determination of the distribution costs and model formulation and results.

The goal of this study is to optimize the distribution network of a hazardous material for a company that operates in the north of Mexico City, allowing the use

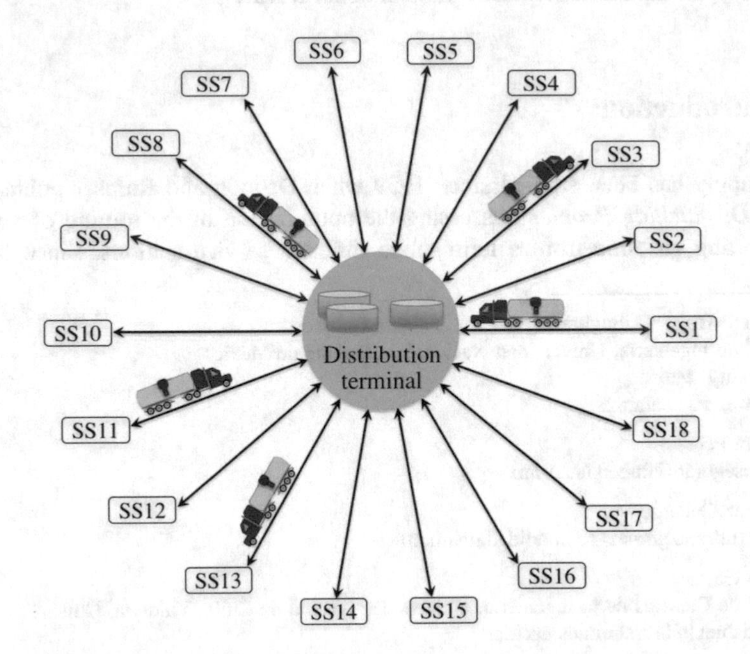

Fig. 1 Representation of the study problem

of a heterogeneous fleet, through a mathematical programming model and simulation.

As specific objectives, the following can be mentioned:

- Collection of information on and statistical analysis of the present state of the fuel supply system in the Azcapotzalco territorial delegation.
- Estimation of fuel demands and distribution costs for the service stations in the Azcapotzalco region.
- Construction of a mathematical programming model to be able to obtain a good solution for fuel ordering quantity, periodicity and the type of truck to be used, considering the collected and estimated information in the previous steps.
- Selection and simulation of different scenarios representing possible critical situations.
- Evaluation of the proposed scenarios by determining the corresponding performance measures, to be able to define possible improvements that can be implemented in the fuel supply system.

2 Theoretical Background

The optimization of vehicle routing and scheduling problems has been studied extensively in specialized literature. This kind of problems aim at establishing the best possible way to distribute products and goods from an origin node to a destiny node, considering changes in the network structure, satisfying the customer demands and minimizing the total costs. This cost is usually expressed in terms of transport costs, inventory costs, opportunity costs, investment, and location-allocation costs.

2.1 Designing Distribution Networks

The models for designing distribution networks are composed of several sub-problems to be optimized. The main ones are: location, allocation, routing, and inventory; different models result when variables are static or dynamic, deterministic or stochastic, discrete or continuous, among others.

The design of the distribution network considers different types of decisions, as for example the location of the elements of the network, fleet dispatching, client and provider assignment, inventory and routing management [8]. Each of these decisions can be optimized independently or jointly. For example, the *vehicle routing problem* (VRP) combines the decisions of selecting the best route and client assignment with homogeneous or heterogeneous fleet [15, 31, 49]. The *location routing problem* (LRP) combines the decision of locating and assigning clients to distribution terminals [35, 45, 90].

Min et al. [57] present the origin and evolution of LRP problems, including different mathematical formulations; they present different LRP classifications based on the number of deposits to locate, demand variations, vehicle number and capacities, distance between nodes, time restrictions or the form of the objective function. Routing and locating models for real-life problems are reported by several authors; see for example [1, 5, 6, 9, 10, 12, 14, 24, 29, 30, 33, 38, 44, 50, 52–54, 56, 60, 62, 67, 69, 72, 73, 75, 87–89].

Problems studying inventory control and vehicle routing jointly are known as *inventory routing problems* or IRP [23, 35, 42, 45, 90]. IRP problems are closely related to *vendor managed inventory* (VMI) problems, having the following characteristics: inventory levels are monitored by the vendor, which decides order quantity and moment, and if shortage of stock is allowed.

At present, models have been developed that consider at the same time decisions on localization, routing and inventories [1, 13]. However, the high complexity involved in solving the complete problem with a sole algorithm originated the formulation of models that solve the problem in stages, in order to find a good solution in the smallest possible computer time. These methodologies involve exact or heuristic algorithms to solve the required decisions. For example, this is the case in the study presented by Flisberg et al. [27] where an exact solution algorithm is proposed to obtain vehicle flows whereas the TABU search method is used in a second step to find optimal routes towards the clients. The use of matheuristics for solving different types of vehicle routing problems, making use of mathematical programming models in a heuristic framework, is assessed in the interesting review presented by Archetti and Speranza [3].

Different kinds of transportation networks include direct shipping, milk runs, crossdocking and tailored networks [18]. The direct shipping network delivers products from suppliers to their customers and is the one used in this study.

2.2 Models for Designing Distribution Networks for Hazardous Materials (DDNHM)

Since the publication of *The Truck Dispatching Problem* [22], several studies have been published on the optimization and simulation of fuel supply; see for example [77] or [58].

According to Winkler [91], the fuel distribution process consists of three steps. The first step includes the distribution from the extraction and/or production plant to the storage terminal, the second step corresponds to the transport of the fuel from the storage terminal to the retail customers (in this case the service stations), while the third and last step corresponds to the distribution to the final client, being cars and/or trucks in this case.

The project presented in this study focuses on fuel distribution in the second stage, that is, from the storage terminals to the retail customers, service stations or petrol stations. In this stage, distribution is carried out by fixed-capacity trucks, as specified in corresponding regulations.

The work of Çetinkaya et al. [16] shows that fuel truck dispatch policies for stock replacement can be carried out regarding two metrics, based on quantity or time. Results showed that truck dispatch based on required quantity provides higher savings in transport costs. In this study, truck dispatching is thus planned based on quantity and using the EOQ inventory model.

Chopra [17] considers the parameters associated with the designing of the distribution network to be directly related to the customer's necessities and the costs needed to implement the network. The first of them include the response time, the variety and availability of offered products, post-sales services, etc. The latter involve the holding costs, transport costs, costs of physical installations and the associated cost of the information system used.

The study by Flisberg et al. [27], mentioned before, presents a truck dispatching problem where daily routes of woodworking trucks deliver to a combination of clients using heterogeneous fleet and taking multiple planning horizons trough mathematical programming and TABU search.

An analysis of literature in the field shows that one of the heuristic algorithms more frequently used to solve the optimization of distribution networks is GRASP (see for example [26, 70]) in combination with mathematical programming. Table 1

Table 1 Most relevant studies for the optimization of distribution networks

Title	Author	Model
A bi-objective GRASP algorithm for distribution of oil products by pipeline networks	Sousa et al. [83]	GRASP
A GRASP heuristic for the mixed Chinese postman problem	Corberán et al. [20]	GRASP
A heuristic for minimizing inventory and transportation costs of a multi–item inventory–routing system	Sombat [81]	EOQ, GRASP, IRP
A reactive GRASP and path relinking for a combined production–distribution problem	Boudia et al. [11]	GRASP
Heuristics for the bi-objective path dissimilarity problem	Martí et al. [55]	GRASP
Model and algorithm for an inventory	Shen et al. [78]	GRASP, IRP
The vehicle routing problem with conflicts	Hamdi-Dhaoui et al. [36]	GRASP, VRPC, ILS, ELS
GRASP with path relinking for the two-echelon vehicle routing problem	Crainic et al. [21]	VRP, GRASP
A GRASP for real-life inventory routing Problem: application to bulk gas distribution	Dubedout et al. [25]	GRASP, IRP
A GRASP ELS for the vehicle routing problem with basic three-dimensional loading constraints	Lacomme et al. [48]	VRP, GRASP
GRASP with VLSN for an inventory-routing problem	Sombat [82]	GRASP, IRP, VLSN, EOQ

shows the most relevant studies that optimize the DDM for different products, specifically hazardous materials.

In this study, the GRASP heuristic was initially explored as the solution method for the optimization model; however, due to the specific nature of the problem where ordering quantity is limited by the used storage tank sizes, feasible solutions are only very small proportion of all possible solutions. As unfeasible solutions increase drastically when the size of the problem increases, the GRASP heuristic would not be time efficient in this study. Still, it was considered the basis of a problem tailored heuristic.

2.3 DDNHM Model Construction

As presented by Chopra [17], the basic components of a DDN model are:

- Localization of the network elements
- Inventory management
- Fleet design
- Vehicle routing

Reyes et al. [71] propose the development of a distribution network in three phases: diagnostics of the distribution system, design of the logistic network and implementation of the network. Each of these phases includes a series of steps, as shown in Table 2.

Table 2 Procedure to construct a distribution network

Phases	Steps
PHASE I: Diagnosis of the distribution system	Step 1: Inventory of the existing equipment
	Step 2: Obtaining information on the current organization of the distribution system
	Step 3: Graphical description and map analysis of the territory of the study object
	Step 4: Description of the existing route
	Step 5: Feasibility study
	Step 6: Temporal analysis of the distribution system
	Step 7: Analysis of the demand by segment and customers
	Step 8: Cost analysis
PHASE II: Design of the logistic network	Step 9: Description of the proposed route
	Step 10: Analysis of the feasibility of the design
	Step 11: Development of the information system
PHASE III: Implementation of the network	Step 12: Implementation of the new logistic network
	Step 13: Measurement and analysis

2.4 Inventory Management and Optimization

Inventory management is defined as the inventory planning and control carried out to meet competitive priorities of the organization [47]. Taha [85] states that the inventory problem consists of keeping in stock just enough articles to satisfy fluctuations of the demand, based on an inventory policy that answers the question of how much and when to order.

Different models have been presented for the optimization of inventories, including models based on dynamic programming [4], linear programming models [41], non-linear programming models [76] and geometric models [46]. Dynamic programming of inventories is based on the minimization of production, retention or holding costs [28]. The Wagner-Whitin algorithm is a classical dynamic programming model that minimizes the fixed ordering and linear procurement and holding costs, over a finite horizon, providing good results [37]. Non-linear programming models to mathematically optimize inventories are proposed by [2, 43, 46, 76]. These models look for the optimal ordering quantity by optimizing the EOQ model.

To know the behaviour of the demand it is necessary to analyse it statistically and know if it is deterministic or stochastic [92]; a suitable criterion is the variation coefficient (VC) introduced by Silver and Peterson [80]. The VC is determined by Eq. (1), where σ is the standard deviation and μ the mean value of the demand.

$$CV = \frac{\sigma}{\mu} \quad (1)$$

If its value is below or equal to 0.2, the data has a low dispersion with respect to the mean value, indicating that the demand can be considered to be deterministic. In the opposite case, it is stochastic. For stochastic demands, a goodness-of-fit test must be carried out to determine the corresponding type of distribution [2].

Taha [85] distinguishes four types of cost related to inventory problems, being the acquisition cost, preparation or ordering cost, retention or holding cost and the stockout cost (see Table 3).

Table 3 Types of inventory costs

Cost type	Definition
Acquisition cost	Unitary price of an inventory product
Ordering cost	Fixed charge due to placing an order, regardless of the ordered quantity
Holding cost	Costs due to having a certain level of existence during a specific time-period; these include the opportunity cost of the inverted money, the storing cost (rental fees, heating, illumination, refrigeration, security etc.), depreciations, taxes, insurance fees, deterioration and obsolescence [59]
Stockout cost	Penalty incurred when the company runs out of a product of the inventory. It includes the loss of income, production disruptions, transaction costs to replace inventory and loss of customer's goodwill

Different methods of inventory replenishment exist; their application depends on the used inventory model, as the demand is the leading factor for replenishment. Examples of procurement systems are given by the Wilson equation (EOQ), the Wagner-Within dynamic programming procedure and the Silver-Meal heuristic, being the EOQ model one of the most extendedly used, as it can be adapted for both deterministic and stochastic demands [4, 79]. In this study, the EOQ model will be applied.

2.5 Simulation

An optimization model is useful to establish the best possible way to distribute products and goods from an origin node to a destiny node; however, in real-life situations observed parameters and or variables change constantly, in which case the proposed schedule must be adjusted. A simulation analysis that compares two different possible scenarios is a cost-efficient and cheap way to decide for one of two future options before these changes take place.

Simulation is the process of reproducing the features, appearance and behaviour of a real system. It is based on three ideas: (1) Imitate, with a mathematical model, a real situation, (2) Study the model's operative characteristics and the system's expected properties making analogies within the simulation model; and (3) Make conclusions and take actions in the system based on the results obtained in the model [37].

Simulation studies are developed in three research levels: descriptive, exploratory and explanatory. Different authors apply the methodology proposed by Law [51], consisting of the steps shown in Fig. 2; see for example [19]. After formulating and planning the study, the data is collected and the model can be constructed. If the model is shown to be valid for the study system, it is implemented in a computer program and should be verified. Experimentation is done for different scenarios of interest; finally, output data is analysed and interpreted.

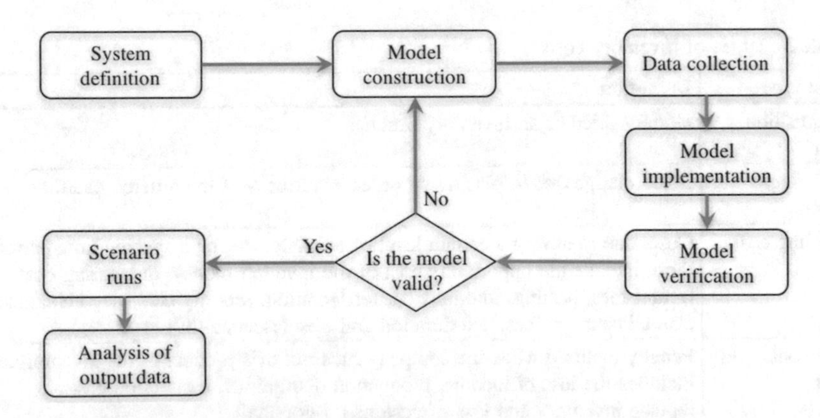

Fig. 2 Simulation methodology, adapted from Law [51]

3 Methodology

The problem addressed in this study is a Design Distribution Network problem where the optimized variables are inventory level, optimal ordering quantity, fleet size and vehicle quantity, as well as order scheduling, corresponding the latter to fleet dispatching or management; a one year planning horizon is used.

Specialized literature describes problems that optimize the previous concepts as inventory routing problems with fleet dispatching. This study presents a special case where each vehicle coming from the storage terminal visits only one client (service station), as the distributed fuel cannot be discharged into fragmented batches due to legal regulations. Reducing the fuel discharge to only one service station by vehicle, the problems seems to be simplified considerably; however, even this special case remains to have an important combinatorial of solutions and therefore stays highly complex.

In the presented case study, both service stations and distribution terminal pertain to the same company, so no out of stocks are considered. Supplied quantities are governed by the vehicle size and required filling level, causing small residuals to exist for technical reasons at the end of the year. These residuals will always be smaller than the truck capacity and are assumed to be transferred without any problem to the next planning horizon. In consequence, demand and supply are assumed constant and always satisfied. According to the information above, this study designs a distribution network with the lowest possible operational costs considering constant fuel demand without stock disruption, heterogeneous fleet, continuous inventory review policy, and fixed capacity of vehicles and storage tanks.

To solve the problem, a methodology of nine steps was used, including the tailored heuristic solution algorithm and mathematical programming.

Step 1: At first, a *data collection* was carried out to obtain existing sales information for the three fuel types considered in the study. Information was found for two service stations; consistency of data was analysed. Storage tank size was obtained for all service stations. Missing information was estimated.

Step 2: The *behaviour of the demand* was analysed for the existing information, including variability, normality and distribution parameters.

Step 3: Based on the previous information, *demand estimation* was done for the rest of the service stations in the study region, considering demographic information and service station characteristics.

Step 4: *Distribution costs* were determined, including holding costs, variable and fixed transport costs and ordering costs.

Step 5: A *mathematical programming model* was developed to describe the truck assignment problem in the study problem.

Step 6: The state space was downsized before solving the model heuristically, so that only *feasible solutions* would be analysed. Truck combinations were restricted to the existing storage tank sizes in each service station. This, in fact, is an optimization step in the solving process, as solution are found in a more efficient and quick way.

Step 7: *Programming* of the linear programming model and its heuristic solution algorithm in R.

Step 8: Determination of specific *simulation scenarios* representing possible critical situations, used to compare different management policies for truck assignment in the fuel distribution.

Step 9: Determination of cost performance measures to *evaluate* the proposed scenarios and definition of possible improvements to be implemented in the fuel supply system.

Figure 3 represents the study methodology graphically.

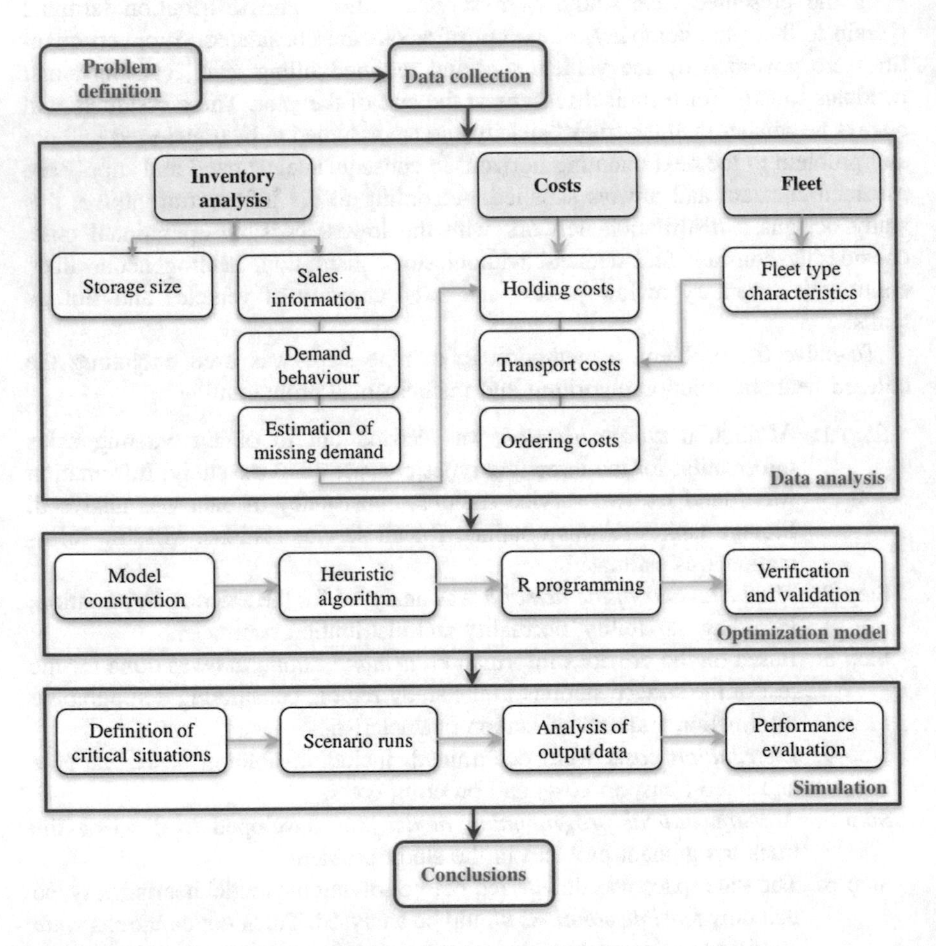

Fig. 3 Study methodology

4 Case Study

4.1 Description of the Problem

In the México City greater area four fuel Distribution Terminals (DT) can be found, being three of them located in Mexico City and the fourth one in the State of Mexico (EDOMEX), from where three types of fuel (gasoline A, gasoline B and Diesel) are distributed towards 371 service stations (SS) [64] in one of the 16 political delegations and some municipalities in the EDOMEX state.

In a pilot phase, this study was carried out in the Azcapotzalco political delegation, located in the northern part of Mexico City (Fig. 4). This delegation has approximately 400 000 inhabitants and one of the DT is located in this area, supplying 18 service stations. Each of the service stations has implemented an inventory review and control system that provides a forecasting method and a weekly ordering schedule into maintain an appropriate service level for fuel consumers, both people and industries.

At present, the Azcapotzalco distribution terminal is using a homogeneous fleet with a 20 m^3 capacity to provision periodically fuel to each of the service stations. The company wants to know if total costs can be minimized when using a heterogeneous fleet and optimizing the supply frequency for the different service stations.

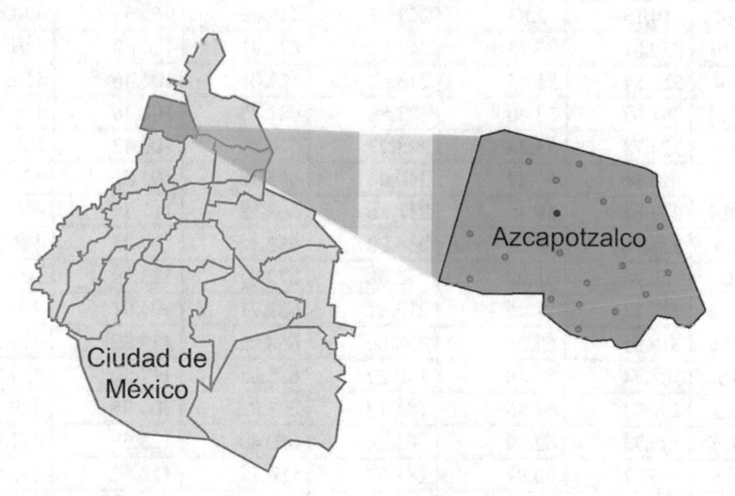

Fig. 4 Study area. Adapted from: http://www.mapa-mexico.com/Mapa_Ubicacion_Azcapotzalco_Mexico_DF.htm

4.2 *Data Collection*

4.2.1 Monthly Sales Information

Monthly sales information was available from January 2014 until October 2015 for service stations SS1 and SS6; this information is presented in Table 4.

As can be seen in Table 4, SS6 sells about double the quantity of fuel for each of the three fuel types. In both service stations, the most sold fuel is gasoline A, which determines about 53% of the total sales. This information is consistent with information reported by INEGI [39] from where it can be determined that average sales per service station for the same period in the Mexican republic were respectively 271.26, 61.27 and 332.54 for gasoline A, gasoline B and diesel. The specific quantity sold in each service station depends on its size and correspondingly on the number of hoses installed for each type of fuel.

In service station SS6 two values were missing for the diesel sales; a simple average of the two closest values was used to estimate these missing values.

Table 4 Monthly sales for SS1 and SS6, January 2014–October 2015

	SS1			SS6		
	Gasoline A (m^3/month)	Gasoline B (m^3/month)	Diesel (m^3/month)	Gasoline A (m^3/month)	Gasoline B (m^3/month)	Diesel (m^3/month)
Jan 2014	333.16	50.86	210.46	663.25	102.92	427.79
Feb 2014	314.15	52.85	203.07	618.09	96.85	430.44
Mar 2014	339.81	55.73	242.77	679.91	110.69	516.37
Apr 2014	319.84	51.15	213.92	622.61	104.36	454.36
May 2014	344.65	57.20	232.58	687.75	104.36	490.70
Jun 2014	324.78	53.33	228.97	648.80	106.42	NA
Jul 2014	322.36	54.64	240.94	635.28	105.56	502.76
Aug 2014	332.63	59.46	212.36	662.18	112.38	469.20
Sep 2014	325.78	56.52	216.90	652.32	107.25	456.31
Oct 2014	339.96	61.07	232.38	679.04	113.79	504.48
Nov 2014	318.11	57.96	213.37	632.71	104.01	450.75
Dec 2014	318.61	63.26	208.06	666.96	119.38	418.25
Jan 2015	306.74	56.88	190.22	636.68	117.50	434.75
Feb 2015	299.33	54.56	181.12	597.82	106.95	429.72
Mar 2015	329.73	62.70	201.74	650.40	123.67	471.84
Apr 2015	305.77	60.82	181.87	586.13	118.42	495.90
May 2015	325.38	65.44	199.77	621.87	125.79	501.20
Jun 2015	325.12	65.03	191.08	614.57	121.96	488.31
Jul 2015	332.40	65.26	198.83	629.44	129.38	506.51
Aug 2015	322.23	69.93	177.31	625.63	137.74	479.99
Sep 2015	331.59	67.29	147.94	637.38	147.12	494.85
Oct 2015	357.04	74.91	192.00	678.59	151.80	NA
Average	325.87	59.86	205.35	642.15	116.74	471.22

4.2.2 Behaviour of the Demand

The service stations are franchises of the same company that owns the fuel supplier (DT), so they are supposed never to run out of stock. In consequence, the monthly demand for both stations SS1 and SS6 are matched to the monthly sales presented in Table 4.

Similar behaviour is observed for fuel demand at different service stations, despite differences in quantities sold (Fig. 5). Of the three fuel types, only gasoline

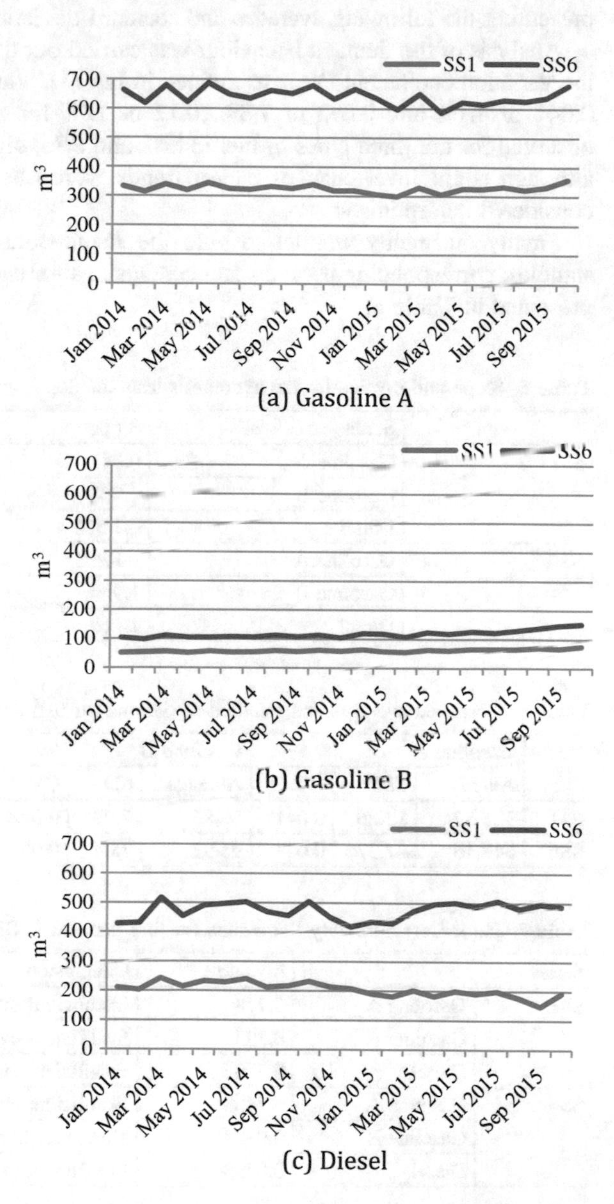

Fig. 5 Behaviour of the demand. **a** Gasoline A, **b** gasoline B, **c** diesel. January 2014–October 2015

B is presenting a marked trend in both service stations (Fig. 5b). This was corroborated determining the corresponding linear regression models and hypothesis tests to verify if the slope is statistically significant. Calculations in R show slope and p-values as presented in Table 5.

As can be expected from graphical results presented in Fig. 5b, gasoline B demand has a statistically significant positive slope in both service stations, while diesel has a statistically significant negative slope or slight level change in service station SS6. To carry out normality tests, the corresponding tendency was removed, presenting the following averages and standard deviations (Table 6).

Analysis of the demand behaviour was carried out through the determination of the variation coefficient (VC), as defined in Eq. (1). Variation coefficients between 0.041 or 4.1% and 0.073 or 7.3% (0.12 or 12% for non-corrected values) were observed for the three types of fuel in SS1 and SS6; all values were below 0.2 so, although slight level changes and/or trends were observed, the demand can be considered deterministic.

Finally, normality was tested with the Jaque-Bera statistic adjusted to small samples; corresponding p-values and conclusions for each of the demand series are presented in Table 7.

Table 5 Slope and p-value for the hypothesis tests on slope significance

	Series	Slope	p-value
SS1	Gasoline A	0.041	0.928
	Gasoline B	0.863	4.85e-08
	Diesel	−2.570	0.0001
SS6	Gasoline A	−1.281	0.169
	Gasoline B	1.964	9.63e-08
	Diesel	1.265	0.224

Table 6 Corrected averages and standard deviations for fuel demand, SS1 and SS6

	Gasoline A			Gasoline B			Diesel		
	Average	SD	CV	Average	SD	CV	Average	SD	CV
SS1	325.87	13.26	0.041	50.86	2.96	0.058	226.09	16.11	0.071
SS6	642.16	27.37	0.043	95.92	7.04	0.073	473.12	30.37	0.064

Table 7 Jaque-Bera normality test results for fuel demand in SS1 and SS6

Series		p-value	Conclusion
SS1	Gasoline A	0.778	Insufficient evidence to reject normality
	Gasoline B	0.395	Insufficient evidence to reject normality
	Diesel	0.5705	Insufficient evidence to reject normality
SS6	Gasoline A	0.7468	Insufficient evidence to reject normality
	Gasoline B	0.0327	Normality is rejected at a 5% level
	Diesel	0.05386	Insufficient evidence to reject normality

All series can be considered to have a normal distribution, unless gasoline B in SS6, for which the hypothesis test was rejected at a 5% significance level. However, even for this type of gasoline, normality is accepted at a 3% level.

4.2.3 Demand Estimation

As the first step in the determination of the demand for the other 16 SS, Voronoi diagrams [34, 40] were implemented to delimit the areas to be supplied [65] for each of the service stations; they capture information on the proximity of a set of points P decomposing the plane in convex polygonal regions. AutoCAD tools were used to define these areas, whereas INEGI [39] information was used to obtain the corresponding number of inhabitants. Land use classification of the service stations was obtained from SEDUVI [74].

Figure 6 represents respectively the political divisions in Azcapotzalco (a) and the corresponding Voronoi polygons (b). The red dots indicate the location of the service stations.

Table 8 shows the resulting Voronoi area for each service station, as well as the corresponding number of inhabitants and land use type.

The Voronoi diagram method supposes that customers will get their fuel supplies in the establishment closest to their domicile. However, as an important difference exists in inhabitants registered in residential and industrial areas, the number of inhabitants resulted not to be a suitable measure to determine the demand; using it as a proportionality coefficient to estimate the demand in each SS, industrial areas would have an artificially low demand as a low number of inhabitants can be expected. On the other hand, land use analysis indicates that both SS where information exists are located in areas classified as mixed residential.

An additional proportionality coefficient was needed, so SS1 and SS6 demands were compared regarding the number of hoses installed for each type of fuel; the results are presented in Table 9. The standard deviations for gasolines A and B can be considered statistically equivalent in both service stations, being mean demand per hose slightly lower for gasoline A in SS1 with respect to SS6. The demand of gasoline B per hose can be considered statistically equivalent in both stations. The standard deviation for the diesel demand is almost two times higher in SS1 with regard to ES6, while the mean diesel demand per hose is also higher in SS1. This higher variability in diesel demand can be explained by the proximity of ES6 to the industrial areas, where a more constant diesel demand is expected. The previous analysis shows differences in the demand per hose in both service stations; however, values are of the same order of magnitude, so the number of hoses for each type of gasoline installed in the service station in combination with the information presented in Table 9 will be used to estimate demands in the other 16 service stations.

The number of serving hoses for each type of gasoline and service station was obtained from information provided by PROFECO [68], visual inspection in a field visit and/or photographical analysis in Google Street View [32]; the results are

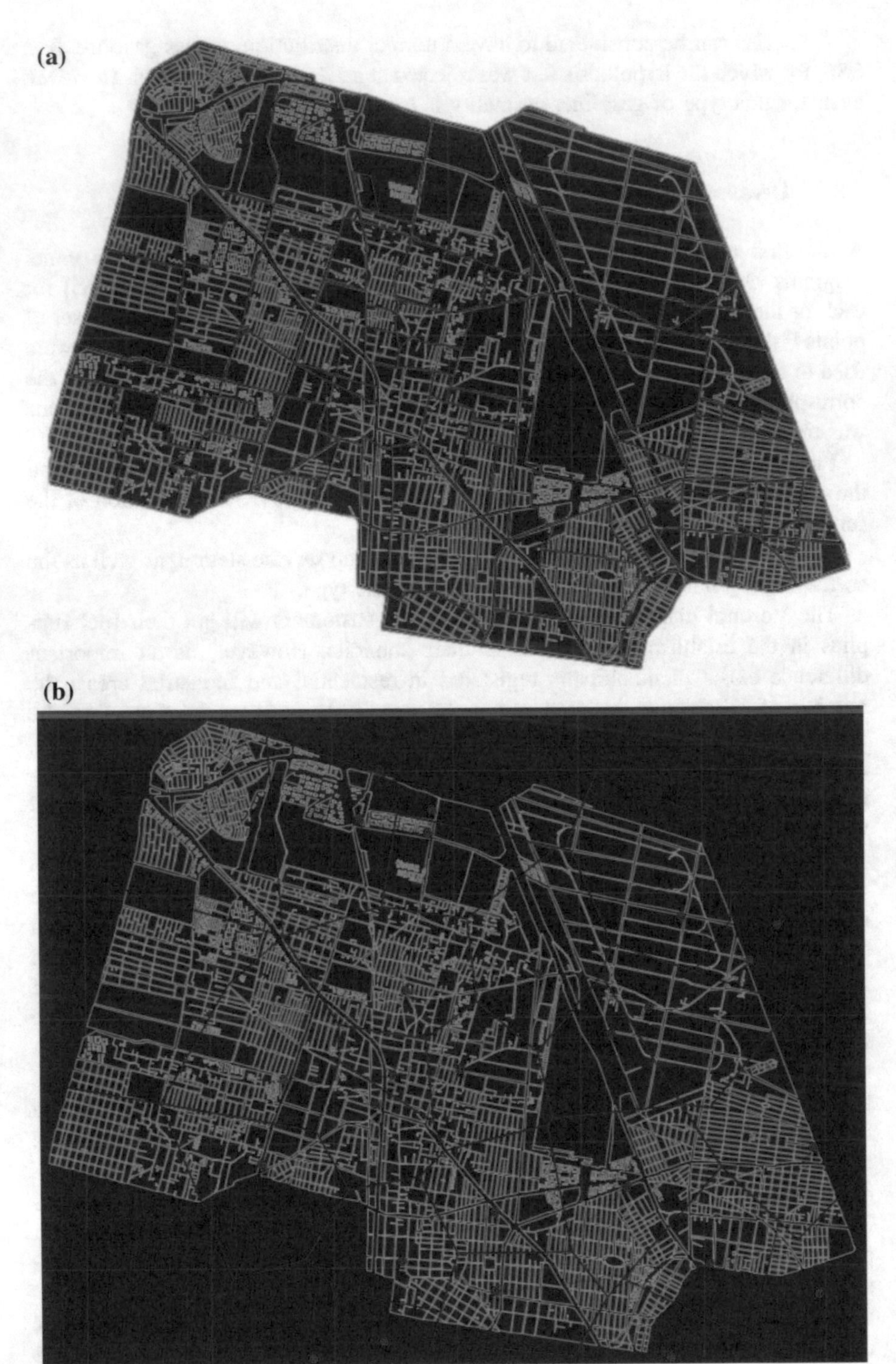

Fig. 6 Azcapotzalco political delegation. **a** Political divisions and **b** corresponding Voronoi polygons

Table 8 Voronoi area, number of inhabitants and land use type for each service station

Service station	Voronoi area	Nr of inhabitants	Land use
SS 1	1.88	11534	Mixed residential
SS 2	1.18	1598	Industrial
SS 3	1.399	10501	Residential
SS 4	1.435	14856	Mixed residential
SS 5	2.068	29838	Residential
SS 6	1.469	348	Mixed residential
SS 7	0.536	6302	Mixed residential
SS 8	1.975	2650	Residential/commercial
SS 9	2.262	429	Industrial
SS 10	1.322	18603	Industrial
SS 11	1.987	16248	Residential
SS 12	2.1	36381	Residential
SS 13	1.017	25008	Mixed residential
SS 14	2.12	14190	Residential
SS 15	1.42	30845	Residential
SS 16	1.145	16257	Residential
SS 17	1.164	10796	Industrial
SS 18	0.881	3800	Industrial

Table 9 Number of hoses and corresponding demands (l/hose) for SS1 and SS6

		Gasoline A	Gasoline B	Diesel
SS1	Number of hoses	6	6	2
	Average demand (l/hose)	54.06	9.86	102.99
	Standard deviation (l/hose)	2.21	1.06	11.43
SS6	Number of hoses	11	11	5
	Demand (l/hose)	58.22	10.46	94.24
	Standard deviation (l/hose)	2.49	1.32	6.25
Average estimated demand (l/hose)		56.14	10.16	98.62

presented in Table 10. Values equal to 0 correspond to service stations that do not sell the corresponding gasoline.

Multiplying the average demand per hose (Table 9) by the number of hoses (Table 10) for each type of gasoline and service station, the estimated demand can be found (Table 11). The demand of SS2 calls the attention; it is considerably higher than the demand in the other service stations due to its location in the biggest industrial area of Azcapotzalco.

Table 10 Number of hoses for each type of gasoline and service station

Service station	Gasoline A	Gasoline B	Diesel
SS 1	6	6	2
SS 2	16	16	18
SS 3	8	8	0
SS 4	13	13	0
SS 5	8	8	0
SS 6	11	11	5
SS 7	10	10	0
SS 8	12	12	4
SS 9	16	16	0
SS 10	8	8	3
SS 11	8	6	2
SS 12	16	12	4
SS 13	12	12	3
SS 14	8	8	0
SS 15	16	16	0
SS 16	12	12	7
SS 17	12	8	4
SS 18	8	8	8

Table 11 Estimated demand for each type of gasoline and service station

Service station	Gasoline A (m^3)	Gasoline B (m^3)	Diesel (m^3)
SS 1	324.39	59.14	205.98
SS 2	898.27	162.54	1,775.13
SS 3	449.14	81.27	0
SS 4	729.85	132.07	0
SS 5	449.14	81.27	0
SS 6	640.42	115.07	471.22
SS 7	561.42	101.59	0
SS 8	673.71	121.91	394.47
SS 9	898.27	162.54	0
SS 10	449.14	81.27	295.86
SS 11	449.14	60.95	197.24
SS 12	898.27	121.91	394.47
SS 13	673.71	121.91	295.86
SS 14	449.14	81.27	0
SS 15	898.27	162.54	0
SS 16	673.71	121.91	690.33
SS 17	673.71	81.27	394.47
SS 18	449.14	81.27	788.95

4.2.4 Storage Tank Size

As the ordering quantity for each gasoline type and service station depends on the size of the storage deposit, deposit sizes were obtained for the 18 service stations in study. Table 12 presents the corresponding information. Values indicating (*) correspond to estimated values, considering the estimated demand. Values of 0 indicate the station does not sell diesel.

Note that the presented values correspond to the tank size, but not to the maximum tank capacity. Due to security regulations, fuel volume in the storage tank should be maximum 90% of its nominal capacity. All storage tanks have an overfill valve installed [61, 66].

4.3 Determination of the Distribution Costs

The costs that must be considered in the distribution network include holding costs, transport costs and ordering costs.

Table 12 Storage tank size for each type of gasoline and service station

Service station	Gasoline A (m^3)	Gasoline B (m^3)	Diesel (m^3)
SS 1	50	40	40
SS 2	80	50	120(*)
SS 3	40	40	0
SS 4	120	60	0
SS 5	50	50	0
SS 6	80	40	40
SS 7	80	60	0
SS 8	100	100	60(*)
SS 9	100	100	0
SS 10	50	40	0
SS 11	60	60	60
SS 12	50	40	40
SS 13	100	100	60(*)
SS 14	100	100	0
SS 15	160	80	0
SS 16	200	80	60
SS 17	100	50	60(*)
SS 18	100	50	60(*)

4.3.1 Holding Costs

Benitez [7] considers among the holding costs rates for physical storage, return on capital detained in stock, insurance, transport, manipulation and distribution of material and finally obsolescence of the material in stock. Tawfik and Chauvel [86] consider holding costs are generally between 14 and 36% of the mean valuation of the stocked products. In the case of fuel service stations, the material is discharged directly in the storage tanks, so no intern transport costs must be considered. On the other hand, the obsolescence concept is not applicable in fuel supply. Accordingly, in this study the holding cost, $C_{h_{ik}}$, is considered as a 20% rate of the cost required to acquire the average monthly demand, being the latter half the ordered demand. Considering a purchase cost, C_{p_i}, of 10 MXN/l, the holding cost for fuel i in service station k is:

$$C_{h_{ik}} = 20\% \cdot C_{p_i} \cdot \frac{Q_{ik}}{2} \quad \forall\, i,\, k \tag{2}$$

where

$C_{h_{ik}}$ Holding cost for fuel i in SS k, MXN/month
C_{p_i} Purchase cost for fuel i, MXN/m^3
Q_{ik} Demanded quantity of fuel i in SS k in each order, m^3

4.3.2 Transport Costs

Transport costs include fixed and variable components. Variable components are directly proportional to the distance between origin and destiny, in addition to taking into account the type of merchandise transported and its weight and volume. These costs change depending on the road type, and if the transport is long range or short range. Fixed costs include purchase costs of the fleet, salaries, driving licenses, insurance, installations for maintenance workshops and parking lots, taxes and recovery of financial capital. Information provided in a report presented to the Ministerio de Transportes y Telecomunicaciones in Chile [84] suggest that fixed costs are about 125% of the fuel cost. No specific information was found for Mexico; as an approximation, the average obtained in the above transport report will be used in this analysis.

Considering that each fuel truck is supplying only one service station in a round-trip, variable fuel costs per trip were determined for three types of trucks j, having capacities of respectively 20, 45 and 60 m^3, as follows:

$$C_{vijk} = \frac{2d_k}{R_j \cdot \eta_j} \cdot c_D \quad \forall\, i, j, k \tag{3}$$

where

C_{vijk} Variable transport cost for fuel i, truck type j and SS k, MXN/order
d_k Distance between the DT and SS k, km
R_j Fuel consumption rate of truck type j, km/l
η_j Performance efficiency of truck type j, %
c_D Required fuel cost for the trip, MXN/l

Table 13 shows the specifications considered for each of the transporting units. The trucks run on diesel; for the diesel cost, a value of 13.77 MXN/l was used (diesel cost in México in August 2016).

For the fixed transport costs, average values of distance, fuel consumption, efficiency and trip number were considered for each service station, in accordance with the demand obtained in Table 11 and the tank capacity of the service station (Table 12), resulting in:

$$C_{f_k} = 1.25 \frac{2\overline{d}_k}{\overline{R}_j \cdot \overline{\eta}_j} \overline{n} \cdot c_D = 1.25 \frac{2(5)}{(2.42)(0.7)} 623 \cdot c_D = 63300 \text{ MXN/year} \qquad (4)$$

The average number of trips per year, $\overline{n}$, was determined including information on required trips for the three fuel types sold in each of the service stations. The amount determined by Eq. (5) is for the whole service station and has to be divided by the number of trips carried out per year in each service station, n_k, to obtain the fixed cost per trip; n_k depends on the obtained supply schedule.

Considering both fixed and variable transport costs, the total transport cost can be determined thus by:

$$c_T = C_{f_k} + \sum_{j=1}^{3} n_{ijk} \cdot C_{v_{ijk}} = \frac{63300}{n_k} + \sum_{j=1}^{3} n_{ijk} \cdot \frac{2d_k}{R_j \cdot \eta_j} \cdot C_D \qquad (5)$$

where n_{ijk} is the number of trips carried out per order for fuel i, truck type j and SS_k.

Table 13 Data sheet for the 20, 45 and 60 m^3 capacity used by the transporting company

	Truck type		
Specifications	20 m^3	45 m^3	60 m^3
Truck type	3C	T3-R2	T3-S2-R4
Minimum fuel consumption rate (km/l)	3.66	ND	ND
Real fuel consumption rate (km/l)	2.95	2.48	1.83
Performance efficiency (%)	0.8	0.65	0.65
Model	Freighliner M2 35k	Freighliner Columbia	Freighliner Columbia
Motor type	MBE4000 de 12.8L EPA 04	Cummins ISX	Cummins ISX
Size of the fuel deposit (l)	189	270	271

4.3.3 Ordering Costs

Concepts used for the determination of the ordering cost include the salary of the personnel that intervene in the ordering process and fixed costs as electricity, telephone, computer use, security clothing and equipment, wheel shims for the tank truck, fire extinguishers and measuring equipment to check the fuel quality.

Regarding the personnel cost, two people are considered to have to be present at the time of discharging, including the person in charge of the service station during the first part of the fuel discharge. A salary of 10 000 MXN/month is considered for the employee, while the station manager has a higher salary but must only be present part of the time. The charge and/or discharge of a 20 m^3 tank truck takes between 30 and 45 min, but time must be added for operations like connection and disconnection of the discharge hoses, security revisions of equipment, quality measuring of the material to discharge, leading to an estimate of 1 h for the complete operation [63]. Considering a finite truck fleet, transport times to and from the DT and recharging times, a maximum of 4 trips per day is considered. In addition to the discharging personnel, a secretary with a monthly salary of 10 000 MXN is considered to dedicate 1 h of her time to each order. Considering 6 weekly working days per week or 25 per month, and 8 daily working hours, a salary of 50 MXN per hour and a total salary cost of 150 MNX per emitted and supplied order.

Fixed ordering costs apportioned per order are assumed to ascend to the same amount, giving a total of 300 MNX per order emitted and per service station.

4.4 *Mathematical Optimization Model*

4.4.1 Model Formulation

Defining the indexes, decision variables and employed parameters, a mathematical model can be developed which can be used for the determination of a good solution for the ordering quantity in each service station, in addition to the type of truck that minimizes the objective function. The model is based on mixed integer programming, with linear restrictions but a non-linear objective function.

In addition to the previously defined variables (see Sect. 4.3), the following indices, variables and parameters are used in the model (Table 14):

Table 14 Indexes and additional variables and parameters used in the model

Indexes		Model parameters and decision variables	
i	Fuel type (i = 1, 2, 3)	C_j	Truck capacity for truck j [m^3]
j	Truck type (j = 1, 2, 3)	O_{ik}	Number of orders for fuel i in SS k
k	Service station (SS) number (k = 1, 2, …, 18)	D_{ik}	Yearly demand for fuel type i in SS k [m^3]
		S_{ik}	Storage tank size for fuel type i in SS k [m^3]

The objective function minimizes total distribution costs. It is a function of the ordered quantity Q_{ik} for fuel i and service station k, which corresponds in each case to the sum of the number of ordered trucks of each type j multiplied by the capacity of the truck, C_j (Eq. 6). Note the factor of 0.9, which indicates that the fuel vessel should be filled to approximately 90% of its rated capacity; this restriction is imposed by corresponding safety regulations (see for example [61]) to avoid accidents due to overload and/or fuel leaks.

$$Q_{ik} = \sum_{j=1}^{3} n_{ijk} \cdot 0.9 \cdot C_j \quad \forall\, i, k \tag{6}$$

The distribution costs C_{ik} for fuel i and service station k (Eq. 7) are calculated as the sum of holding costs and transport costs as defined by Eqs. (2) and (5). The transport cost in Eq. (5) was determined considering the number of fuel trucks n_{ijk} in one order, so it must be multiplied by the number or orders for that fuel and service station.

$$C_{ik} = 20\% \cdot C_{p_i} \cdot \frac{Q_{ik}}{2} + O_{ik} \cdot \left(C_{f_k} \sum_{j=1}^{3} n_{ijk} \cdot C_{v_{ijk}} \right) \quad \forall\, i, k \tag{7}$$

To obtain the total cost (Eq. 8), the above costs are summarized for the three fuel types in each service station k and this quantity is increased with the ordering cost. If the total cost is to be minimized, orders for the different fuel types in a specific service station should be concurrent. Assuming concurrency, the number of orders for service station k in an annual planning horizon equals the order number of the most frequently ordered fuel. The latter depends on both the demand and storage capacities of the fuels.

$$C_T = \sum_{k=1}^{18} \left(C_{o_k} \cdot \max_{\forall i} [O_{ik}] + \sum_{j=1}^{3} C_{ijk} \right) \quad \forall\, i, k \tag{8}$$

The constraints of the model are the following:

- Each truck only supplies one service station in each trip.
- Security constraint: fuel trucks must not be overloaded; they are assumed to be charged at 90%. This restriction is considered in the formulation of the ordered quantity (Eq. 6).
- A single order is considered for any combination of truck and fuel types arriving at a service station on a specific day.
- The demand is always satisfied; only a remnant smaller that the ordering quantity can exist and will be transferred to the next planning horizon. As mentioned before, this is a direct consequence of the specific restrictions in loading capacity in fuel transport and containers vessels. For each iteration, the

number of orders corresponds to the demand divided by the ordered quantity and the following restrictions must be fulfilled:

$$O_{ik} = \frac{D_{ik}}{Q_{ik}} \quad \forall\, i, k \tag{9}$$

- An unlimited fleet is considered.
- Capacity constraint: the total ordered quantity for fuel i and service station k for all types of truck in each order cannot exceed the capacity of the corresponding storage tank:

$$Q_{ik} \leq S_{ik} \quad \forall\, i, k \tag{10}$$

  The tank is assumed to be at its minimum level at the time of ordering.
- No negativity constraint: all physical quantities should be positive.

Finally, the optimization model was implemented in the R programming language, being the leading open-source tool in data analysis. Its main advantages are that it is platform independent, open-source, free and very flexible and straightforward to use. It can handle big amounts of data due to its power and efficient calculations. In addition, R allows integration with other languages as C/C++, Java or Phyton and has packages allowing to integrate the optimization model within a user-friendly interface.

4.4.2 Proposed Solution

Greedy randomized adaptive search procedure (GRASP) is a metaheuristic technique for combinatory optimization. The technique consists of an iterative process, where each iteration is composed of two stages: in the first stage, a feasible solution is constructed, while the second stage consists of a local search in the neighbourhood of the previous solution [70].

The proposed model is highly combinatorial; considering the maximum storage tank capacity of the fuel involved in the problem (200 m^3), a maximum of 66 different truck combinations can be found for each fuel type and service station, giving a total of 1.8×10^{98} combinations. Only a very small amount of them correspond to feasible truck combinations, as a certain combination cannot deliver more fuel than is possible to receive in the storage tank.

Instead of determining the objective function for all possible combinations and discarding those that violate the storage capacity restriction, the number of feasible truck delivery combinations is determined for each storage tank size. As, due to security reasons, fuel vessels should be loaded at a maximum of 90% and only fixed storage tanks exist (see Table 16), in this study the 45 m^3 truck is assumed to be loaded with a maximum of 40 m^3 of fuel to simplify calculations. As an example of how feasible combinations were determined, Table 15 presents all possible

Table 15 Feasible combinations for a 80 m^3 storage tank

Combination	Number of 20 m^3 trucks	Number of 45 m^3 trucks	Number of 60 m^3 trucks	Total quantity delivered (m^3)
1	1	0	0	20
2	0	1	0	40
3	2	0	0	40
4	0	0	1	60
5	1	1	1	60
6	3	0	0	60
7	0	2	0	80
8	1	0	1	80
9	2	1	0	80
10	4	0	0	80

Table 16 Number of feasible combinations for the storage tanks used in the problem

Storage tank size (m^3)	Number of feasible combinations
40	3
50	3
60	6
80	10
100	15
120	22
160	40
200	66

combinations to deliver a maximum of 80 m^3. This amount can be supplied with two 45 m^3 trucks, one 20 m^3 and one 60 m^3 truck, two 20 m^3 trucks and one 45 m^3 truck or, finally, four 20 m^3 trucks. However, since the storage tank must not necessarily be filled completely, there exist other combinations where the total quantity supplied is less than 80 m^3. For this example, there are a total of 10 feasible combinations.

The number of feasible combinations in this problem for commercial storage tanks in Mexico are given in Table 16.

Selecting only feasible delivery combinations, the solution space was reduced to 5.75×10^{38}, which is only a fraction of the original problem search space.

To obtain a solution, at each iteration a feasible delivery combination was randomly selected for each type of fuel and service station. For this combination, total distribution costs are determined. If the solution obtained is better than the best one from previous iterations, it is stored as the best solution. If not, it is discarded. The algorithm stops at a fixed number of iterations, or when the solutions are not improving at a given number of iterations.

4.5 *Simulation*

To show how simulation can be used as a tool to assess managerial decisions, we analysed if it is convenient to consider a heterogeneous fleet to deliver the fuel orders, instead of the current homogeneous fleet. A planning horizon of one year was considered.

4.5.1 Scenario Definition

To be able to analyse if the inclusion of tank trucks with a higher capacity (45 and 60 m^3 respectively), will minimize total distribution costs, the model was set up for two situations:

- If the present scheme of homogeneous fleet is considered (only 20 m^3 trucks), the 54 storage tanks in the 18 service stations will be supplied only with these trucks. As the maximum storage size is 200 m^3, maximum ten feasible truck combinations exist. For instance, the 80 m^3 storage tanks can be supplied with one, two, three or four 20 m^3 trucks (combinations 1, 3, 6 or 10 in Table 15). If the supply in all fifty four storage tanks is considered at the same time, a total of 2.2×10^{22} feasible combinations exist.
- When 45 and 60 m^3 trucks are included, 5.75×10^{38} feasible combinations exist, as explained in Sect. 4.4.2.

Simulation conditions:

- Considering the parameters included in the model (for example, unlimited number of trucks and drivers) for the present study, the distribution cost at a certain service station does not depend on information at other service stations. For this reason, the optimization of the above scenarios can be carried out at each service station independently to increase the algorithm's efficiency. The independency between service stations can be lost, of course, if more information becomes available in a later stage and for example resources are shared between them.
- A total of 100 000 iterations per simulation and 10 repetitions were carried out for each scenario. Running time was about 25 s in a MacBookPro 2 GHz for each run. Analysis of the repetitions suggested that 100 000 iterations was enough to come to a good solution.

4.5.2 Experiments and Discussion

Solving the model for the first scenario where only 20 m^3 trucks are programmed, a minimum total distribution cost of 1 980 076 MXN was obtained.

Table 17 Truck allocation scheme proposed by the model

SS	20 m³ trucks			45 m³ trucks			60 m³ trucks		
	Gasoline A	Gasoline B	Diesel	Gasoline A	Gasoline B	Diesel	Gasoline A	Gasoline B	Diesel
1	0	0	0	1	1	1	0	0	0
2	0	0	0	0	1	0	1	0	2
3	0	0	0	1	1	0	0	0	0
4	0	1	0	3	1	0	0	0	0
5	0	0	0	1	1	0	0	0	0
6	0	0	0	2	1	1	0	0	0
7	0	0	0	2	0	0	0	1	0
8	0	0	0	1	1	0	1	1	1
9	0	0	0	1	1	0	1	1	0
10	0	0	1	1	1	0	0	0	0
11	0	0	0	0	0	0	1	1	1
12	0	0	0	1	1	1	0	0	0
13	0	0	0	1	1	0	1	1	1
14	0	0	0	1	2	0	1	0	0
15	2	0	0	0	2	0	2	0	0
16	0	0	0	0	2	0	3	0	1
17	0	0	0	1	1	0	1	0	1
18	0	0	0	1	1	0	1	0	1

The model was rerun for the scenario that considers a heterogeneous fleet (20, 45 and 60 m³). For this case, a minimum distribution cost of 1 733 585 MXN was found, showing an improvement of minimum 12.4% with respect to the current costs. Since up to now no optimization rules have been applied, the current distribution costs can still be higher than the 1 980 076 MXN obtained in the simulation scenario. In other words, the inclusion of tank trucks with more capacity seems to decrease distribution costs considerably.

The best truck allocation scheme found by the model is presented in Table 17. The proposed allocation scheme shows preference towards trucks with a higher capacity, which is consistent with the conclusion that the use of a heterogeneous fleet can cut distribution costs.

The corresponding ordered quantity, the number of orders in the yearly planning horizon and the order frequency (in days) can be found in the Table 18.

If the company is not willing to buy trucks with other capacities on a short term (vehicles can be substituted for example only when their useful life is over), the proposed model can still be used to optimize the fuel distribution with the current homogeneous fleet, as this study showed the following:

- Transport costs seem to be an important portion of the total distribution cost. This is suggested by the fact that bigger trucks are preferred.

Table 18 Ordered quantity, number of orders and order frequency proposed by the model

SS	Ordered quantity (m^3)			Annual orders			Order frequency (days)		
	Gasoline A	Gasoline B	Diesel	Gasoline A	Gasoline B	Diesel	Gasoline A	Gasoline B	Diesel
1	36	36	36	110	20	70	3.3	18.3	5.2
2	54	36	108	203	55	200	1.8	6.6	1.8
3	36	36	0	152	28	0	2.4	13.0	–
4	108	54	0	83	30	0	4.4	12.2	–
5	36	36	0	152	28	0	2.4	13.0	–
6	72	36	36	109	39	160	3.3	9.4	2.3
7	72	54	0	95	23	0	3.8	15.9	–
8	90	90	54	92	17	89	4.0	21.5	4.1
9	90	90	0	122	22	0	3.0	16.6	–
10	36	36	18	152	28	200	2.4	13.0	1.8
11	54	54	54	102	14	45	3.6	26.1	8.1
12	36	36	36	304	42	134	1.2	8.7	2.7
13	90	90	54	92	17	67	4.0	21.5	5.4
14	90	72	0	61	14	0	6.0	26.1	–
15	144	72	0	76	28	0	4.8	13.0	–
16	162	72	54	51	21	156	7.2	17.4	2.3
17	90	36	54	92	28	89	4.0	13.0	4.1
18	90	36	54	61	28	178	6.0	13.0	2.1

- For the same reason, it can be cheaper to supply bigger and less frequent orders; obviously, considering the maximum capacity of the storage tank for each type of fuel. As mentioned before, in this study an unlimited existing homogeneous fleet is considered. Restrictions in the number of available trucks can change the outcome of the model.
- The fuel with the highest demand (in most cases gasoline A) seems to govern the ordering scheme, suggesting that it is possible to make the planning in stages and adjust the reordering schedule of the less requested gasolines based on the optimal ordering schedule for gasoline A. More simulation runs should be carried out to revise this assumption.

In conclusion, the proposed scenario of including trucks of different capacity showed to be less costly than the current situation in which a homogeneous fleet is used; the tool presented in this study can be used for the optimization of the allocation and delivery scheme with the current fleet or when other vehicles with different capacity are included, as well as for different "what-if?" questions raised by the management of the company.

5 Conclusions

The methodology applied in this chapter corresponds to a mathematical programming model with a tailored heuristic solution, originally based on the GRASP algorithm, to optimize total distribution costs in a fuel distribution network. The model provides estimates of the fuel quantity to order, the best type of truck to carry out the supply, as well as the ordering schedule that minimizes the associated costs of distribution and inventory. Subsequent simulation of several scenarios related to critical situations provides a cheap, flexible and quick way to assess different managerial decisions.

Scenarios that were analysed include the selection of a homogeneous versus a heterogeneous fleet. The current homogeneous fleet was not proven to be the most cost-effective option. In addition, the model is an interesting tool to learn more about the posed supply problem, as for example the preference of supplying bigger quantities on a less frequent basis.

With the present model, what-if analysis can easily be carried out on questions as for example:

- What if the fuel company decides to construct a new service station?
- What if the fleet is limited? In which case should new trucks be purchased?
- What if in the future more companies (and thus different fuel terminals in the same region) start to operate?
- Is a bigger storage tank needed for some fuels?
- In the last months of 2016, the price of diesel increased by approximately 25% due to political and economic instability in Mexico. Does this affect the optimal selection of the fleet? If this raise in diesel cost would persist, would the previous conclusions remain valid?

The flexibility of R to program this kind of optimization model makes it very easy to include more advanced features or extend the problem to a larger spatial scale. Programming in R gives very fast answers, so it should not be a problem to consider, among other, more service stations, political divisions, truck types or cost concepts. Even unexpected situations such as traffic problems due to major maintenance or construction roadworks in a heavily congested city such as Mexico can be evaluated, for example by considering an "equivalent distance" for detours in the determination of variable transport costs. With these relatively simple adaptations, the nature of "what if?" questions which can be posed is very extensive.

As several variables, such as demand or fuel price, are stochastic in nature, future investigations may include determining the corresponding behaviour with a probability density function; in this case, the quantity to be ordered will be determined based on these probability functions and it may be necessary to change the solution strategy. More efficient solution strategies can be considered in the future to find a solution more quickly in these complex situations.

Finally, it should be noted that, due to the specific nature of the problem studied in this chapter, an existing heuristic was not necessarily the best option to find a quick and good solution. Although today there are very powerful mathematical tools and computers to solve an operational research problem without worrying too much about the required computing resources, common sense indicates that the very nature of optimization and engineering prefers to apply simpler strategies, based on previous knowledge of the problem, if they achieve a less intensive use of resources.

References

1. Ambrosino, D., Sciomachen, A., & Scutellà, M. (2009). A heuristic approach based on multi-exchange techniques for a regional fleet assignment location-routing problem. *Computers and Operations Research, 36,* 442–460.
2. Andrade, R., Tello, D., & Romero, O. (2015). *Modelo matemático para la administración de inventario en una microempresa del rubro alimenticio, caso de aplicación.* Ingeniería, negocios e innovación1, 7–21.
3. Archetti, C., & Speranza, M. G. (2014). A survey on matheuristics for routing problems. *EURO Journal on Computational Optimization, 2,* 223–246.
4. Arrelid, D. Y., & Backman, S. (2012). *How to manage and improve inventory control. A study at AB Ph Nederman & Co for products with different demand patterns.* Lund, Lund University. Master thesis.
5. Arvidsson, N. (2013). The milk run revisited: A load factor paradox with economic and environmental implications for urban freight transport. *Transportation Research Part A, 51,* 56–62.
6. Benavides, F. (2012). Planificación de movimientos aplicada en robótica autónoma móvil. Tesis para Maestría en Informática: Universidad de la República Uruguay.
7. Benítez, R. (2012). *Influencia de los costos de mantenimiento en la toma de decisiones* (p. 13). La Habana: Centro de Inmunología Molecular.
8. Benítez, R., Segura, E., & Lozano, A. (2014). Inventory service-level optimization in a distribution network design problem using heterogeneous fleet. In *17th International Conference on Intelligent Transportation Systems (ITSC), Qingdao, China* (pp. 2342–2347).
9. Bilgen, B., & Ozkarahan, I. (2007). A mixed-integer linear programming model for bulk grain blending and shipping. *International Journal of Production Economics, 107,* 555–571.
10. Birgisson, G., & Lavarco, W. (2004). An effective regulatory regime for transportation of hydrogen. *International Journal of Hydrogen Energy, 29,* 771–780.
11. Boudia, M., Louly, M., & Prins, C. (2007). A reactive GRASP and path re-linking for a combined production-distribution problem. *Computers and Operation Research, 34,* 3402–3419.
12. Bruns, A., & Klose, A. (1995). An iterative heuristic for location-routing problems based on clustering. In *Proceedings of the 2nd International Workshop on Distribution Logistics, The Netherlands* (pp. 1–6).
13. Campbell, A., & Savelsbergh, M. (2004). A decomposition approach for the inventory routing problem. *Transportation Science, 38,* 488–502.
14. Caramia, M., Dell'Olmo, P., Gentili, M., & Mircandani, P. B. (2007). Delivery itineraries and distribution capacity of a freight network with time slots. *Computers and Operations Research*, *34*, 1585–1600.
15. Caric, T., & Gold, H. (2008). *Numerical analysis and scientific computing.* InTech.
16. Cetinkaya, S., Mutlu, F., & Lee, C. Y. (2006). A comparison of outbound dispatch policies for integrated inventory and transportation decisions. *European Journal of Operational Research, 171,* 1094–1112.

17. Chopra, S. (2001). *Designing the distribution network in supply chain.* Kellogg School of Management, Northwestern University.
18. Chopra, S., & Meindl, P. (2001). *Supplier chain management—Strategies, planning, and operation.* Upper Saddle River, New Jersey: Prentice Hall.
19. Contreras, E., & Silva, J. (2014). Simulación del proceso de logística inversa de envases y empaques vacíos de plaguicidas en la unidad de riego pantano de Vargas de Boyacá-Colombia. En: VII Simposio Internacional de Ingeniería Industrial: Actualidad y Nuevas Tendencias. Lima, Perú. 689 p.
20. Corberán, A., Martí, R., & Sanchis, J. (2002). A GRASP heuristic for the mixed Chinese postman problem. *European Journal of Operational Research, 142,* 70–80.
21. Crainic, T., Mancini, S., Perboli. G., & Tadei, R. (2012). GRASP with path re-linking for the two-echelon vehicle routing problem. CIRRELET.
22. Dantzig, G., & Ramser, J. (1959). The truck dispatching problem. *Management Science, 6,* 80–91.
23. Desrochers, M., Lenstra, J., & Savelsbergh, M. (1990). A classification scheme for vehicle routing and scheduling problems. *European Journal of Operational Research, 46,* 322–332.
24. Du, T., Wang, F., & Lu, Y. (2007). A real-time vehicle-dispatching system for consolidating milk runs. *Transportation Research Part E, 43,* 565–577.
25. Dubedout, H., Dejax, P., Neagu, N., & Yeung, T. (2012). A GRASP for real-life inventory routing problem: Application to bulk gas distribution. *9th International Conference on Modeling, Optimization & Simulation, Bordeaux, France.*
26. Festa, P., & Resende, M. (2009). An annotated bibliography of GRASP—Part I: algorithms. *International Transactions in Operational Research, 16,* 1–24.
27. Flisberg, P., Lidén, B., & Rönnqvist, M. (2007). A hybrid method based on linear programming and tabu search for routing of logging trucks. *Computers and Operations Research, 36,* 1122–1144.
28. Flores, I. (2015). Programación Dinámica. Universidad Nacional Autónoma de México, Facultad de Ingeniería. Class notes.
29. Ghafoori, E., Flynn, P., & Feddes, J. (2007). Pipeline vs. truck transport of beef cattle manure. *Biomass and Bioenergy, 31,* 168–175.
30. Gillet, B., & Johnson, J. (1976). Multi-terminal vehicle-dispatch algorithm. *Omega, 4,* 711–718.
31. Golden, B. L., Raghavan, S., & Edward, A. (2010). The vehicle routing problem: latest advances and new challenges. *Operations Research/Computer Science Interfaces Series.*
32. Google Street View. https://www.google.com.mx/maps/@40.396764,-3.713379,6z.
33. Gribkovskaia, I., Halskau, O., Laporte, G., & Vlcek, M. (2007). General solutions to the single vehicle routing problem with pickups and deliveries. *European Journal of Operational Research, 180,* 568–584.
34. Guth, N., & Klingel, P. (2012). Chapter 15. Demand allocation in water distribution network modelling—A GIS-based approach using Voronoi diagrams with constraints. In M. A. Bhuiyan (Ed.), *Application of geographic information systems.*
35. Halper, R. (2011). The mobile facility routing problem. *Transport Science, 45,* 413–434.
36. Hamdi-Dhaoui, K., Labadie, N., & Yalaoui, A. (2011). The vehicle routing problem with conflicts. *IFAC Proceedings, 44,* 9799–9804.
37. Heizer, J., & Render, B. (2008). *Dirección de la producción y operaciones* (8ª ed.). Madrid: Pearson Education.
38. Helgesen, Ø., Havold, J., & Nesset, E. (2010). Impacts of store and chain images on the ‘‘quality–satisfaction–loyalty process’’ in petrol retailing. *Journal of Retailing and Consumer Services, 17,* 109–118.
39. INEGI. (2016). Cuentame - número de habitantes en la Ciudad de México. http://cuentame.inegi.org.mx/monografias/informacion/df/poblacion/. Consultation date: 04 mayo 2016.
40. Iwaszko, T. (2012). *Généralisation du diagramme de Voronoï et placement de formes géométriques complexes dans un nuage de points.* Diss. Université de Haute Alsace-Mulhouse.

41. Janssens, G., & Ramaekers, K. (2011). A linear programing formulation for an inventory management decision problem with a service constraint. *Expert Systems with Applications, 7,* 7929–7934.
42. Kapur, K. (1970). Mathematical methods of optimization for multi-objective transportation systems. *Socio-Economic Planning Sciences, 4,* 451–467.
43. Kasthuri, R., & Seshaiah, C. (2013). Multi-item EOQ model with demand dependent on unit price. *Applied and Computational Mathematics, 2,* 149–151.
44. Knust, S., & Schumacher, E. (2011). Shift scheduling for tank trucks. *Omega, 39,* 513–521.
45. Köksalan, M., & Süral, H. (1999). Efes beverage group makes location and distribution decisions for its malt plants. *Interfaces, 29,* 89–103.
46. Kotb, K., Genedi, H., & Zaki, S. (2011). Quality control for probabilistic single-item EOQ model with zero lead time under two restrictions: A geometric programming approach. *International Journal of Mathematical Archive, 2,* 335–338.
47. Krajewski, L., Ritzman, L., & Malhotra, M. (2008). *Administración de Operaciones* (8ª ed.). Pearson Educación: México D.F.
48. Lacomme, P., Toussaint, H., & Duhamel, C. (2013). A GRASP x ELS for the vehicle routing problem with basic three-dimensional loading constraints. *Engineering Applications of Artificial Intelligence, 26,* 1795–1810.
49. Laporte, G. (2009). Fifty years of vehicle routing. *INFORMS, 43,* 408–416.
50. Laporte, G., Gendreau, M., Potvin, J., & Semet, F. (2000). Classical and modern heuristics for the vehicle routing problem. *International Transactions in Operational Research, 7,* 285–300.
51. Law, A. M. (2015). *Simulation modeling and analysis* (5th ed.). New York: McGraw-Hill. 804 p.
52. Lenstra, J. K., & Rinnooy-Kan, A. H. G. (1981). Complexity of vehicle routing and scheduling problems. *Networks, 11,* 221–227.
53. Li, J., Borenstein, D., & Mirchandani, P. (2006). Truck scheduling for solid waste collection in the City of Porto Alegre, Brazil. *Omega, 36,* 1133–1149.
54. Lizotte, Y. (1987). Truck and shovel dispatching rules assessment using simulation. *Mining Science and Technology, 5,* 45–58.
55. Martí, R., et al. (2011). Multi-objective grasp with path-relinking. AT&T Labs Research Technical Report.
56. Melkote, S., & Daskin, M. S. (2001). Capacitated facility location/network design problems. *European Journal of Operational Research, 129,* 481–495.
57. Min, H., Jayaraman, V., & Srivastava, R. (1998). Combined location-routing problems: a synthesis and future research directions. *European Journal of Operational Research, 108,* 1–15.
58. MirHassani, S. (2008). An operational planning model for petroleum products logistics under uncertainty. *Applied Mathematics and Computation*, *196*, 744–751.
59. Mora, L. (2008). Gestión Logística Integral. Bogotá: Ecoe Ediciones.
60. Muyldermans, L., & Pang, G. (2010). On the benefits of co-collection: Experiments with a multi-compartment vehicle routing algorithm. *European Journal of Operational Research, 206,* 93–103.
61. MWLAP. (2002). *A field guide to fuel handling, transportation and storage* (41 p.). Ministry of water, Land and Air Protection, British Columbia, Canada.
62. Navarro, A. (2007). *Planificación de redes de distribución: aproximación vía clustering, diagramas de Voronoi y búsqueda tabú*. Tesis para optar al grado de Magíster en Ciencias de la Ingeniería. Santiago, Pontificia Universidad Católica de Chile. Escuela de Ingeniería.
63. NOM-012-SCT-2-2014. (2014). Publicado en el Diario Oficial de la Federación, México. http://www.dof.gob.mx/nota_detalle.php?codigo=5368355&fecha=14/11/2014.
64. NOM-EM-001-ASEA-2015. (2015). Publicado en el Diario Oficial de la Federación, México. http://dof.gob.mx/nota_detalle.php?codigo=5418780&fecha=03/12/2015.
65. Okabe, A., Boots, B., & Sugihara, K. (1992). *Spatial tessellations: concepts and applications of Voronoi diagrams*. New York: Willey.

66. Pemex. (2008). Manual de operaciones de la Franquicia PEMEX. Versión 2008-1. http://www.ref.pemex.com/files/content/02franquicia/sagli002/controlador358e.html?Destino=sagli002_01.jsp.
67. Pootakham, T., & Kumar, A. (2010). A comparison of pipeline versus truck transport of bio-oil. *Bioresource Technology, 101,* 414–421.
68. PROFECO. (2016). Reporte gasolina. http://200.53.148.113/qqg/?page_id=13. Consultation date: November 12, 2016.
69. Raviv, T., & Kaspi, M. (2012). The locomotive fleet fueling problem. *Operations Research Letters, 40,* 39–45.
70. Resende G. C., & Ribeiro C. C. (2013). Chapter 11. GRASP: Greedy randomized adaptive search procedures. In E. K. Burke & G. Kendall (Eds.), *Search methodologies: introductory tutorials in optimization and decision support techniques* (2nd ed., pp. 287–312). New York: Springer.
71. Reyes, E., Tamayo, Y., & Leyva, M. (2011). Procedimiento para el diseño de redes de distribución logística. Contribuciones a la economía Julio. s.p.
72. Ronen, D. (1988). Perspectives on practical aspects of truck routing and scheduling. *European Journal of Operational Research, 35,* 137–145.
73. Russell, R., Chiang, W., & Zepeda, D. (2008). Integrating multi-product production and distribution in newspaper logistics. *Computers and Operations Research, 35,* 1576–1588.
74. SEDUVI. (2016). Programa delegacional de desarrollo urbano en Azcapotzalco. Zonificación y norma de ordenación. Secretaría de Desarrollo Urbano y Vivienda, CDMX. http://www.data.seduvi.cdmx.gob.mx/portal/docs/programas/programasdelegacionales/PLANO-DIVULGACION_PDDU_AZCAPOTZALCO.pdf.
75. Segura, E. Carmona, R., & Lozano, A. (2013) A heuristic for designing inventory-routing networks considering location problem and heterogeneous fleet (pp. 1–18). In *13th World Conference on Transport Research.*
76. Segura, E., Wellens A. G., Tello Gaete D. A., Gómez Hernández A. A., & Rojas Mojía N. L. (2016). Optimización de una red de distribución de un material peligroso al norte de la ciudad de México. En: XIX Congreso Panamericano de Ingeniería de Tránsito, Transporte y Logística, México.
77. Shah, N. (1996). Mathematical programming techniques for crude oil scheduling. *Computers and Chemical Engineering, 20,* 1227–1232.
78. Shen, Q., Chen, H., & Chu, F. (2008). Model and algorithm for an inventory routing problem in crude oil transportation. *Journal of Advanced Manufacturing Systems, 7,* 297–301.
79. Silver, E., & Meal, H. (1973). A heuristic for selecting lot size for the case of a deterministic time varying demand rate and discrete opportunities for replenishment. *Production and Inventory Management, 14,* 64–74.
80. Silver, E., & Peterson, R. (1985). *Decision systems for inventory management and production planning*. New York: Willey.
81. Sombat, S. (2004). A heuristic for minimizing inventory and transportation costs of a multi–item inventory–routing system. *Symposium: Research Operations Annual, 2548,* 59–66.
82. Sombat, S. (2013). GRASP with VLSN for an inventory—routing problem. *Proceedings of the International Conference on Tourism, Transport and Logistics, Paris, France.*
83. Sousa, E., Ferreira, E., Goldbarg, M., & Navarro, T. (2010). A biobjetive GRASP algorithm for distribution of oil products by pipeline networks. *Rio Oil & Gas Expo and Conference.*
84. Steer D. G. (2011). Análisis de costos y competitividad de modos de transporte terrestre de carga interurbana: Informe final. Preparado para el Ministerio de Transportes y Telecomunicaciones in Chile, 162 p.
85. Taha, H. (2012). *Investigación de Operaciones* (9ª ed.). Pearson Education: México D.F.
86. Tawfik, L., & Chauvel, A. M. (1992). *Adminstración de la producción* (404 p.). Mc Graw-Hill.
87. Ulusoy, G. (1985). The fleet size and mix problem for capacitated arc routing. *European Journal of Operational Research, 22,* 329–337.

88. Villegas, J., et al. (2013). A matheuristic for the truck and trailer routing problem. *European Journal of Operational Research, 230,* 231–244.
89. Wang, X., & Regan, A. (2002). Local truckload pickup and delivery with hard time window constraints. *Transportation Research Part B, 36,* 97–112.
90. Wesolowsky, G. O., & Trustcott, W. G. (1975). The multiperiod location-allocation problem with relocation of facilities. *Management Science, 22,* 57–65.
91. Winkler, J., Henning, B., & Marsosudiro, P. (1992). Project Summary. The liquid and gaseous fuel distribution system. United States Environ-mental Protection Agency. Air and Energy Engineering, EPA/600/S2-91/057.
92. Winston, W. (2004). *Investigación de operaciones. Aplicaciones y Algo-ritmos* (4ª ed.). México: International Thomson Editores.

Printed in the United States
By Bookmasters